T0301854

ROTATION SENSING WITH LARGE RING LASERS

Ring lasers are commonly used as gyroscopes for aircraft navigation and attitude control. The largest ring lasers are sensitive enough that they can be used for high resolution inertial rotation sensing of the Earth in order to detect tiny perturbations in the Earth's rotation caused by earthquakes or global mass transport. This book describes the latest advances in the development of large ring lasers for applications in geodesy and geophysics using the most sensitive and stable devices available. Chapters cover our current knowledge of the physics of the laser gyroscope, how to acquire and analyze data from ring lasers, and what the potential applications are in the geosciences. It is a valuable reference for those working with ring lasers or using the data for applications in geodesy and geophysics, as well as researchers in laser physics, photonics, and navigation.

K. ULRICH SCHREIBER is a professor at the Technical University of Munich. He has more than 30 years of research experience in the technology of space geodesy, in particular satellite and lunar laser ranging, ring laser development, and optical time transfer. He received the Huygens Medal for Instrumentation from the European Geosciences Union in 2016.

JON-PAUL R. WELLS is a professor of physics at the University of Canterbury in Christchurch, New Zealand. His research interests include large ring laser gyroscopes, optical interferometry, and laser spectroscopy of inorganic solids.

ROTATION SENSING WITH LARGE RING LASERS

Applications in Geophysics and Geodesy

K. ULRICH SCHREIBER

Technical University of Munich

JON-PAUL R. WELLS

University of Canterbury

Shaftesbury Road, Cambridge CB2 8EA, United Kingdom

One Liberty Plaza, 20th Floor, New York, NY 10006, USA

477 Williamstown Road, Port Melbourne, VIC 3207, Australia

314–321, 3rd Floor, Plot 3, Splendor Forum, Jasola District Centre,
New Delhi – 110025, India

103 Penang Road, #05–06/07, Visioncrest Commercial, Singapore 238467

Cambridge University Press is part of Cambridge University Press & Assessment,
a department of the University of Cambridge.

We share the University's mission to contribute to society through the pursuit of
education, learning and research at the highest international levels of excellence.

www.cambridge.org
Information on this title: www.cambridge.org/9781108422550

DOI: 10.1017/9781108524933

First published 2023

A catalogue record for this publication is available from the British Library.

Library of Congress Cataloging-in-Publication Data
Names: Schreiber, Ulrich (Geodesist), author. | Wells, Jon-Paul, author.
Title: Rotation sensing with large ring lasers : applications in geophysics
and geodesy / K. Ulrich Schreiber and Jon-Paul R. Wells.
Description: Cambridge : Cambridge University Press, 2022. |
Includes bibliographical references and index.
Identifiers: LCCN 2022041589 (print) | LCCN 2022041590 (ebook) |
ISBN 9781108422550 (hardback) | ISBN 9781108524933 (epub)
Subjects: LCSH: Lasers. | Rotation sensors. | Optical gyroscopes.
Classification: LCC QC688 .S37 2022 (print) | LCC QC688 (ebook) |
DDC 621.36/6–dc23/eng20221121
LC record available at https://lccn.loc.gov/2022041589
LC ebook record available at https://lccn.loc.gov/2022041590

ISBN 978-1-108-42255-0   Hardback

To the pioneers of our large ring laser adventure: Hans Bilger, Richard Falke, Morrie Poulton, Clive Rowe, Manfred Schneider and Geoffrey Stedman.

Contents

Preface

At the beginning there is always a dream: "Wouldn't it be nice to just go to the basement and read the instantaneous Earth rotation signal straight from a laser gyro, instead of waiting for several days for a computed result from a global network of GNSS receivers and VLBI telescopes?"[1] This dream, expressed by my supervisor Professor Manfred Schneider in 1991, was the short and challenging job description that I obtained when I inherited the *Ring Laser project* in my early postdoc years at the Technical University of Munich. Luckily, I was too young to understand the immense difficulties behind this request, and luckily again, I had the opportunity to find the ring laser group at the University of Canterbury, who already had a crudely working but promising prototype. When we teamed up, we *only* had eight more orders of magnitude of precision to go, and we had a lot of optimism too, which turned out to be an essential asset. Words like *accuracy* and *stability* did not even enter our minds in those early days, and that explains why we have continued on this thorny road of sensor development for so long. After three decades of struggle on this bumpy road, it is time to look back and to summarize our experience.

Large ring lasers are very suitable sensors for the precise monitoring of the rotation of the Earth. Their potential long-term stability, size, high sensitivity and rigid mechanical properties suggest themselves for terrestrial deployment. Ring lasers were developed through the 1970s, when the potential substitution of a complex mechanical rotating mass assembly against a simple laser cavity became a realistic expectation. Commonly, these gyros are single longitudinal mode helium–neon lasers operated at a wavelength of 632.8 nm in aircraft navigation applications. Laser gyroscopes used for inertial navigation usually have an area <0.02 m^2, corresponding to a perimeter of 30 cm or less. With a bandwidth of ≈1.8 GHz for the available laser transition in neon, this ensures single longitudinal mode operation. The typical sensitivity of such devices is around 5×10^{-7}rad/s/ $\sqrt{\text{Hz}}$, and the drift is as low as

---

[1] GNSS: global navigation satellite system; VLBI: very long baseline interferometry.

0.0001°/h. This performance level is fully sufficient for navigational requirements but falls short by several orders of magnitude for most geophysical applications. Since the scale factor and hence the sensitivity increases with the area enclosed by the two counter-propagating laser beams, upscaling has been the logical route to sensor improvements. However, the value of large ring laser gyroscopes was initially met with a great deal of skepticism, mostly because it was regarded as doubtful that the operation of a single longitudinal laser mode could be obtained as the cavity free spectral range (FSR) decreased considerably. As it turned out, this was easily achieved.

Frequency pulling from two weakly coupled oscillators with neighboring frequencies can cause frequency synchronization. In ring lasers this effect, known as *locking*, can cause the beat note to disappear, even though the laser is physically rotating. The first large ring laser gyroscope to unlock on Earth rotation alone was the 0.748 m^2 Canterbury-I ring laser C-I, situated in Christchurch, New Zealand [207]. It was a planar, essentially square geometry defined entirely by dielectric *super* mirrors having a nominal reflectance of 99.9985%. In fact, it was the advances in dielectric mirror coating technology that made upscaled ring lasers a viable technology. The choice of a square ring was primarily made to optimize the signal-to-noise ratio but also because of the expectation of reduced backscattering for mirrors used at a 45° angle of incidence. In this early design, the required thermo-mechanical stability was achieved by placing the mirrors on super-invar holders, themselves attached to a 1 m^2 Zerodur plate. Yielding a Sagnac frequency of 76 Hz, this device only rotation sensed for short periods. Tiny leaks in the viton vacuum seals caused the gas mix to degrade quickly. The initial location in a high rise building and the early generation super-mirrors employed caused further problems. Ultimately, C-1 was shifted to an old wartime bunker in the Christchurch suburb of Cashmere, the Cashmere Caverns facility. Its operation was a tremendous step forward, and the technical advances employed in this early prototype were essential for the state-of-the-art ring lasers in use today. In merely 15 years, ring lasers operating on a single longitudinal laser mode increased in size from ~0.02 m^2 to an astounding 834 m^2. In fact, they could be made significantly larger; however, geometrical instabilities and increased losses at the mirror surfaces from larger beam spots do not make this an attractive option at the time of writing. Further, the mode competition processes that govern the start-up time of large ring lasers (i.e., the time the sensor takes to settle into mono-mode operation) may become prohibitively long.

Ring laser-based Sagnac interferometers measure any non-reciprocal effect, which gives rise to a difference in the respective optical path lengths of the counter-propagating laser beams within the cavity. However, the HeNe ring laser gyroscopes discussed here are best suited to the measurement of physical rotations, externally imposed on the monument of the device. Rotation sensing gyroscopes, which utilize the Sagnac effect in the optical domain, essentially fall into two categories: passive and active. Fiber optic gyroscopes are the most prominent passive optical Sagnac interferometers, while ring laser gyroscopes represent the group of active Sagnac devices. The latter group provides the most sensitive and most stable class of gyroscopic devices to date. Of course, alternatives, such as atom interferometry, have the often stated (intrinsic) advantage that atomic masses are much greater than the photon mass. As such they have very high potential as gyroscopes, although they do not (as yet) compete with advanced, large scale ring laser technology.

Highly sensitive rotation sensors have many applications. These range from robotic guidance and inertial navigation systems for a variety of vehicles, aerospace and military hardware; measurement of high order (non-reciprocal) optical effects in condensed phase systems, such as chiral liquids; through to very demanding high resolution measurements in seismology, geodesy and geophysics and even tests of fundamental physics. The vast array of applications necessitate a correspondingly wide range of different sensor types and specifications to satisfy these demands. For the required stable operation for applications in geodesy and geophysics specifically, the fundamental observable, the Sagnac frequency, δf, is strongly influenced by three factors:

- *Scale factor*: The variability of the sensor geometry and effects from laser functions (such as dispersion, laser gas aging and backscatter coupling) are reflected in the measurement quantity mostly as a slowly changing bias.
- *Sensor orientation*: The alignment of the normal vector of the sensor with the Earth's rotation vector as a function of time is critical. Pressure loading around the sensor site, varying wind loads, ground water variations, microseismic activity and solid Earth tides are readily visible.
- *Variations in the Earth's rotation*: This is our very small measurement quantity of prime interest, and it does not exceed a value of $\delta\omega \approx 6 \times 10^{-12}$ rad/s. For periodic diurnal and semidiurnal geophysical signals, these contributions can be well isolated, but aperiodic trends are hard to discriminate from (variable) orientational issues and scale factor instabilities.

Although δf is the only direct measurement quantity obtained from a Sagnac interferometer, one can treat these three contributors independently. The inclusion of auxiliary operational parameters helps in the identification of these signal sources. High resolution tiltmeters, together with an appropriate model for atmospheric attraction, based on gridded regional atmospheric pressure values from a meteorological service [118], for example, allow the estimation of variations in orientation but not the orientation itself. The measurement of perimeter variations through the self-referenced estimation of the FSR allow us to track scale factor variations. The scale factor itself is determined with the help of an optical frequency standard, such as an iodine referenced HeNe laser or an optical frequency comb. Sensor drift caused by changes in backscatter coupling can be determined from the continuous observation of the beam powers and the backscatter amplitudes of the two counter-propagating laser beams. Unfortunately, most of the available auxiliary measurements cannot be easily related to a single error mechanism, so that the correction of the interferometric measurements of a ring laser remains a very involved process.

This book summarizes approximately 30 years of laboratory experience on the development and operation of large HeNe ring lasers. In the beginning, C-I provided the rotation rate of the Earth with a stability of about 10%. Roughly one decade later, we successfully obtained ocean loading effects and the solid Earth tides for the first time, and shortly after that, the continuous detection of diurnal polar motion became a common observable, followed by the detection of very long period signals, like the Chandler wobble in 2012. Today our flagship gyro 'G' routinely operates at a resolution around and below 1 part in 10^8, and the most pressing questions are not primarily *sensitivity* as in the early days, but *stability* and *accuracy*. Large ring lasers are highly coherent optical interferometers, not unlike gravitational wave antennas. Despite some significant differences, they share a number of common problems. While gravitational wave antennas need to be well isolated from the body of the Earth, ring lasers need to be rigidly attached or, better put, *strapped down* to Earth.

On the other side, very subtle non-reciprocal effects between the two counter-propagating laser beams immediately cause a significant offset in the measured Earth rotation rate, which in most cases is slowly changing with time. As a consequence, we estimate, that the accuracy of G with respect to the estimated rotation rates is still only around a level of 1 ppm. The difficulty of giving an exact number here starts already with the problem of fixing the true orientation in terms of latitude of this single component gyro in its strapped down location in the underground ring laser facility of the Geodetic Observatory Wettzell (GOW) in Germany. The stability of G can finally be compared to the sensitivity once the laboratory is in thermal equilibrium, the pressure vessel closed and the optical frequency in the ring cavity actively stabilized. So one can see that this field of development is highly advanced but far from complete. However, we believe it is important to collect and document all our experience to date in this monograph, so that other groups with an interest in the application of large laser gyros have an easier start than we had. We have arranged the book as follows:

- There is a brief review of some important early experiments in rotation sensing by optical interferometry in Chapter 1,

- We take a short stop in Chapter 2 to look at some fundamentals of HeNe-based laser gyroscopes. This does not replace the recommendation to consult the classical work of [11, 40, 236] for the details of the laser theory involved.
- Chapter 3 constitutes the central part of this book. It discusses all the important aspects for the construction and operation of large ring laser gyros and provides a description of a number of very subtle error sources and not so obvious side effects.
- The successful operation of a large gyro does not stop at the design and the properties of the instrument alone. Aspects of data processing and the necessary auxiliary sensor components are also an essential part of gyro operation. Furthermore, we also need to take a look at the geophysical quantities of interest, their magnitude and their spectral range, in order to match the sensor performance to the application. All this is contained in Chapter 4.
- This is followed by a brief discussion of alternative rotation sensing concepts. Chapter 5 puts our work into perspective of the entire field of high resolution inertial rotation sensing, before
- Chapter 6 finally summarizes where large strapped down ring laser gyroscopes have contributed over the years and where we expect to be in the future.

As authors we have shared our effort. Jon-Paul Wells has written Chapters 1 and 2, while Ulrich Schreiber has written Chapters 3–6. Apart from a full account of our activities over the years, we hope that we have also managed to provide a profound overview of the existing large body of literature in this research field, both from an instrumental point of view and with respect to the application. Although the ring laser has come a long way already, there are still a number of challenges ahead. Absolute rotation sensing still requires several orders of magnitude of improvement in order to become a viable technique for tests in fundamental physics. A still more improved sensor stability is another item on this wish list. Due to the large inertia of the Earth, most geodetically relevant signals have signatures in the nHz regime, corresponding to periods of months and years. From that point of view, we are still lacking a consistent sensor fusion of large gyroscopes, the global navigation satellite system (GNSS), and the very long baseline interferometry (VLBI) technique. We hope that this book encourages more activities in that direction.

Acknowledgments

The development of the large ring lasers presented in this monograph was made possible through the collaboration between Satellite Geodesy Research Unit of the Technical University of Munich (Germany), University of Canterbury, Christchurch (New Zealand) and the Federal Agency of Cartography and Geodesy, Frankfurt

(Germany). We also gratefully acknowledge the strong contribution from the Ludwig Maximilian University Munich (Germany). Their input was essential for the development of the field of rotational seismology for which the ring laser technology has become a major measurement tool. University of Canterbury research grants, contracts of the Marsden Fund of the Royal Society of New Zealand and also several grants from the German Research Foundation (DFG) within the research group FOR584 and individual grants SCHR 645/2-2, SCHR 645/2-3, SCHR 645/6-1, SCHR 645/6-2 and GE 3046/1-1 are gratefully acknowledged. The GEOsensor was funded under the program GEOTECHNOLOGIEN of BMBF and DFG.

Special thanks go to our colleagues Hans Bilger, Jacopo Belfi, Nicolo Beverini, Athol Carr, Steven Cooper, Angela Di Virgilio, Robert Dunn, Richard Falke, Yuri Filatov, Andre Gebauer, Urs Hugentobler, Robert Hurst, Heiner Igel, Thomas Klügel, Jan Kodet, Graeme MacDonald, Morrie Poulton, Rüdiger Rodloff, Clive Rowe, Wolfgang Schlüter, Manfred Schneider, Dmitry Shabalin, Geoffrey Stedman, Robert Thirkettle, Alexander Velikoseltsev, Joachim Wassermann, Brian Wybourne and Jie Zhang. Finally we must pay tribute to the hard work of the postgraduate thesis students in both Germany and New Zealand who have contributed so much to the project over the last 25 years.

1

Pre-history of Large Ring Lasers

1.1 The Sagnac Effect and Some Early Experiments

The reader is doubtless familiar with the 1881–1887 Michelson–Morley experiments in which Albert Michelson, together with Edward Morley, famously attempted (repeatedly with ever greater accuracy; first in Potsdam and subsequently in Cleveland) to prove the existence of the aether wind – the supposed medium that permeated space allowing for the propagation of light. Ultimately, they failed to observe any fringe shift in their L-shaped interferometer – a finding in direct conflict with accepted scientific theory at that time and which ultimately culminated in the development of the theory of general relativity. The early history of the Sagnac effect is also intimately associated with that intensive search for the aether around the end of the 19th century.

Following on from the Michelson and Morley experiment, a fundamental question of the time was whether the Earth dragged the aether along with it, as it moved through space. In order to investigate such a possibility, the English physicist Sir Oliver Lodge carried out careful experiments [7, 139, 140] involving rapidly rotating disks in order to induce a dragging of the aether. Both Michelson and Lodge firmly believed in the aether theory – Lodge going so far as to believe that the spirit world existed within it. It is therefore something of an irony that it was their experiments that constituted the primary evidence that it did not exist. Lodge's 1883 and 1887 experiments were serious undertakings involving one meter diameter steel disks spun on the vertical axis of an electric motor at speeds of up to 3000 rpm. The light path of an interferometer then passed between the beams, the idea being that if the aether were dragged along with the spinning disks, this would show up as a fringe shift in the resultant interference pattern. As with the Michelson and Morley experiments in the preceding decade, a null result was obtained.

A further experiment by George Sagnac demonstrated the absence of a *"whirling of the ether"* for a 20 m perimeter vertical ring within the error limits of $\frac{1}{1000}$ th fringe, a conclusion from which was taken that no radial velocity gradient with non-zero curl existed within these error limits [7]. Furthermore, it is interesting to note that Franz Harress, a young scientist in Jena (Germany), who used a fast rotating glass prism ring (12.5 rev. per sec.) to investigate the dispersion properties of glass, came up with an observed fringe shift of

$$\Delta = \frac{2lq}{\lambda c},$$ (1.1)

with q the angular velocity of the rotating body, l the light path inside the glass prisms, λ the optical wavelength and c the velocity of light in a vacuum [120]. For the simplified case of a circular structure, $l = 2\pi r$ and $q = 2\pi r\omega$, so that Eq. 1.1 becomes

$$\Delta = \frac{8\pi A}{\lambda c},$$ (1.2)

with A the enclosed area. This equation turns out to be what is today known as the *Sagnac Equation*. However, Harress and his colleagues did not recognize its significance. Since they assumed aether was dragged along with the prism ring, they misinterpreted his measurement result as the drag coefficient. Incidentally, it is noteworthy that this drag coefficient could not be derived with this kind of apparatus, and the correct interpretation of this signal as a Sagnac fringe shift was excluded implicitly by the assumption of dragged aether. In retrospect, his measurements were more precise than those of George Sagnac, who was the first to correctly combine the theoretical expectations with an experiment. He generated a coherent beam of light, which he guided around a contour with a predetermined area of 0.086 m². The entire apparatus was then rotated with a frequency of approximately 2 Hz [173, 174]. With the help of a beam splitter and several mirrors, he managed to generate two counter-propagating beams, passing them around the same optical path. He observed a shift in the interferogram of 0.07 ± 0.01 fringes and found that the measured shift was directly proportional to the rate of rotation. Further, he established that the effect that now bears his name does not depend on the shape of the optical circuit or the center of rotation. His technical skill in building the instrument with sufficient mechanical stability such that no bending of optical components under the substantial centrifugal forces had an impact on his measurements has to be admired. Finally, we remark that his observation, referred to as the *Sagnac effect* today, would require an aether at rest and was in contradiction with Michelson's findings. Again, it is an irony of history to note that G. Sagnac performed his experiment in order to prove the existence of the aether. Furthermore, his experiment can be described either by the theory of relativity or by a classical aether theory, so that it is not possible from this experiment to decide which of the theories is right

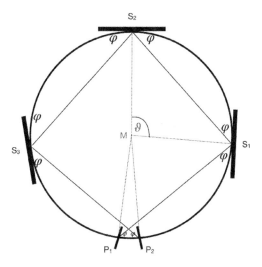

Figure 1.1 Schematic reproduction of the thought experiment by Max von Laue, used in his proof that an experiment, such as the historical Sagnac experiment, produces the same result, irrespective of the assumed theory [228].

or wrong [129]. As a result of these collective experiments, the aether theory was given up.

In 1911, Max von Laue presented a thought experiment [228] on the propagation of light around a rotating circular body (see Figure 1.1), which he solved by geometric considerations in the framework of special relativity, for the case of a preferred frame theory, and in classical electrodynamics. In all cases he obtained the same expression for the experienced difference in propagation time between counter-propagating light beams, namely

$$\Delta \tau = \frac{4A}{c^2} \Omega \sin \varphi, \tag{1.3}$$

which corresponds exactly to the result of George Sagnac. A is the area, c the velocity of light, Ω the experienced rate of rotation and φ the latitude in Eq. 1.3. A full description of the Sagnac effect is based on general relativity [61]; however, as worked out by von Laue, a classical interpretation yields the same result [95, 105]. A simplistic derivation is possible by considering the physically limiting case of a circular beam path of radius R with counter-propagation beams contained within [149]. In the absence of an externally imposed rotation, we may write for the round trip path length L for either beam and their time travel time t:

$$L = 2\pi R \tag{1.4}$$

$$t = \frac{L}{c} = \frac{2\pi R}{c}. \tag{1.5}$$

If the body of this hypothetical interferometer is subject to a physical rotation in the clockwise sense and has an angular velocity of Ω, we now write for the clockwise sense of propagation:

$$L_{cw} = 2\pi R + \Omega R t_{cw} \tag{1.6}$$

$$t_{cw} = t + \frac{\Omega R t_{cw}}{c}. \tag{1.7}$$

With the aid of Eq. 1.5, we get

$$t_{cw} = \frac{2\pi R}{c} + \frac{\Omega R t_{cw}}{c} \tag{1.8}$$

and therefore

$$t_{cw} = \frac{2\pi R}{c - \Omega R}. \tag{1.9}$$

Similarly we may write for the counter-propagating beam travelling in the counter-clockwise direction:

$$L_{ccw} = 2\pi R + \Omega R t_{ccw} \tag{1.10}$$

$$t_{ccw} = \frac{2\pi R}{c + \Omega R}. \tag{1.11}$$

Thus, the path difference is

$$\Delta L = c(t_{cw} - t_{ccw}) = 2\pi R \left[\frac{1}{c - \Omega R} - \frac{1}{c + \Omega R} \right] \tag{1.12}$$

$$= 2\pi R c \frac{2\Omega R}{c^2 - \Omega^2 R^2} = \frac{4\pi R^2 \Omega}{c} \frac{1}{\left[1 - \frac{R^2 \Omega^2}{c^2} \right]} \tag{1.13}$$

$$\approx \frac{4\pi R^2 \Omega}{c}. \tag{1.14}$$

The observed phase difference can then be written as

$$\delta\phi = 2\pi \frac{\Delta L}{\lambda} = \frac{8\pi^2 R^2 \Omega}{\lambda c} = \frac{8\pi A \Omega}{\lambda c}. \tag{1.15}$$

Including the dependence upon the orientation of the interferometer body to the rotation axis, we have:

$$\delta\phi = \frac{8\pi A}{\lambda c} \mathbf{n} \cdot \mathbf{\Omega}, \tag{1.16}$$

where **n** is the normal vector upon A. Equation 1.16 relates the obtained phase difference to the rate of rotation of the entire apparatus and can be interpreted as the gyroscope equation [208]. Fiber optic gyros (FOG) are modern embodiments of this kind of optical gyroscope. Because glass fibers with a length of several hundred meters are used, the scale factor can be made very large by winding the fiber into a coil, and the rotational sensitivity is therefore much larger than for Sagnac's experiment. With L the length of the fiber and R the radius of the resultant coil, one obtains

$$\delta\phi = \frac{4\pi LR}{\lambda c}\mathbf{n} \cdot \mathbf{\Omega}. \tag{1.17}$$

The rotation rate of the Earth alone would have generated a fringe shift of $\approx 0.4 \times 10^{-6}$ on Sagnac's historic instrument, which is well outside the range of sensitivity of the comparatively small apparatus. Based on Sagnac's experiment, it was possible to estimate the size required for an instrument capable of resolving an angular velocity of ≈ 50 μrad/s, which corresponds to the Earth's rotational velocity as it would be experienced at mid-latitude. The goal of course was to measure a very small, nearly constant angular velocity. The experiment was not intended to prove the existence of the Earth's rotation as such.

1.2 The 1925 Michelson, Gale and Pearson Experiment

As early as 1904, Michelson laid forth the conceptual essence of a large scale, closed path, interferometric measurement of the relative motion of the Earth and the aether [147], as well as a method to calibrate the proposed device. The idea was put into practice at Clearing, west of Chicago, in 1925 – the area currently home to Chicago Midway airport. It is very noteworthy that the effort was still done in pursue of the aether. For this mammoth experiment, 12 inch water pipes were utilized to create an evacuated, rectangular optical path 2010 ft by 1113 ft (approximately 613×339 m) [148]. The pipes were joined to cast-iron corner boxes, with a degree of mechanical decoupling provided for by flexible joints. As the entire beam path was evacuated, screw and lever systems were used to align the mirrors. By all accounts [201], Michelson was ill at the time, and the experiment proceeded through the efforts of Henry Gale and Fred Pearson. Figure 1.2 shows a design draft of this experiment. Highly reflecting mirrors were positioned at D, E and F, while the plates at A, B and C acted as beamsplitters via the thin optical coatings applied. The critical feature of this experiment was that the rotation to be measured (the Earth's rotation) could not be switched off or reversed in its sense of rotation. Michelson and Gale had to prove that any observed fringe shift was indeed a measurement quantity and not an artifact generated from the finite thickness of the beamsplitter

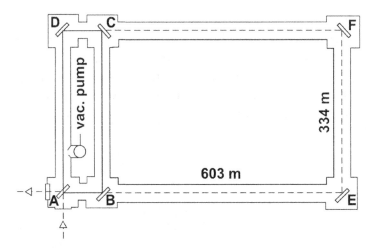

Figure 1.2 Schematic of the 1925 Michelson and Gale experiment. The optical interferometer had a length of 603 m and a width of 334 m. (Reprinted with permission of AIP Publishing from [182], ©2022, American Institute of Physics)

or multiple reflections in the interferometer itself. Thus the set of fringes generated by the circuit ABCD was used as a calibration point, since the area enclosed by this section was insufficient to generate a measureable fringe shift. The main circuit ADEF generating the measurement quantity, with the rotation rate of the Earth at the location of Clearing (Illinois, USA) generating a shift of 0.23 fringes, measured with an uncertainty of no more than 0.005 fringes. This corresponds to a measurement error of only 2%. This was a huge experimental feat.

1.3 The Advent of Lasers: The 1963 Macek and Davis Ring Laser

It is arguably the case that by the 1950s, optics as a field had stagnated somewhat. This changed radically when a technique for coherent light amplification by stimulated emission was found, aside from anything else giving breath and form to both non-linear optics and quantum optics. On the afternoon of May 16, 1960, Theodore Maiman demonstrated the first working laser, made using a gain medium consisting of a 2 cm rod of ruby (Al_2O_3 doped with trivalent chromium – the optically active, dopant ion) [144]. Shortly thereafter followed lasers utilizing a gaseous discharge which converted electrical energy directly into optical energy, as opposed to the optically pumped ruby laser [108].

It appears that the first idea of using the Sagnac effect in a resonant cavity to measure rotation dates back to the late 1950s with the work of Clifford Heer, as documented in a patent disclosure dated October 7, 1959 [89, 90], although the work of Adolph Rosenthal at a comparable time [167] must surely be acknowledged.

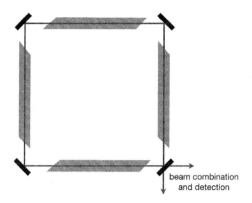

beam combination
and detection

Figure 1.3 A design sketch of the first laser gyro at around 1963. Four large gain tubes with Brewster windows were placed around the sides of the square cavity, in order to compensate for the significant losses of the mirrors. The system operated on an infrared transition at $\lambda = 1.153$ μm [142].

In the account given by MacKenzie [143], Warren Macek of the Sperry Gyroscope Company was in the audience when Rosenthal gave his presentation at the 1961 annual meeting of the Optical Society of America [167]. Macek was a member of an optics group started at Sperry as early as 1957 and already had ambitions to build rotation sensing devices using lasers. By all accounts, internal proposals of this nature were not well met by the Sperry management, and perhaps as a consequence, what was to ultimately become the world's first laser gyroscope was put together with pre-existing plasma tubes and associated equipment. Figure 1.3 depicts a block diagram of this very first laser gyroscope. High quality mirrors were difficult to obtain, and therefore three dielectric mirrors and one curved gold mirror were used to form the cavity (the gold mirror having been coated by a relative [89, 143]). Macek's one meter square gyroscope was constructed from individual sealed plasma tubes with Brewster windows at each end. Radio frequency excitation fueled the gain medium in a cavity designed to support amplification at 1.15 μm on the $2s_2 \longrightarrow 2p_4$ transition of neon in a helium–neon gain medium. Having 20 intra-cavity surfaces, rotation sensing was only achieved on a mechanical turntable, a feat first achieved in late december 1962 (a mere two years after the first demonstration of the HeNe laser itself – see Chapter 2). Beat notes were observed between 1 and 40 kHz, limited by the available equipment.

Unlike earlier Sagnac interferometers, such as the Michelson, Pearson and Gale ring [148], the ring laser gyroscope converts phase information measured via a fringe shift into frequency, since the fractional path length change experienced under the influence of an externally imposed rotation is equal to the fractional frequency shift experienced by the counter-propagating beams in a resonant cavity.

The measurement of a frequency splitting comes with a concomitant increase in sensitivity of many orders of magnitude. The technological significance of their achievements were not lost on Macek and his co-workers, as the final paragraph of their seminal 1963 paper [142] makes clear: "The principle demonstrated in this experiment may be utilized for rotation rate measurement with high sensitivity over an extremely wide range of angular velocities. Such sensors would be self-contained, requiring no external reference."

Following the demonstration of the laser gyroscope, research programs in the USA, Soviet Union, France and the UK were initiated. Of these, most notable was the long term effort by researchers at Honeywell, which included Frederick Aronowitz and Joseph Killpatrick. The objective at Honeywell was commercial success in the medium grade market for military and commercial aircraft. To achieve this, they came up with the, now familiar, triangular configuration drilled into a solid block of quartz (which evolved into a glass ceramic construction of firstly Cer-Vit and latterly Zerodur). In their 'monolithic' design, the laser discharge (which filled the entire cavity) was excited via a single cathode and double anode structure. In addition, they introduced the mechanical dither approach to overcome backscattering from the intra-cavity mirrors, which prevents small gyros from rotation rate sensing at low input rates. In fact, a mechanically dithered gyroscope developed at Honeywell in 1964 was probably the first laser gyro to detect the Earth's rotation. By 1966, Honeywell could claim the first flight test of a laser gyro system. Ring laser gyroscope-based inertial navigation systems require 0.01°/h accuracy and went into large scale civilian airline service in 1981. The modern market for commercial navigational gyroscopes is dominated by companies such as Honeywell, Northrop–Grumman, Thales, iXBlue and Sagem, who supply Airbus and Boeing, among other aircraft manufacturers. The overall gyro market is valued at well over 1.5 billion US dollars at the time of writing, albeit that this includes all sensor types, not just ring laser gyros.

1.4 Passive Optical Gyroscopes

Another approach for precise interferometric measurement of rotation is the externally injected excitation of the TEM_{00} cavity mode of a ring resonator experiencing rotation. This method was explored by Ezekiel and Balsamo in 1977 [71], since it was believed at the time to be free from the frequency lock-in issue as well as deleterious effects associated with an intra-cavity gain medium, such as bias drift and scale factor variations. A block diagram of the setup is shown in Figure 1.4. In their experiment, the beam of an external, single frequency HeNe laser was divided in two by a beamsplitter. Each beam was offset in frequency by an acousto-optical modulator to match the respective cavity modes of the clockwise and counter-clockwise senses

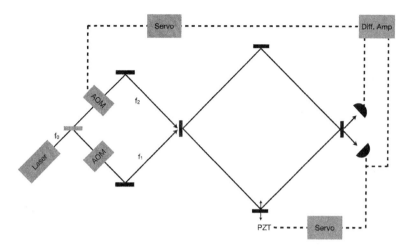

Figure 1.4 A simplified schematic of the externally excited passive resonator gyroscope of Ezekiel and Balsamo [71].

of propagation of a square Fabry–Perot cavity, and the polarization was aligned with one of the polarization axes of the resonator. Both offset frequencies were derived from a stable low noise radio frequency RF oscillator, so that the effective frequency jitter was caused by the laser source and was therefore identical for both counter-propagating beams and hence cancels out in the interferogram when the beat note Δf is evaluated to obtain the rate of rotation Ω. In the ideal case, Ω is strictly proportional to Δf and scales with the area enclosed by the light path. By dithering the resonance frequency of the cavity, a feedback loop was used to lock the resonator to the external laser beam injected into the cavity in the clockwise direction. A second feedback loop then adjusted the offset of the second laser beam to hit the resonance of the cavity in the counter-clockwise direction.

Such an experiment, although conceptually simple, comes with numerous hurdles in practice. Free space, external injection requires very careful alignment and mode matching of the beam from the external laser source to the passive cavity, to ensure the avoidance of high order transverse modes (a particular issue as the passive cavity gets larger). Furthermore, laser gyros rely on common mode rejection to achieve their sensitivity; again meticulous alignment is required to ensure and maintain coincident counter-propagating beam paths. Probably the most serious drawback of passive gyros is the fact that the cavity resonance is of the order of several hundred Hz wide. Active ring laser gyros, in comparison, generate a linewidth In the regime of 10 μHz. A consequence of this is that passive cavities with external excitation generally have a lower resolution and poorer common mode noise rejection. For an instrument size of 0.49 m^2, a sensitivity limit of about

2.4×10^{-8} rad/s for 100 seconds of integration was achieved [175]. Although this concept was expected to be free of lock-in behavior, it was later shown that the lock-in effect was similar to the effect seen in active ring laser gyroscopes, however only shifted to the two feedback loops [237].

Fiber ring interferometers were first reported in 1976 as low loss, single mode fibers became available [223, 224]. The significance of single mode fibers is that different modes have different effective path lengths and propagation velocities; thus a multi-mode fiber will not maintain the coherence required for a high contrast Sagnac interference pattern. By the early 1980s, research at both Stanford and MIT led to reports of fiber optic gyroscopes for fiber lengths up to 900 m [22, 23, 51, 199]. Both HeNe and semiconductor diode (GaAs) lasers were used as external sources in the various measurements, the best results achieving a rotational sensitivity of around $0.1°/$h for 30 seconds of averaging (which is of the order of $5 \times 10^{-3}\ \Omega_E$) with near-photon noise-limited behavior [51]. By 1982, fiber optic gyroscopes having a sensitivity suitable for inertial navigation requirements had been developed [130]. Single mode fiber resonators were also developed in that year [211], and subsequently, Ezekiel's group at MIT reported the development of a passive fiber-optic ring resonator [146] constructed from 3.1 m of fiber formed into a ring and closed off with an evanescent wave coupler. A single frequency HeNe laser operating at 632.8 nm was used as the externally injected source, which was split into two beams, and either beam was frequency shifted by acousto-optical modulators as in the 1977 Ezekiel and Balsamo free space ring resonator experiment discussed above. The influence of backscatter was minimized through the use of phase modulation of one of the input beams. These initial measurements achieved a rotational sensitivity of $0.5°/$h for one second of averaging.

1.5 Large Gyroscope Experiments in the Early 1980s

The early suggestions to measure general relativistic precessions using gyroscopes date back to shortly after Einstein advanced the theory of general relativity in 1915. In the 1930s Blackett examined the prospect of constructing a laboratory gyroscope for such a purpose, concluding that with the technology of the time it was not possible [69]. With the advent of satellite technology, the situation had changed, and renewed proposals were put forward by Schiff [179] and others. The timely coincidence of proposals to measure relativistic effects with gyroscopes and the development of the laser, and subsequently the laser gyroscope, did not go unnoticed, as the Ezekiel and Balsamo experiment [175] (Section 1.4) makes quite clear. In 1982, The Frank J. Seiler Research Laboratory (FJSRL) of the USAF Academy in Colorado Springs commenced a substantial program to develop a passive resonant ring laser gyroscope (PRRLG), building on the work of Ezekiel and Balsamo [71]

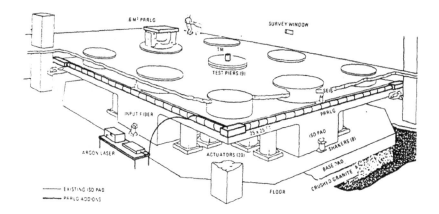

Figure 1.5 Schematic representation of the FJSRL passive resonant ring laser gyroscope project. (Reprinted from the final report on Optical Rotation Sensors JSRL-TR-86-0002 [169])

and motivated by Balsamo's arrival at FJSRL. While the primary objective of this program was clearly military in nature, a secondary objective was to consider the use of large gyroscopes for the study of geophysical phenomena as these were of interest to the Air Force Geophysical Laboratory and the Defense Mapping Agency Geodetic Survey Squadron.

The FJSRL approach was that of the passive ring resonator having the excitation source external to the cavity, closely following the Ezekiel and Balsamo experiment. This appears to have been motivated by the erroneous belief that such an approach avoids the lock in phenomenon, since the laser is external to the cavity. The concept was to initially develop a 0.62 m^2 feasibility model followed by a larger ring of approximately 60 m^2, a sketch of which is shown in Figure 1.5. Both would be placed on a large isolation test pad, providing seismic stability to the ring body but also to provide known input signals for calibration purposes [29, 202, 203]. This program was exceedingly ambitious for the time, having the objective of achieving performance at a resolution of 10^{-10} of the Earth's rotation. The isolation platform itself was 7.62 meters square and constructed from steel-reinforced concrete supported by 20 (pressurized air fed) pneumatic actuators. The excitation source was to be initially a stabilized, single frequency HeNe laser, to be replaced by an argon laser (having an expected linewidth of less than 3 kHz) in due course for the higher photon count. Light from the excitation source was to be fed to the ring via optical fiber, thereby isolating the cavity from any mechanical disturbance induced by the argon laser, which requires high flow rate, water cooling. The large PRRLG was intended to occupy the full perimeter of the iso-pad, with Zerodur structurally stabilized arms, and included active perimeter control. The optical beam path was to be evacuated, and significant work went into the appropriate mode matching optics required to minimize excitation of high order transverse resonator modes [14].

The large PRRLG was never completed, its stop order given on November 27, 1985. It appears the project fell victim to the levels of manpower and budget required to complete the task, together with the prerogatives of the Air Force Office of Scientific Research [170] at the time. What is impressive is the legacy left behind in the literature and mostly resulting from the 20 or so support projects conducted by other institutions, such as the work conducted at the University of New Mexico on quantum detection limits, relativistic experiments and gravitational wave detection, to give one example. It is further noted that two of the scientists involved in the FJSRL experiment had a hand to play in the large ring laser experiments beginning in the late 1980s in New Zealand. These were Hans Bilger, who was involved in the early experiments on ring lasers sized 1 m^2 and below, and Robert Dunn, who was involved in the 'Ultra-G' big ring program at the very beginning of the 21st century. Both of these scientists brought very considerable expertise to the big ring, active gyro programs that followed the FJSRL experiment.

2

Aspects of Helium–Neon-Based Laser Gyroscopes

To paraphrase Javan [107] himself, it was "The date, twelve of december nineteen sixty, and the time, 4:20 pm…" that the first continuous wave optical maser was experimentally realized at AT&T Bell Laboratories in Murray Hill, New Jersey. It was in 1961 that Javan, Herriott and Bennett [108] published their seminal work, demonstrating the first laser utilizing a gaseous gain medium, based around a helium and neon mixture, in *Physical Review Letters*. The original device operated at 1.15 μm, with operation on the now familiar 632.8 nm red transition achieved the following year. The rest, as they say, is history, with millions of red helium—neon (HeNe) lasers having been produced over the years for applications in metrology, scientific research, teaching laboratories and a vast number of commercial applications – such as their early implementation in barcode scanners [50]. Solid state lasers, in particular variants of the semiconductor diode laser, have replaced the HeNe laser (indeed, gas lasers more generally) in very many applications. However for metrology, and more specifically interferometry, the helium–neon laser remains the optical source of choice [88]. To some degree, this is true for the very same reasons that made the HeNe laser successful in the first instance – high quality temporal and spatial coherence, good beam quality and divergence, and visible wavelengths suitable for high quality, low noise detection systems.

For the purposes of gyroscopic measurement, the helium–neon system possesses several other highly desirable characteristics. It has a comparatively low turbulence discharge, excellent spectral purity and a negligible number of scattering centers per unit length of gain (in stark contrast to *any* solid state medium), and it can be excited using a transverse radio frequency electric field, which eliminates (minimizes) the axial flow of the constituents of the gain medium. Finally, it is the ability to manipulate the isotopic composition of the atomic neon ensemble within the gain medium that proves to be critical, as we will see in Section 2.5.

2.1 The Helium–Neon Gain Medium and Laser Oscillation

Both helium and neon are elements from group 0 of the periodic table, otherwise known as the noble gases. These elements have full valence shells, and as a consequence neither atomic helium nor neon engage in the formation of molecules. In the helium–neon laser, a mixture of these chemically inert atoms is used to achieve population inversion in the neon atoms via resonant energy transfer from metastable excited states of helium. Direct electronic excitation of neon atoms in the absence of helium is possible; however, efficient excitation of the states required to achieve optical emission is significantly assisted by the addition of helium. In the HeNe laser an electrical discharge, generated by strong electric fields between an anode and cathode, ionizes the helium gas. The ionized helium atoms then recombine with the electrons, and some of the helium atoms end up in metastable states via excitation by those electrons, which occupy the high energy tail of the kinetic energy distribution. Helium has a multitude of excited states of which precisely two (2^3S and 2^1S) can be referred to as metastable – meaning their radiative transition probability is something very close to zero, and they cannot easily decay back to the ground state ($^1S_0$) via the emission of a photon. Thus the helium excitation process follows

$$\text{He}(^1S_0) + e \longrightarrow \text{He}(2^1S, 2^3S). \tag{2.1}$$

The helium atoms can, however, decay while undergoing a collision with other atoms, if the conditions are appropriate. Thus the excited helium atoms are the energy drivers for the neon atoms, between whose energy levels a population inversion can be generated. It might be tempting to think that this process has something to do with the kinetic energy of the helium atoms; however, even at typical plasma temperatures of around 400 K, this is not even close to the energy required. Energy is transferred from the helium atoms to the neon atoms during the period that the two species are in close spatial proximity and given that there is a near resonance between the energy eigenstates of the respective atoms. The helium to neon energy transfer pathways are as follows:

$$\text{He}(2^1S_0) + \text{Ne}(^1S_0) \longrightarrow \text{He}(^1S_0) + \text{Ne}(2p^55s) + \Delta E_1, \tag{2.2}$$

$$\text{He}(2^3S_1) + \text{Ne}(^1S_0) \longrightarrow \text{He}(^1S_0) + \text{Ne}(2p^54s) + \Delta E_2, \tag{2.3}$$

where the energy mismatches ΔE_1 and ΔE_2 are of the order of kT and are converted into atomic motion.

There are many possible resultant transitions from which laser action has been derived, spanning the visible through to the far infrared [6]. The highest gain transition is the $3s_2 \longrightarrow 3p_4$ transition at 3.39 μm (see Fig. 2.1); however, this is firmly in the infrared and generally considered to be at too long a wavelength to be useful in

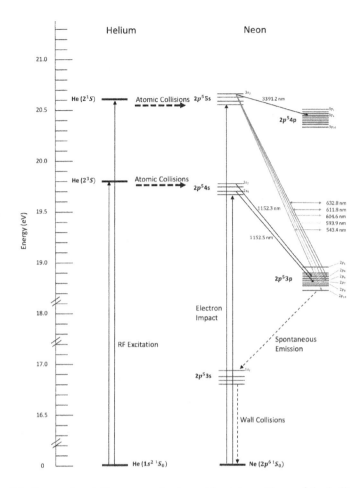

Figure 2.1 Energy level diagram and relevant laser transitions of the helium–neon gain medium.

laser gyroscopes. Table 2.1 lists the four transitions that have been successfully used in laser gyroscopes and their associated properties. The most commonly used wavelength in commercial gyroscopes and large ring laser gyroscopes is the $3s_2 \longrightarrow 2p_4$ 632.8 nm transition, which also has the highest gain of any transition in the visible. We note that the gain available on the 3.39 μm transition is approximately 45 times that at 632.8 nm [88]. Both transitions share the same upper (metastable) state, which can be problematic if due care is not taken. The precise gain and gain difference, among the various neon transitions, are dependent upon the dimensions of the gain tube, the gas ratios used, the neon isotopic composition and the excitation density – thus the values given in Table 2.1 should be taken as indicative values only.

Table 2.1. *Neon transitions used in laser gyroscopes*

Transition	Wavelength (nm)	Gain (% m^{-1})	^{20}Ne-^{22}Ne Isotope Shift (MHz)
$2s_2 \longrightarrow 2p_4$	1152.3	8	261 ref [214]
$3s_2 \longrightarrow 2p_4$	632.8	10	875 ref [48]
$3s_2 \longrightarrow 2p_6$	611.8	2	
$3s_2 \longrightarrow 2p_{10}$	543.4	0.5	1000 ref [75]

Laser operation in the near infrared around 1.15 μm occurs as a doublet transition for pure neon. In the absence of helium, direct rf excitation of neon proceeds via

$$\text{Ne}(2p^{6\,1}S_0) + e \longrightarrow \text{Ne}(2p^5 5s) + e \tag{2.4}$$

$$\text{Ne}(2p^{6\,1}S_0) + e \longrightarrow \text{Ne}(2p^5 4s) + e. \tag{2.5}$$

Simultaneous laser oscillation then occurs on the $2s_2 \longrightarrow 2p_4$ and $2s_4 \longrightarrow 2p_7$ transitions at 1152.3 and 1152.5 nm, respectively. The addition of even small quantities of helium (with partial pressures around 0.07 mbar) quenches the $2s_4 \longrightarrow 2p_7$ transition by elevating the population of the $2p_7$ level.

The short wavelength visible transitions of the helium–neon gain medium have, since the 1980s, been proposed for applications such as lithography, illumination sources for surgical lasers or even laser gyroscopes [88, 125]. While tunable wavelength HeNe lasers covering these wavelengths have been available from companies such as Research Electro-Optics for some time, the implementation of short wavelength neon transitions in gyroscopes is far more recent. There are several challenges. First, for operation on the shortest wavelength 543.4 nm $3s_2 \longrightarrow 2p_{10}$ transition, the gain is very low. This necessitates the use of long discharge tubes with the accompanying prospect of time varying, longitudinal variations in the available gain. Second, and with respect to the 611.8 nm $3s_2 \longrightarrow 2p_6$ transition, the presence of the nearby $3s_2 \longrightarrow 2p_7$ and $3s_2 \longrightarrow 2p_8$ transitions at 604.6 and 593.9 nm limits the excitation densities that can be applied for operation in the absence of the weaker high energy transitions, which tend to generate mode competition. However, it is probably worth noting that gyroscopic operation has been achieved in some form on all three of these transitions. Finally, we note that the $3s_2 \longrightarrow 2p_3$ $3s_2 \longrightarrow 2p_5$ transitions at 635.1 and 629.3 nm have been isolated and operational in large ring cavities, although gyroscopic operation was never attempted. Doubtless, this would be successful if attempted, although there is no particular advantage in doing so.

In very nearly all large ring lasers, the gain medium fills the entire laser cavity, regardless of its volume. Excitation of the gain medium occurs within a pyrex capillary having a diameter in the 4–6 mm range and a length appropriate to the particular laser system or neon transition employed. Relative to commercial linear laser systems or aircraft gyros, such lasers are overpressured in order to remain in the weak saturation limit (see Chapter 3). Thus the helium to neon ratios employed vary from a factor of 10 anywhere up to a factor of 50 (10–20 times larger than their commercial counterparts). Excitation (predominantly of the helium atoms due to their prevalence in the gaseous discharge) occurs via a capacitively coupled, radio frequency transverse electric field using external electrodes.

2.2 The Sagnac Effect in a Ring Laser Cavity

In a laser cavity, the condition for coherent amplification requires zero transverse electric field at the mirrors and that an integer number of wavelengths fit inside the cavity. As such, for a bidirectional ring laser subject to an externally imposed rotation, the counter-propagating laser modes will become frequency non-degenerate. The magnitude of the frequency shift experienced by each beam will be proportional to the input rotational velocity as either beam attempts to fit into the rotating cavity. This is because the optical path length varies during the time of flight of the light, when the laser body is subject to a rotation. The fractional optical frequency shift is proportional to the fractional change in optical path length per cavity round trip:

$$\frac{\Delta v}{v} = \frac{\Delta P}{P}. \tag{2.6}$$

Thus the problem is essentially obtaining an expression for ΔP. The most general case of an optical path, having arbitrary shape and rotating about an arbitrarily positioned axis, is treated in ref. [105] (less general derivations are given elsewhere, see for example [95]). Following [105] and with reference to Figure 2.2, we can express the condition for coherent amplification as follows:

$$n\lambda = \oint \mathbf{dl} = P, \tag{2.7}$$

where P is the perimeter. We note that the time required to traverse an infinitesimal path element $\mathbf{dl}$ is $\mathbf{dl}/c$, and the associated velocity component along the path at any point in the cavity is given by

$$v_{\text{path}} = (\mathbf{\Omega} \times \mathbf{r}) \cdot \hat{i}, \tag{2.8}$$

with $\hat{i}$ being the unit vector tangential to the path at a point with Cartesian co-ordinates x, y, z. Therefore we may write,

$$\Delta P = \oint_{\text{ring}} [(\mathbf{\Omega} \times \mathbf{r}) \cdot \hat{i}] \frac{\mathbf{dl}}{c} = \frac{1}{c} \oint_{\text{ring}} (\mathbf{\Omega} \times \mathbf{r}) \cdot \mathbf{dl}, \tag{2.9}$$

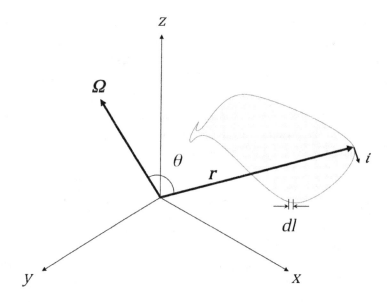

Figure 2.2 A ring laser of arbitrary shape, rotating with an angular velocity Ω.

where $\mathbf{dl} \equiv \hat{i}\, dl$. By Stokes' theorem,

$$\oint_{\text{ring}} (\boldsymbol{\Omega} \times \mathbf{r}) \cdot \mathbf{dl} = \iint_{\text{area}} [\nabla \times (\boldsymbol{\Omega} \times \mathbf{r})] \cdot \mathbf{dA}. \qquad (2.10)$$

Therefore

$$\Delta P = \frac{1}{c} \iint_{\text{area}} [\nabla \times (\boldsymbol{\Omega} \times \mathbf{r})] \cdot \mathbf{dA}. \qquad (2.11)$$

Since $\nabla \times (\boldsymbol{\Omega} \times \mathbf{r}) = 2\boldsymbol{\Omega}$ for a purely rotational field, we have

$$\Delta P = \frac{1}{c} \iint_{\text{area}} 2\boldsymbol{\Omega} \cdot \mathbf{dA} = \frac{2}{c} \boldsymbol{\Omega} \cdot \mathbf{A}. \qquad (2.12)$$

Then we obtain

$$\frac{\Delta \nu}{\nu} = \frac{\Delta P}{P} = \frac{2}{cP} \boldsymbol{\Omega} \cdot \mathbf{A}, \qquad (2.13)$$

and therefore

$$\Delta \nu = \frac{2\nu}{cP} \boldsymbol{\Omega} \cdot \mathbf{A} = \frac{2}{\lambda P} \boldsymbol{\Omega} \cdot \mathbf{A}. \qquad (2.14)$$

Finally then, we obtain an expression for the beat note detected by the beam combining optics external to the ring laser, which is

$$\delta f_{\text{Sagnac}} = 2\Delta\nu = \frac{4}{\lambda P}\boldsymbol{\Omega}\cdot\mathbf{A} = \frac{4A}{\lambda P}\boldsymbol{\Omega}\cdot\hat{\mathbf{n}}. \tag{2.15}$$

This formulation allows us to define a scale factor $(4A/\lambda P)$ and an orientational dependence for the magnitude of the measured beat frequency, which depends upon the relative alignment of the rotational axis and a unit vector pointing normal to the plane of the ring laser body. An important point to note is that it is the assumption of a constant $\boldsymbol{\Omega}$ that yields the independence of the Sagnac effect from the center about which the laser is rotated. For a comprehensive, if somewhat mathematically terse, review of the Sagnac effect, the reader is referred to Post (1967) [162].

2.3 Cavity Stability

The ring oscillator starts to lase when the gain in the laser medium exceeds the experienced losses in the resonator and the beam path retraces itself after completing one loop around the cavity. From a practical point of view, it is also necessary to select the radius of curvature of the mirrors of a given size for a cavity in a way that beam steering effects from mirror misalignments do not make the adjustment of the cavity unduly difficult. The cavity stability conditions of a symmetric N-mirror cavity have been studied by [28] and revisited later by [49]. The cavity stability criteria have to be satisfied both in the in-plane and the out-of-plane directions. A common approach for this is the application of the ABCD matrix formalism, where a lens L of focal length f, a mirror M of radius of curvature R, and a open distance D of length d are described as

$$L(f) = \begin{pmatrix} 1 & 0 \\ -1/f & 1 \end{pmatrix}, \ M(R) = \begin{pmatrix} 1 & 0 \\ -2/R & 1 \end{pmatrix}, \ D(d) = \begin{pmatrix} 1 & d \\ 0 & 1 \end{pmatrix}. \tag{2.16}$$

For a complete discussion we refer to the description given in [49]. The stability criterion for a complete laser cavity $\mathcal{L}$ with a total ABCD matrix $M(\mathcal{L})$ then requires that the trace $A + D$ of the matrix $M(\mathcal{L})$ has a modulus of less than 2:

$$|A + D| = |\chi M(\mathcal{L})| \leq 2. \tag{2.17}$$

A similar approach for the misalignment, derived in [28], provided the condition that

$$\kappa_N(\gamma) \equiv (-1)^N \det D_N + 2 \tag{2.18}$$

also must have a modulus of less than 2:

$$|\kappa_N(\gamma)| \leq 2. \tag{2.19}$$

For the design of a practical system, it is important to avoid the boundary value of 2 for both cavity stability and misalignment, since the incipient cavity instability becomes serious as the matrix D_N becomes singular. In the early days of the large ring laser project, we often used combinations of two flat and two curved mirrors, both in a configuration where the flat mirrors were either in an adjacent or a diagonal configuration. In all cases we obtained the best performance when the cavity design was as symmetric as possible with respect to the gain section. This was of particular importance for the large rectangular cavities UG-1 and UG-2. Today, essentially all of our current lasers have either a square or an equilateral triangular cavity and employ spherical mirrors of the same radius of curvature (ROC) at all corners. As a general design rule in our systems, the ROC usually corresponds to the length of the arm. For a practical and intuitive cavity design in more complex situations, we suggest the hands-on procedure given in [164].

2.4 Astigmatism and TEM Modes in the Ring Laser

When a spherical mirror is viewed from an angle other than zero incidence, its radius of curvature appears different in plane than out of plane. Hence two parallel beams parallel to the plane of incidence see a different focal length compared to two parallel beams lying in a plane perpendicular to the plane of incidence. These phenomena give rise to astigmatism in ring lasers, which leads to elliptical beams [84]. Geometrical arguments lead to the results

$$R_i = R\cos\theta \quad \text{in plane} \tag{2.20}$$

$$R_o = \frac{R}{\cos\theta} \quad \text{out of plane.} \tag{2.21}$$

In order to find the lasing frequencies of all TEM modes, we require that the phase of a mode satisfies the boundary condition that in a full circle around the cavity, the phase evolves through an integer number of cycles. For a linear laser we find the Hermite–Gaussian functions that define the mode $U(x, y, z)$ by taking the phase of the mode $\phi(z)$ and substituting into the boundary condition

$$\phi(z_2) - \phi(z_1) = n\pi, \tag{2.22}$$

where z_1 and z_2 define the locations of the mirrors. For a linear laser with no astigmatism and mirrors with radius of curvature $R_{1,2}$, this is found to give

$$\nu = \frac{c}{2L}\left(n + (m + l + 1)\frac{\phi(z_2) - \phi(z_1)}{\pi}\right) \tag{2.23}$$

$$= \frac{c}{2L}\left(n + (m + l + 1)\frac{\cos^{-1}\sqrt{g_1 g_2}}{\pi}\right), \tag{2.24}$$

where $g_{1,2} = 1 - L/R_{1,2}$. Now in the astigmatic case, we have to evaluate the Hermite–Gaussian modes for the in-plane and the out-of-plane components with their own respective radii of curvature. The cavity mode is given by $U(x, y, z) = u_m(x)u_l(y)e^{-ikz}$, where the x-axis lies in plane. The Hermite–Gaussian functions in a single dimension are taken from [204, p. 686], so for example we have

$$u_m(x) = \left(\frac{2}{\pi}\right)^{\frac{1}{4}} \frac{e^{-i(m+\frac{1}{2})\phi(z))}}{\sqrt{2^m m! \, \omega(z)}} H_m\left(\frac{\sqrt{2}x}{\omega(z)}\right) e^{\left(-\frac{ikx^2}{2R(z)} - \frac{x^2}{\omega^2(z)}\right)}. \tag{2.25}$$

The following definitions are used in the above expression:

$$\omega(z) = \omega_0 \sqrt{1 + \left(\frac{z}{z_R}\right)^2} \tag{2.26}$$

$$R(z) = z + \frac{z_R^2}{z} \tag{2.27}$$

$$\phi(z) = \tan^{-1}\left(\frac{z}{z_R}\right). \tag{2.28}$$

From this one can find $U(x, y, z)$ and that

$$\phi(z) = \left(m + \frac{1}{2}\right)\phi_i(z) + \left(l + \frac{1}{2}\right)\phi_o(z) + kz. \tag{2.29}$$

along the axis of the laser cavity. We then find the lasing frequency:

$$\nu = \frac{c}{2L}\left(n + \left(m + \frac{1}{2}\right)\frac{\phi_i(z_2) - \phi_i(z_1)}{\pi} + \left(l + \frac{1}{2}\right)\frac{\phi_o(z_2) - \phi_o(z_1)}{\pi}\right). \tag{2.30}$$

Now if the radii of curvature of the two curved mirrors can be taken as equal, then we may use the identity $\phi(z_2) - \phi(z_1) = \cos^{-1} g$ and express the final lasing frequency as

$$\nu = \frac{c}{2L}\left(n + \left(m + \frac{1}{2}\right)\frac{\cos^{-1} g_i}{\pi} + \left(l + \frac{1}{2}\right)\frac{\cos^{-1} g_o}{\pi}\right), \tag{2.31}$$

where g_i and g_o represent the g value for the in-plane and out-of-plane values of R.

2.5 Hole Burning, Saturation, Cross- and Self- Saturation

In this section we look at the laser beam excitation and the contribution of the active laser gain medium to the experienced null-shift of the laser gyroscope. An excellent treatment of the behavior of the HeNe laser gain medium in a ring cavity can be found in the 1971 monograph *The laser gyro* by F. Aronowitz [11], which in turn is based on the semiclassical self-consistent formalism developed by W. Lamb [126]. As opposed to a linear cavity, where the standing wave patterns of the two counter-propagating beams are constrained to be equal, the two laser modes in a ring cavity are independent of each other, which means that they can oscillate on different frequencies and with different amplitudes. The result is a set of coupled differential equations, which describe the time evolution of the amplitudes and frequencies of the two counter-propagating laser modes, assuming single mode operation for each sense, here given in the form as derived in [16]:

$$\dot{I}_1 = \frac{c}{P}\left[\alpha_1 I_1 - \beta I_1^2 - \theta_{12}I_1 I_2 + 2r_2\sqrt{I_2 I_1}\cos(\psi + \epsilon)\right], \tag{2.32}$$

$$\dot{I}_2 = \frac{c}{P}\left[\alpha I_2 - \beta I_2^2 - \theta_{21}I_1 I_2 + 2r_1\sqrt{I_1 I_2}\cos(\psi - \epsilon)\right], \tag{2.33}$$

$$\dot{\psi} = \omega_s + \sigma_2 - \sigma_1 + \tau_{21}I_1 - \tau_{12}I_2$$
$$- \frac{c}{P}\left[r_1\sqrt{\frac{I_1}{I_2}}\sin(\psi - \epsilon) + r_2\sqrt{\frac{I_2}{I_1}}\sin(\psi + \epsilon)\right]. \tag{2.34}$$

The dimensionless intensities are defined as

$$I_i = \frac{|\mu_{ab}|^2}{2\eta^2\gamma_a\gamma_b}E_i^2, \tag{2.35}$$

where μ_{ab} is the electric dipole matrix element between the laser states, with γ_a and γ_b the decay rates of the upper and lower energy states, namely $a = 3s^2$ and $b = 2p^4$. It is given by

$$\mu_{ab} = \sqrt{\pi\epsilon_0\frac{\lambda^3}{(2\pi)^3}\hbar A_{ik}}, \tag{2.36}$$

where A_{ik} is the radiative decay rate between the two energy states of neon. Furthermore, we have the relative value of the homogeneous line broadening $\eta = \gamma_{ab}/\Gamma_D$, which is composed of the radiation decay rate and the collision induced rate to the Doppler broadening Γ_D. For the HeNe laser systems treated in this book, we have $\Gamma_D = \sqrt{2k_B T_p/m_{Ne}}/\lambda$, with k_B the Boltzmann constant, T_p the plasma temperature and m_{Ne} the atomic mass of neon. The term ω_s in Eq. 2.34 represents the difference in the eigenfrequencies of the two beams in the cavity due to the experienced rate of rotation, and ψ corresponds to the instantaneous phase difference. The coefficients α_i, β_i, θ_{ij} and τ_{ij} are the Lamb coefficients, defined in terms of the complex plasma dispersion function as elaborated in [11, 226]. The real part accounts for the dispersion of the active medium, and the imaginary part is proportional to the population inversion

$$Z(\xi) = 2i \int_0^\infty e^{-x^2 - 2\eta x - 2i\xi x} dx, \tag{2.37}$$

where $\xi = (\omega - \omega_0)/\Gamma_D$ describes the detuning from line center, normalized to the Doppler width, for each beam, and η is defined as above. For a HeNe gas laser in the Doppler limit, values of $\eta \ll 1$ are typical, and Eq. 2.37 can then be approximated by

$$Z_I(\xi) \approx \sqrt{\pi} e^{-\xi^2} - 2\eta \tag{2.38}$$

and

$$Z_R(\xi) \approx -2\xi e^{-\xi^2}. \tag{2.39}$$

The Lamb coefficient α in the Eqs. 2.32 and 2.33 describes the relationship of the laser gain (G) and the losses (μ) in the gyroscope:

$$\alpha_i = G \frac{Z_I(\xi_i)}{Z_I(0)} - \mu_i. \tag{2.40}$$

The β term in turn accounts for the gain saturation of each laser beam on itself as a consequence of hole burning, while θ_{ij} accounts for the gain saturation effect of each beam on the respective other beam:

$$\beta_i = G \frac{Z_I(\xi_i)}{Z_I(0)} \tag{2.41}$$

$$\theta_i = G \frac{Z_I(\xi_i)}{Z_I(0)} L(\xi), \qquad L(\xi) = \frac{1}{1 + (\xi/\eta)^2}. \tag{2.42}$$

The parameters σ_i and τ_{ij} in Eq. 2.34 incorporate the effect of the oscillation frequency pulling away from the eigenfrequencies and mode pushing as a result of hole burning in the dispersion curve:

$$\sigma_i = \frac{f_{FSR}}{2} G \frac{Z_R(\xi_i)}{Z_I(0)}, \qquad \tau_{ij} = \frac{f_{FSR}}{2} \frac{\xi}{\eta} G \frac{Z_I(\xi_i)}{Z_I(0)} L(\xi), \tag{2.43}$$

where $f_{FSR} = c/P$ is the free spectral range of the cavity, and P the perimeter circumscribed by the laser beam. The amplitudes r_i and the relative phase of the backscattered light usually present the most dominant bias for our measurements. From the point of view of the actual laser excitation, Eq. 2.34 provides the three main contributions, listed in Table 2.2. The first term presents the desired rotational signal, the second term is the differential frequency pulling, which causes an apparent scale factor reduction. The terms related to τ_{ij} cause mode pushing effects and are associated with hole burning in the laser gain curve. The last term is depends on

Table 2.2. *The main signal contributions to the observed interferometer beat note.*

Contribution	Equation 2.34
Sagnac frequency	$\Delta f = \frac{4A}{\lambda P} \mathbf{n} \cdot \mathbf{\Omega}$
Scale factor correction	$(\sigma_2 - \sigma_1)$
	$\tau_{21} I_1 - \tau_{12} I_2$
Backscatter	$\frac{c}{P} \left[r_1 \sqrt{\frac{I_1}{I_2}} \sin(\psi - \epsilon) + r_2 \sqrt{\frac{I_2}{I_1}} \sin(\psi + \epsilon) \right]$

coupling effects between the two laser beams, caused by mutual scatter losses at the mirrors.

All practical HeNe laser gyroscopes operate on an isotopic mixture of neon, which consists of 0.1 hPa ^{20}Ne and 0.1 hPa ^{22}Ne. Since the two isotopes cause a shift of the laser transition of 885 MHz, this avoids mode competition and means that the two laser beams are independent of each other. Both laser beams pass along the exact same optical path, so one would expect the intensities of the two laser beams to be identical. However, this condition of equal beam powers has never occurred in any of our many ring laser gyros. We always observe that the intensities are different by a small amount ($0.2\% < \delta I < 5\%$). This slight asymmetry could have several causes, which are still not well understood. Birefringence of the mirror coatings in the presence of a small amount of Faraday rotation is the most obvious candidate, which would introduce the necessary non-reciprocal properties. The plasma for the laser excitation is another possible candidate, and a polarization rotation effect was indeed reported in the early literature [11]. The IBS mirrors in all of our instruments are a factor of 10 more lossy for p-polarization than for s-polarization. However, since the plasma region of the G ring laser is thoroughly shielded by Mu-metal (permalloy), and the Earth's magnetic field is the only source for magnetism near the gyro, a Faraday rotation together with the polarization selecting mirrors, acting in a non-reciprocal way on both laser beams, is highly unlikely but cannot be entirely ruled out until the real cause is identified. The null-shift caused by non-reciprocal loss is contained in the cross saturation terms and can be commonly addressed as $\tau \Delta I$. In principle it should be possible to derive a correction value from the exploitation of the intensity measurements of each beam, according to

$$\Delta f_0 = \frac{f_{\text{FSR}}}{2} G \frac{\xi}{\eta} \frac{Z_I(\xi)}{Z_I(0)} L(\xi) \Delta I. \qquad (2.44)$$

However, there are a few purely experimental obstacles in the way, which make the estimation of the unperturbed laser intensity for each beam a delicate task. Differences in the contamination from residual background light on each of the detectors, subtle differences in the transmission of the applied spectral filters, and non-identical amplifier gain in the detectors make the correct estimation of the DC component of the intensity difficult. A common experience for the estimated null-shift from the G ring laser is that this term mostly causes an offset with compara-tively little variability. For example, we have observed an offset of about -300 µHz (≈ 1 part in 10^6) and a slow trend of ± 3 µHz.

The null-shift bias represents an odd function, where a value of zero is obtained only at the line center. The backscatter amplitudes in G are usually unequal by a factor of two and vary slowly over time, mostly driven by a temperature-induced change in the separation distance of the individual mirrors. This drift can also be observed under tight active control over the integral length of the perimeter, albeit at a lower rate. The values slowly change between 2 and 10 mV relative to the DC part of the intensity of about 1 V for both beams, and it is a typical situation that r_1 increases when r_2 decreases and vice versa, although not at the same rate. The detrimental effect of backscatter coupling, however, reduces as the cavity size increases, due to the then smaller dihedral acceptance angle for the scattered light. In the G ring laser, the uncorrected backscatter induced drift may amount to 90 µHz per day, even if the entire perimeter is actively controlled to be stable to ± 60 kHz. For UG-2, the magnitude of the frequency pulling was estimated to 40$\pm$22µHz [96].

2.6 Polarization

A high quality factor Q is the most important property of the ring laser cavity. Therefore it is of paramount importance to minimize the number of optical surfaces inside the laser resonator. The best performing laser gyros do not have Brewster windows, and neither are there any mode selection devices, such as etalons. The only surfaces of interaction are the thin film coatings on the mirror substrates. Proper operation, however, is only obtained when the polarization of the laser beams in the cavity remains stable. All of our gyros exhibit beams in (sagittal) s-polarization. This is achieved by the properties of the mirror coatings. There is a difference of at least a factor of 10 in transmission loss of the dielectric thin films. Typical specification values for the s-polarization are of the order of 8 ppm total loss at a 632.8 nm wavelength. This includes scatter transmission and absorption, where the transmission loss is typically between 0.2 and 1 ppm. In this mirror design, the (parallel) p-polarization experiences losses that are an order of magnitude larger, and this is enough to discourage laser oscillation in any other polarization than s reliably.

All this assumes for a square cavity that the alignment is entirely planar and that the mirrors are non-gyrotropic. In a non-planar (square) ring, the polarization becomes circular as the polarization is rotated out of the s direction in response to the fold angle of the sensor geometry [30]. Imperfections in the mirror coatings are the main cause of the experienced anisotropy. A variable non-ideal thickness of the layers in the complete coating stack is enough to cause this effect. For very small angles of misalignment, this results in a situation where birefringence of the mirror coatings remains the dominant cause for a small ellipticity in polarization and hence increased losses. From the calculations in [30] for the misalignment alone, one would expect a high sensitivity of the gyro polarization with respect to even a small cavity non-planarity. However, this is not observed. A deliberately introduced tilt of one corner by 3 mrad did not produce any detectable change in the performance of the gyro. Neither indicator, namely ringdown time and observed exit polarization, changed within the resolution of the measurements. In agreement with [30], we conclude that mirror birefringence is the dominant effect in our real world gyros. This conclusion is further supported by the observation that a ring laser gyro can be biased noticeably when one of the cavity mirrors is subjected to a strong magnetic field. We will revisit this aspect in Section 4.11.

3

Large Scale Helium–Neon Gyroscopes

3.1 General Remarks

The application of sensing the small variable part of the Earth's rotation Ω_E requires sensors that operate very stably in the regime of low angular velocities. The Earth rotates at 15°/h, which corresponds to 72.7 µrad/s. Variations of the Earth's rate ($\delta\Omega_E$) occur at a level below 1 prad/s or, expressed differently, $\delta\Omega_E/\Omega_E \leq 10^{-8}$. With the advent of lasers, it was possible to increase the resolution of Sagnac interferometers substantially. In particular the transition from a phase to a frequency measurement provided for a vast improvement in sensor sensitivity. Placing a laser gain medium inside the closed light path established by four mirrors converts the apparatus into a traveling wave resonator with a square cavity [142]. Along with the lasing condition, which requires the perimeter to contain an integer number of wavelengths $P = i \cdot \lambda$, one obtains the ring laser equation

$$\delta f = \frac{4A}{\lambda P}\mathbf{n} \cdot \mathbf{\Omega}, \tag{3.1}$$

with δf the beat frequency, λ the wavelength of the optical beam, $\mathbf{n}$ the normal vector to the plane of the laser and $\mathbf{\Omega}$ the rate at which the entire apparatus rotates with respect to the universe. This beat frequency is usually referenced to as the *Sagnac frequency*. In the classical approach one can depict the effective co-rotating optical resonator as slightly larger compared to the anti-rotating resonator. Therefore, both optical beams are shifted by the same small amount but in different directions away from the optical frequency that both beams would have if the apparatus were at rest. This path length difference, caused by the Earth's rotation, for the co-rotating and anti-rotating 'cavities' turns out to be exeedingly small. As an example, for a 16 m² large ring cavity at the Geodetic Observatory Wettzell (latitude: 49.16° north), one obtains with $\Delta p = \pm\frac{2\omega A}{c}$ a value of 5.85 × 10⁻¹² m [191]. It is interesting to note that Macek and Davies in their 1963 paper [142] speculate on the possibility of measuring the "*Earth orbital rate around the sun*," while the detection of the

Earth's rotation itself, although not yet achieved at the time when their successful ring laser operation was communicated, appeared to be already a given at this early stage of ring laser development. Retrospectively the detection of the orbital motion of the Earth seems very natural and straightforward as geophysical signals like polar motion have a sidereal periodicity [185], while effects from atmospheric tides are related to the solar day [184]. All these geophysical signals present natural markers that leave their traces in the observations of the ring laser.

3.2 Hot Cavity versus Cold Cavity

The rotation sensing ability of an optical Sagnac Interferometer is entirely determined by the traveling wave oscillator cavity. The laser beams, either generated in this cavity or injected from the outside and subsequently locked to a suitable cavity resonance, are the only elements required to evaluate the rotational motion experienced by this (*rigid*) cavity. The governing parameters defining the ultimate sensitivity for rotational motion are the *area* enclosed by the laser beams and the optical *losses* of the cavity. The fundamental limit of the sensitivity of a laser gyro $\delta\Omega$ relates inversely to the enclosed area and the cavity quality factor Q, which is defined by the cavity losses. A fuller treatment of sensor resolution is given in Section 3.6.5.

Another consequence of these very general sensor properties is the principal equivalence of the active and passive laser gyro concept, since the measurement method only determines how the given cavity properties are interrogated. The effective technical realization finally sets the ultimately achievable resolution limit. These gyro properties are very general and also apply to fiber optic gyros. Therefore it becomes a matter of choice, which of the available technologies are eventually used to make these cavity properties available to the outside world. Furthermore, the sensor performance finally achieved depends critically on the applied measurement technology and in particular on the right balance of desirable and less favorable properties of the chosen concept of interferometric cavity resonance interrogation. An example explains this more clearly. Active gyros require a laser gain medium to be present in the cavity, which modifies the cavity properties in some way. As a consequence the refractive index changes, and the gas, present all around the cavity, may introduce Fresnel drag, Langmuir flow or uncontrollable turbulences. Generally, this is not desirable. Passive gyros, on the other hand, work with an empty, evacuated, cavity. However, the externally injected spectrally much wider laser source has to excite the same cavity mode of oscillation for both senses of propagation, which in turn is very hard to achieve at high spectral resolution. While this is only one single aspect out of a whole chain of critical technological difficulties that come with each sensor concept, it ultimately means that the best working solution for long term stable high resolution rotation sensing will critically depend on how well the unavoidable downsides of each of the respective technologies can be contained.

Active and passive ring cavities use the same type of mirrors, so losses from absorption, scatter and transmission are identical in both concepts. It also means that systematic errors from backscatter coupling are the same for both concepts. We will discuss this in more detail in Section 3.9.4. Mechanical cavities, circumscribing areas of $16\,\text{m}^2$ or more, designed to be long term stable for the detection of variations of the effective changes of cavity length to ≈ 60 zm (zepto-m) and below, have to be set up as optical interferometers with an extremely high degree of reciprocity. This holds true for gravitational wave antennas [1] as well as for large Sagnac interferometers, regardless of whether they are active or passive by concept.

Every physical realization of a rotation sensor has to fulfill the same requirement and this again illustrates that with all the problems of mounting such devices onto the crust of the Earth being equal, they may be achievable by one sensor concept but not by another. So where are the differences, which one day may decide which of the concepts will perpetuate? Passive gyros sample an entirely empty cavity, while practical working high resolution ring lasers have to deal with a low pressure gas mixture and the non linear dynamics of laser radiation excitation. HeNe ring laser gyroscopes, on the other hand, exhibit a laser linewidth well below 1 mHz in cavities that have Q factors in excess of $Q > 10^{12}$, while the cold cavity resonance of the identical cavity is of the order of $\Delta \nu_c \approx 265$ Hz, from which the center has to be determined (see Eq. 3.10 and discussion in Section 3.6.5). In order to reach the required resolution, we have to demand absolute reciprocity of the beam path, a condition that is fairly natural for an active laser and has yet to be demonstrated for a passive gyro. The locking scheme of the external laser source has to maintain the center of the cavity resonance accurately to a very small fraction of the actual laser linewidth. Apart from early work in the late seventies of the last century [71, 175], there are two more recent projects reported in the literature, one by the LIGO group in Pasadena [121] and another by the Center of Gravitational Experiments in Wuhan [137]. Here we will briefly review the available passive concepts before we concentrate on active ring laser gyros.

3.3 Passive Gyro Concepts

In a very simplified approach, one can set up a passive gyroscope by taking a coherent cw laser source, splitting the beam into two parts and shifting the optical frequency with an electro optical modulator (EOM) of each beam until it hits a cavity resonance. The beams are then injected in the cavity, both in the clockwise and counter-clockwise directions. The sketch in Figure 3.1 illustrates this concept.

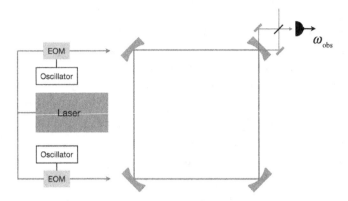

Figure 3.1 The principal layout of a passive ring gyroscope. For simplicity, the mode matching and locking hardware is not shown.

In order to work as a gyroscope, the injected beams have to be matched to a cavity resonance, belonging to either the same or to a neighboring longitudinal cavity mode. In terms of backscatter coupling, it makes no difference if the excited cavity modes have the same longitudinal mode index or not (for details see Section 3.9.4). In practice all recent efforts have chosen to operate on separate laser modes in order to reduce cross-coupling. As a result the crucial common mode bias rejection is lost, while some convenience for the operations is gained. Length variations of the cavity are then no longer compensated. We have explored this mode of operation also with active ring lasers in Section 3.8.2. Due to the currently much higher resolution of the active gyro concept, a better understanding of the limitations of the operation on a separate mode index has been gained. Here we illustrate the effort at the Huazhong University of Science and Technology [137]. The external laser is operated on $\lambda = 1064$ nm in the near-infrared for technical convenience, and the Pound–Drever–Hall method [33] has been applied to lock the mode-matched beams to the respective cavity modes. The detrimental effect of residual amplitude modulation on the injected laser beams has been reduced as much as possible. At one of the corners, the light leakage from both laser beams is recombined to extract the Sagnac beat note biased by 1 FSR due to the choice of the different injection frequencies. For the latitude α of Wuhan at $\alpha \approx 30.5°$ north, the expected beat note is of the order of 35 Hz. Figure 3.2 shows an early result of the obtained time series of the Earth's rotation signal of nearly 30 minutes' duration. Due to some small temperature variations during the period of the measurement, the data has been detrended by fitting a low degree polynomial to the data. At a later stage of the experiment, the cavity has been stabilized to an ultra-stable reference laser, where the residual instability reduces to values of the order of 8×10^{-16}, corresponding

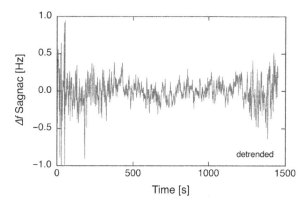

Figure 3.2 An early measurement series from a 1 m^2 passive gyro prototype. The data has been numerically detrended to compensate for temperature effects on the cavity.

to a remaining drift rate of the optical frequency of about 3 kHz/day. Eventually a measurement resolution of 2×10^{-9} rad/s after 1000 seconds of integration time was demonstrated. Although this is a very encouraging start, it does not yet allow speculation about the ultimately achievable long term stability and accuracy of the measured rotation rate. The basic design of the passive gyro development effort of the LIGO group [121] is similar to that in Wuhan and reports an achieved rotation sensitivity of 10^{-7} rad/$\sqrt{\text{Hz}}$ below 1 Hz. Values for the long term stability are not reported. However, the fact that this passive gyro does not operate on the same longitudinal mode index provides a residual length noise contribution that limits the concept, with a common-mode rejection of $v_{\text{FSR}}/v_{\text{opt}} \approx 3.5 \times 10^{-7}$. The same mechanism does apply to an active laser gyro when operated on different longitudinal mode indices and is addressed in Section 3.6.4. It is also noteworthy that the two experiments, introduced here briefly, are both operating in the infrared on a potentially much smaller scale factor than HeNe lasers of a similar size in the visible.

3.4 Prototypes of Early High Sensitive Ring Laser Gyroscopes

Sagnac interferometers can be viewed in different ways. When they are set up such that the two cavities for the counter-propagating beams are as reciprocal as possible, they can be used as highly sensitive sensors for the detection of physical rotation. For Earth observation, in particular when the scale factor is made very large, the detection of polar motion, the Chandler and the Annual wobble, up to the variations of length of day (LoD) at a rate of $\delta\Omega_{\text{LoD}} \approx 7.3 \times 10^{-14}$ rad/s are feasible. Increasing the scale factor even further by making the sensor much larger promises even higher resolution. Over the years this has led to proposals of tests of predictions in fundamental physics, well beyond the regime of the variations of LoD. The first candidates were the precession a gyroscope experiences under the influence of the Lense–Thirring frame dragging effect or the spacetime curvature (geodetic precession) in general relativity [34, 52, 179, 208]. This was motivated by the fact that a vast upscaling of the ring laser gyroscopes became viable, providing a massive boost in sensitivity; however, this was at the cost of mechanical stability, which may in the end limit the resolution of the laser gyroscope. For this application of accurate rotation sensing of geophysical phenomena, the HeNe excited ring laser is currently the system of choice. The size of the plasma is small and well controllable, and the system exhibits only a very small number of non-reciprocal effects, which bias the measurement (see Section 3.9.7). On the other hand, there is an ideal frequency shift of about 885 MHz for the most prominent two neon isotopes, namely ^{20}Ne and ^{22}Ne, to support unperturbed bidirectional lasing, and since the required laser gain is small, the amount of turbulence experienced in the excited plasma is expected to be negligible. However, this said, it has not yet been proven that this expectation is also valid for very small low frequency rotation signals with frequencies well below 1 mHz.

The Argon Ion Gyroscope

Apart from actual rotation sensing, Sagnac interferometers can be exploited for the precise measurement of non-reciprocal parts of the refractive index of the medium in the optical path inside the interferometer. These are effects that can be induced by magneto-optical or non-linear optical interaction. The successful observation of magneto-chiral birefringence [225] in a ring laser configuration is a good example of this. Given that the inherent non-reciprocity of the interferometer is well under control, a considerable sensitivity – in excess of seven orders of magnitude – can be expected in principle. However, in most cases the deliberate introduction of a non-reciprocal effect into a ring laser cavity will go together with a corresponding reduction of the cavity Q-factor, which has the adverse effect of a reduction in sensor resolution. The application of argon for the laser gain medium made the detection of the magneto-chiral birefringence possible, while the turbulence inside the laser gain medium inhibited sensitive rotation sensing. In more recent times a considerable amount of work has been performed using femtosecond mode-locked lasers [10, 124] to perform intra-cavity phase interferometry. There is one experiment that reports the application of an argon laser for a rotation sensing interferometer [93]. In the beginning of this chapter, we identified **area** and **losses** as the most important drivers of a ring laser design. At first glance this appears to be counter-intuitive, since the equation for the ring laser sensitivity seems to suggest that the shot noise causes the most significant limitation. An argon laser can easily provide

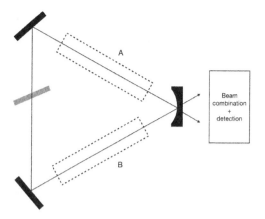

Figure 3.3 Sketch of the argon laser gyro layout. The design was modeled after commercial navigation gyros and was operated in several different configurations, of which one gain section (A) and two symmetrical gain sections (A+B) were the most successful.

100 W of circulating beam power, so the promise was an extremely low shot noise boundary. However, the consequences of this decision were much higher losses due to the requirement of an additional mode selector. Furthermore, there were significant detrimental side effects in terms of a large number of spurious beat frequency components, caused by the plasma turbulence and the Fresnel–Fizeau effect. The construction of the large gyroscope started soon after the mid-1980s. The group in Munich [93] tested a number of different ring cavity configurations, which employed either one or two large gain tubes. The area circumscribed by the beams of a triangular cavity was of the order of 1.4 m^2 and required either an etalon or a prism for mode selection as an additional optical component. A sketch of the best performing setup is shown in Figure 3.3. This led to a significantly increased cavity loss, and the application of the very turbulent argon gain medium gave rise to non-reciprocal frequency fluctuations well in excess of 10 kHz. All the different cavity configurations did not unlock on the rotation rate of the Earth, and the obtained linewidth of the laser was between 400 Hz and 3 kHz, while the expected Sagnac frequency was calculated to be around 100 Hz. Since the experiment was placed on a turntable, the lock-in frequency could be established and varied between 2 and 7 kHz. The high beam power came at the expense of a considerably enlarged line width and a very turbulent gain medium. The large circulating beam power also required additional very lossy filter mechanisms to ensure the required operation at a single longitudinal laser mode. From that point of view, this concept failed in about every aspect that we have identified as favorable in Section 2.1. By analogy with navigational HeNe gyros, it was attempted to balance the gain medium between

two arms of the triangular cavity such that the drag effects would compensate, which turned out to be impossible. These drawbacks outweighed the perceived advantage by far and led to the discontinuation of this approach for high resolution rotation sensing, once the project was completed. At that time, super-mirrors were not even considered for the argon ring.

The Experimental Laser Gyroscope System ELSy

By the mid-1980s, small HeNe gyroscopes had matured sufficiently to make the transition from a laboratory study object to a commercial product. They are still used in large numbers for aircraft and missile attitude control and ship, aircraft and space probe navigation [166]. Compared to applications in geodesy, these devices require a moderate resolution, a high dynamic range and, with the exception of submarine navigation, a moderate measurement stability. Taken from [166], Table 3.1 lists the essential requirements. The *Experimental Laser Gyroscope System* (ELSy) was one concept to go beyond the existing limitations and strive for higher sensitivity and stability. In terms of size, it compared with the commercial products, owing to its background in navigation. However, size was not under consideration for this project. Significant improvements were expected through better geometrical control of the beam path within the cavity in terms of stabilizing the scale-factor, as well as keeping the backscatter contribution to a constant actively controlled minimum. Furthermore, ELSy allowed sensitive control over the positioning of the mode selecting aperture and the DC excited discharge channel, in order to reduce non-reciprocal effects on the counter-propagating laser beams by as much as possible. The laser cavity of ELSy was embedded in a stainless steel recipient, with the helium–neon gas mixture not confined to a separate cell with Brewster windows. These features made the instrument very accessible and allowed many different test configurations in order to minimize the cavity losses and systematic errors. It is interesting to note that the ELSy design was still caught in the prevailing paradigm of the time, that the cavity length of the gyroscope cannot be larger than ≈ 60 cm, in order to ensure that no more than one longitudinal mode is excited by the DC discharge. In the end,

Table 3.1. *Requirements for rotation sensors in navigation applications.*

Parameter	Requirement
Dynamic range	$10^{-6} - 10^{3\circ}$ s^{-1}
Angular resolution	10 µrad
Drift stability	$10^{-3\circ}$ h^{-1}
Scale factor stability	3 ppm

ELSy demonstrated better performance than commercial systems, but the obtained improvement remained well within one order of magnitude.

The Arkansas Ring

An entirely different concept for a sensitive large scale ring laser experiment emerged from the Hendrix College in Arkansas (USA). It is remarkable in several ways. It was built from generally available commercial items, such as a DC-discharge gain tube with Brewster windows on each end. The cavity was formed by two flat mirrors and one curved mirror. The majority of the beam path led through open air, and the structure was triangular [63]. Figure 3.4 illustrates this setup in a simple sketch. There are two different realizations of this concept reported in the literature. The earlier structure had a perimeter of 2.56 m length. The length of the discharge tube was 35 cm, and it was filled with ^{4}He and ^{20}Ne. The presence of a single neon isotope caused a significant difference in beam intensity of about 30% between the two counter-propagating laser beams. The ring was mounted on a turntable and did not unlock on Earth rotation alone, which was estimated to cause a beat note of less than 100 Hz. The frequency difference of the two laser beams had to be larger than 800 Hz to unlock. The later structure was significantly larger, with 13 m length for each side [64]. It was rigidly tied to the ground and showed the Earth's rotation without the requirement of any additional rate bias. Although suffering from a lack of structural stability, this ring laser structure could resolve Earth rotation at the level of $\Delta\omega/\Omega_E \approx 0.1\%$ over an observation time of three hours. A ringdown time for this instrument is not reported. Values of the order of 20 μs can be realistically expected, however, due to the excessive losses incurred from the two Brewster windows and the standard mirrors used, having dielectric coatings with a typical reflectivity of the order of 99.7%.

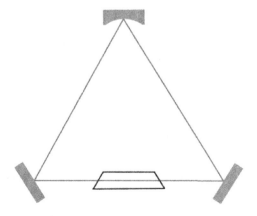

Figure 3.4 The principal layout of the ring laser setup employed by the Arkansas group.

The ring laser activities in Arkansas have contributed two important findings to the development of large ring laser gyroscopes. Overfilling the gain tube of the gyroscope with a total gas pressure of 3.5 mB $\leq$ p $\leq$ 10 mB considerably reduced the detrimental effects from drift-induced laser mode jumps by increasing the homogeneous line broadening. Although the free spectral range in our rings progressively became smaller than 3 MHz, mode jumps occurred only after a drift of the optical frequency of at least 100 MHz in the cavity. Secondly, it found that a stable operation regime of several co-existing phase-locked laser modes allows sensitive gyroscope operation on higher beam power levels, thus decreasing the boundary of the shot noise limit.

The Canterbury Ring Laser Project

The advantages of upscaling the laser cavity to meter-size structures and simultaneously minimizing the cavity losses were pioneered in the Canterbury ring laser project, an effort that started at around the same time as the Arkansas rings and the ELSy system [207]. Although an entirely independent activity, it used the same loss minimizing strategies as ELSy; however on top of that, it also employed large cavities and better super-mirrors, which started to become commercially available at that time. It differed from the argon ring concept in that it used HeNe excitation and gain starvation (operation close to the laser threshold) in order to achieve single mode bidirectional lasing. In contrast to the rings in Arkansas, it used an open gain capillary. Instead of two widely spaced electrodes with a high DC voltage between them, radio frequency (RF) excitation was employed. This technique provided the additional advantage of a significant reduction of gas flow within the gain medium. It practically eliminated systematic errors from Fresnel drag. With RF excitation between two closely spaced antenna studs near the laser gas capillary, the size of the excited plasma could be reduced to a point where the gain dropped below the lasing threshold without quenching the plasma glow. The enclosed area was optimized by using a square cavity geometry, and losses were minimized by allowing no other optical elements in the cavity than the four super-mirrors.

In order to select the desired TEM_{00} laser mode for the operation, the cross-section of the laser excitation capillary was chosen such that it formed an aperture that preferred the desired laser mode by presenting a small loss gradient in radial direction in favor of the TEM_{00} laser mode, which has the smallest mode volume compared to higher order modes. The laser operates around the center of the 633 nm transition on a 1:1 mix of the ^{20}Ne and ^{22}Ne isotopes. With a free spectral range of 86.2 MHz, additional longitudinal laser modes are starved because they experience lower gain than the dominant mode near the center of the transition. The 0.748 m^2 Canterbury-I ring laser (C-I) situated in Christchurch [207] was the first moderately sized planar HeNe ring laser to unlock on Earth rotation alone, exhibiting a Sagnac frequency of 72 Hz. At the height of its performance, it reproduced the Earth's rotation rate within an error margin of 5%. Through the successful demonstration of reasonably stable continuous unlocked operation in a moderately upscaled geometry, together with the reduction of loss generating components inside the laser cavity, the door into the regime of high resolution Earth rotation sensing devices was opened.

3.5 Large Scale Helium–Neon Ring Laser Design

Ring lasers are highly sensitive optical interferometers, which require extreme mechanical stability of the instrumental hardware. It is of paramount importance to obtain a linear relationship between the experienced rate of rotation and the observed unbiased beat note of the interferometer. Following early interpretations [197], one can consider the standing wave pattern of the two counter-propagating traveling monochromatic laser waves as fixed in inertial space. Rotation is then sensed by measuring the rate of change of the nodes and anti-nodes of this space-fixed optical interferogram as the entire apparatus is passing along this vernier of light. The fact that light is applied for this purpose provides the high sensitivity for rotation sensing as well as a very uniform scale. However, in order to extract the true rate of rotation from the gyroscope, we have to guard the rotation sensing device against systematic measurement errors. Three types of errors are typical for laser gyroscopes [11]. In Figure 3.5 the first two error mechanisms are illustrated. A

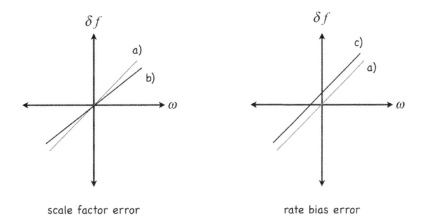

Figure 3.5 Illustration of the main error mechanisms. A scale factor error is present when the experienced rotation rate translates into smaller or larger values for the beat note (b) instead of the true value (a). Rate bias errors (c) occur in the presence of a non-reciprocity, when the beat note does not disappear in the case of zero rotation.

scale factor error can generate a subtle but serious problem. While the relationship between the experienced rotation rate and the obtained beat note is still linear, the input rate (ω) translates into a beat note either below or above the correct value. The most common cause of this error is the angular misalignment (orientation) of the ring laser plane. In order to make this error negligible for the observation of variations of the Earth's rotation rate, the projection of the normal vector of the laser plane must be accurately known to much better than 10 nrad, which is impractical for a single component ring laser gyroscope. The second error type is a rate bias or null-shift error, when the measured rotation rate is offset from the true value. This means that the beat note does not disappear even in the absence of a rotational excitation. This is a very common problem because it is extremely difficult to make the sensor completely reciprocal, and that limits the accuracy that one obtains from the sensor, even when the precision is excellent. Figure 3.6 illustrates the third major error source in optical gyroscopes, namely backscatter coupling. Two weakly coupled oscillators of almost the same frequency tend to synchronize. This is a very common phenomenon, and it occurs for optical frequencies in laser cavities in the same way as in mechanical systems. The beat note disappears for very small

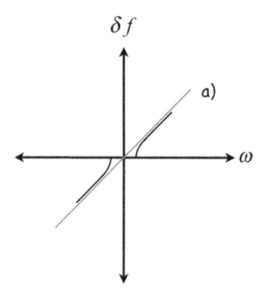

backscatter coupling error

Figure 3.6 Backscatter coupling causes the two counter-propagating laser frequencies to become degenerate. The beat note of the gyro disappears, although the sensor is still turning at a very low rate.

rotation rates and becomes highly non-linear for frequencies slightly larger than the lock-in frequency. The presence of backscatter coupling still causes a bias for frequencies much larger than the lock-in frequency. The problem of backscatter coupling is exacerbated by the fact that the lock-in threshold for a given gyro is not constant. Small variations in the cavity path-length, driven by thermal expansion of the sensor body, change the phase of the coupling and hence the threshold value. In practice we have to constantly evaluate the amount of backscatter and derive a correction for the backscatter contribution continuously. We include these error sources by adding another term to the ring laser equation:

$$\delta f = \frac{4A}{\lambda P} \mathbf{n} \cdot \mathbf{\Omega} + f_{nr}. \tag{3.2}$$

The additional contribution to the Sagnac beat note is caused by a non-reciprocal frequency shift between the two counter-propagating beams, which may change significantly over time.

3.5.1 General Techniques Common to All Lasers

The large HeNe ring lasers covered in this section are designed to maximize the area while simultaneously minimizing losses in the cavity. In order to optimize the sensor design, we can define a worth function $\kappa = A/PN$ and look for the largest value of κ [207]. A is the area of the ring, P the perimeter and N the number of mirrors. The optimum solution is a planar square cavity realized by four super-mirrors (loss < 50 ppm per mirror). Minimizing the losses requires the complete absence of any intra-cavity elements, such as Brewster windows or etalons. As a consequence, the HeNe laser gas is dispersed throughout the entire cavity. The maximum gain of neon is obtained from a partial pressure of $p_{Ne} = 0.2$ hPa and helium with a partial pressure of around $p_{He} = 1.8$ hPa. In order to decouple the two beams in the interferometer as much as possible, we use two isotopes of neon, namely [20]Neon (0.1 hPa) and [22]Neon (0.1 hPa). The isotopic shift of the center of the gain curve is 885 MHz, and this reduces mode competition significantly. On a single isotope ([20]Ne), for comparison, we have observed a difference in brightness of 10:1. In order to make use of the suppression of neighboring longitudinal laser modes by increasing the homogeneous line broadening from atomic collisions, we use total gas pressures in excess of 6 hPa, and the capillary for the actual laser excitation varies between 4 mm in diameter for the smaller rings ($\approx 0.8-2.5$ m^2), 5 mm for middle sized rings ($\approx 12.5-16$ m^2) and up to 6 mm for the really large rings ($\approx 365-834$ m^2). Single longitudinal mode operation is established via the mode starvation technique, in which instruments are operated near the lasing threshold. In order to avoid a systematic bias from Fresnel drag and Langmuir flow effects, all of the ring lasers are excited by radio frequencies in the range of 20–200 MHz. In order to generate stationary cw laser beams, the RF excitation is stabilized with a closed loop locking scheme. The light leakages of the two beams at one mirror are interfered with each other on a non-polarizing beam splitter with the help of two steering mirrors. It is important to minimize the open beam path in order to avoid phase front distortions from turbulence in air. A free space distance of less than 1 cm before the beam splitter is highly recommended. To this end, both the C-II and the G ring laser employ a Kösters prism, which reduces the free space path to about 2 mm.

3.5.2 Monolithic Ring Laser Structures

Ring laser structures for navigational purposes have been traditionally manufactured from solid Zerodur glass ceramic blocks. Triangular constructions have the advantage of a well defined plane of orientation. They provide sufficient stability and angular resolution for attitude control and navigation in a strap-down application. On top of that, they are entirely free of any moving mechanics. The almost zero thermal expansion properties of Zerodur, along with the high mechanical stiffness of these compact structures (effective sensor area: ≈ 0.0023 m^2), is responsible for the great commercial success of these devices. Large ring lasers for applications in geodesy and geophysics generally follow the same basic concept. In order to provide the required six orders of magnitude higher sensitivity, these instruments are vastly larger in size, when compared with their counterparts from vehicle navigation. On top of this, the sensor drift must not exceed 10^{-14} rad/week if reliable estimates of the length of day variations are to be achieved.

To date, two large monolithic ring lasers have been built. These are the 1 m^2 C-II ring laser, located in the Cashmere Cavern in Christchurch (New Zealand), and the 16 m^2 G ring laser (G stands for Grossring) based in an underground laboratory on the Geodetic Observatory Wettzell near the Bavarian–Czech border in southern Germany. Figure 3.7 (left) shows the construction of the C-II prototype [27, 181] housed inside an ambient pressure stabilizing vessel. The body of the gyroscope is made from a square slab of Zerodur, 18 cm thick and 1.20 m on a side. All four corners are bevelled and polished so that discs of ULE (ultra low expansion glass) with optically contacted super-mirrors having total losses of 5–40 ppm (scatter, absorption and transmission loss) can be mounted onto the ring laser body (Figure 3.7 (right)) in order to generate a pre-aligned closed light path. The beam path itself is drilled into the neutral plane of the Zerodur slab, parallel to the sides, so that an area of 1 m^2 is circumscribed by the laser beams. On one side, half way between the mirrors, there is a cut-out in the ring body. A small adjustable capillary with a diameter of 4 mm is placed in this gap. Two electrical loops around it act as an RF-antenna. They are

Figure 3.7 The C-II ring laser under the ambient pressure stabilizing vessel in the Cashmere Cavern, New Zealand (left) and with one corner open during the optical contacting of the super mirrors (right).

used to spark a gas discharge for laser excitation. Another similar cut-out on the opposite site, closed up with a 12 mm wide pyrex tube, was also integrated into the design for future experimental purposes.

There are also two diagonal holes drilled through the entire ring laser structure. A UHV-valve, located in the center above the Zerodur block, seals the cavity off from the environment. This valve can be connected to a turbo molecular pump, in order to evacuate the entire ring cavity to a residual pressure level of about 10^{-6} hPa. Subsequently to pumping, a mixture of ^{4}Helium, ^{20}Neon (0.1 mB) and ^{22}Neon (0.1 mB) can be applied to fill the cavity to 2–10 mbar of total gas pressure. This design has provided a cavity Q of the order of 10^{12} and higher for both the C-II and G-ring laser. The cavity Q, defined as $Q = \omega\tau$ with $\omega = 2\pi f$ and f the optical frequency of the laser, was established from cavity photon decay time (τ) measurements. High values for Q result in a narrow laser linewidth and, equally importantly, a much reduced systematic offset of the Sagnac frequency from its true value from the lock-in effect [3, 11, 236].

Since for all high performing active Sagnac interferometers, the simplest form of operation is on a single longitudinal mode per sense of propagation, they are operated near the laser threshold, employing mode starvation from differential gain between neighboring modes as the selection criterion. The range for a single mode to lase is larger than the free spectral range (FSR), due to homogeneous line broadening from increased gas pressure. The free spectral range (FSR = $\frac{c}{P}$) between 18.75 MHz (G) and 75 MHz (C-II) would otherwise allow many different longitudinal modes to oscillate at higher gain settings. The cross-section of the capillary used for laser excitation is matched to the diameter of the laser beams, such that extra-cavity losses are avoided, in order to keep the Q high but small enough that the TEM$_{00}$ mode with the lowest mode volume is preferred. In the G-ring, the capillary

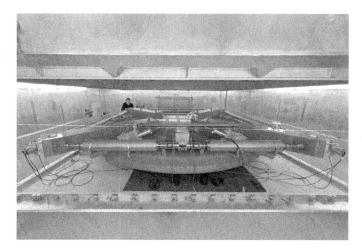

Figure 3.8 The G ring laser during a maintenance period in 2012. (© A. Heddergott of TU-München)

is 5 mm wide, owing to the longer beam path. Figure 3.8 shows the G-ring laser during a maintenance period in 2012.

The G-ring differs from C-II in a few subtle aspects. Since it was not possible to obtain Zerodur in a sufficiently large slab, the G-ring was constructed in a semi-monolithic way on a rigid circular baseplate with a diameter of 4250 mm and a thickness of 250 mm. There are four bars of 2000 mm length, a width of 400 mm and a thickness of 250 mm placed on top in an X-shape configuration spanning an area of 4 m by 4 m, the effective area circumscribed by the laser beams. The end faces of all four bars are drilled and polished such that the four super-mirrors can be optically contacted in the same way as for the C-II ring laser. Both C-II and the G-ring are placed on rigid concrete foundations, which are embedded in the local bedrock. Steel rods firmly tie the respective platforms down to the ground. The top of each platform is made from a highly polished granite slab. A number of small pads of teflon, 0.1 mm thick, were placed between the granite slab and the Zerodur ring laser body. This reduces strain effects to the ring laser body from the different thermal expansion between the granite and the Zerodur material. The G-ring is the most stable ring laser and has the highest *usable* sensitivity of the lasers constructed in our group.

3.5.3 Heterolithic Ring Laser Structures

In order to obtain a stable interferogram from the two laser beams, the cavity length has to be kept constant to within a wavelength. A variable perimeter causes the optical frequency to follow these changes in length. This, however, varies the bias

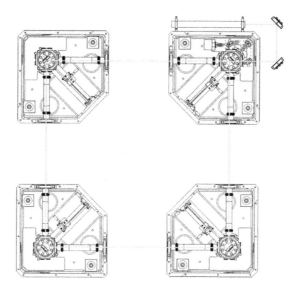

Figure 3.9 The basic construction layout of large heterolithic ring lasers. The mirrors are inside an ultra high vacuum compatible recipient and can be adjusted by long levers, both in plane and out of plane. An external alignment laser is injected to form a closed beam path.

on the beat note from non-reciprocal cavity effects and gives rise to a dispersive shift of the measurements. Therefore the best choice is the construction of ring laser bodies from a low thermal expansion material – Zerodur or ULE glass. However, ring lasers with areas much larger than 16 m^2 cannot be fabricated as monolithic or semi-monolithic Zerodur structures because neither Zerodur nor ULE is available in sufficiently large blocks. Furthermore, the glass ceramic is prohibitively expensive and a delicate material to handle. Figure 3.9 gives an impression of the general ring laser hardware utilized in all our heterolithic ring lasers.

The square laser cavity has 4 turning mirrors, each of which is located inside an adjustable compartment within a solid corner box. All corner boxes are rigidly attached to the floor, which is made from an approximately 30 cm thick steel armed concrete slab. This concrete base, resting on the ground, represents the actual platform whose rotation is monitored. The alignment of the mirrors is facilitated by a folded lever system, which can turn and tilt the mirrors in order to form a closed beam path. Bellows between the mirror compartment and the connecting steel pipes provide the necessary degrees of freedom for the alignment movement. An external HeNe alignment laser ($\lambda = 543$ nm) can be injected as indicated in Figure 3.9 in order to obtain the required closed light path. As indicated in Figure 3.9, a folded lever system allows the alignment of each mirror to within ± 10 seconds of arc, a feature that is common to all heterolithic ring laser structures of our rings. This high

Figure 3.10 Top view of the corner box construction. Behind the mirror compartment, one can see the arrangement for the combination of the two laser beams, composed of two turning mirrors and a beam splitter. The photo-detector is contained in a cylindrical housing. Except for the beam combiner assembly, all corners have an identical structure.

level of alignment resolution is required to ensure lasing from an optically stable cavity. The mirrors themselves are rigidly mounted inside stainless steel corner boxes, which in turn are connected by stainless steel tubes, forming a vacuum enclosure around the laser beam path (see Figure 3.10). In the middle of one side, the steel tubes give way to a small glass capillary of 4 mm diameter and length 10 cm. This is the place where the excitation of the gain medium occurs, as with the monolithic ring laser designs. Figure 3.9 gives an impression of the instrument layout. Since the ring laser is constructed from several individual components, it requires a stable concrete platform or bedrock as a base. The actual area of the ring laser component is not predetermined by the design.

In 1997 it was not generally known whether a 16 m^2 monolithic ring cavity could be operated as a single mode Sagnac interferometer in the same fashion as had been successfully demonstrated with C-II. The approximately four times smaller FSR was perceived as a serious obstacle. In order to investigate the potential of larger gyroscopes, a radical change of concept was therefore necessary. Since stainless steel and concrete have nearly the same thermal expansion coefficient, the basic concept used a UHV-compatible laser cavity made from stainless steel for the laser function and rigidly attached it to one of the massive concrete walls in the Cashmere Cavern, which provided the necessary stable base for the interferometer. Figure 3.11 shows this early large ring laser design [171]. At the time of construction, G-0 was not seriously expected to operate as a ring laser. Since it was designed as a precursor to the G-ring structure, G-0 was only expected to demonstrate single longitudinal mode operation on a cavity much larger than C-II. This was the missing link between

Figure 3.11 The prototype of a large steel and concrete based ring laser called G-0, as constructed in the Cashmere caverns in 1997.

the already working C-II instrument and the newly designed G ring laser in order to justify a 10 million Euro investment at the Geodetic Observatory Wettzell.

However, G-0 performed remarkably well as a Sagnac gyroscope, and as a result, much larger structures, such as UG-1 (366.8 m^2), filling half of the available floor space of the Cashmere facility, and UG-2 (834 m^2), using the entire available perimeter, became viable options. The corner assemblies of UG-1 and UG-2, including the alignment lever system, the laser gain section and the vacuum tube connections, rest on small concrete pedestals around the Cashmere Cavern, with the floor of the cavern acting as the rigid base of the interferometer. Of the two lasers, UG-1 was the more stable device. UG-2 had a tendency to be sensitive to rotations of the corners, yielding in-plane beam movements, a fact attributed to its quite substantial departure from an ideal square [96]. In addition, UG-2 gave a disappointingly low cavity Q of 1.5×10^{12}. This arises from concomitantly greater losses at the mirrors' surfaces as the arm length increases and progressively larger areas of the mirrors are illuminated. Overall, UG-2 represents the high water mark for the ultra large ring laser project (for the time being at least), since above a certain size the cavity Q will not increase unless the mirror manufacturing technology improves substantially. UG-2 was subsequently dismantled in favor of a structure much closer to square. Figure 3.12 gives an impression from the construction of UG-2 on a large number of small concrete piers around the walls of the cavern. When the vacuum pipes were joined, dry nitrogen was flooded through the cavity in order to reduce the intrusion of humid cavern air. Although the UG-2 ring laser provided the highest nominal sensitivity, this could not be exploited, due to the geometric instabilities associated with this large scale heterolithic design.

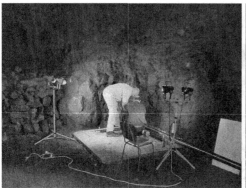

Figure 3.12 Illustration of the heterogeneously constructed UG-2 ring laser. Many small concrete foundations on the floor of the Cashmere cave support the 834 m^2 ring laser structure. The laser is too large to capture on a single photograph due to the layout of the Cashmere caverns. The cavity was vented with boiled off liquid nitrogen during the construction in order to avoid humidity creeping into the cavity.

Because of their high sensitivity, ring lasers are also very suitable sensors for the detection of earthquake-induced ground rotations. A much smaller version (A = 2.56 m^2) of a similar heterolithic design was constructed and placed on a rigid 30 cm thick reinforced concrete slab in order to provide a relatively simple, but still very sensitive device to the seismological community. As there is no real need for long term bias stability, this GEOsensor also uses a steel structure, as shown in Figure 3.13 during the initial installation at the Geodetic Observatory Wettzell. Equation 3.1 suggests and experimental evidence confirms that ring lasers have a very wide and linear frequency response. The complete absence of any moving mechanical parts, such as inertial probe masses and springs, are responsible for this valuable property. Therefore, the transfer function, which maps rotational motion into a beat frequency (Eq. 3.1), can be considered linear with a slope of unity for all practical purposes, which is another great advantage. The high sensitivity of large ring lasers to rotational ground motion has jump-started the field of rotational seismology [99, 157, 213]. The application of ring lasers in seismology is covered in more detail in Section 6.3.

3.5.4 Comparison of Ring Laser Concepts

In order to summarize the basic performance of all important large rings built so far, Table 3.2, an updated and condensed version of earlier published tables [182, 210], lists some relevant physical quantities from these instruments. Apart from the area (A), the quality factor $Q = 2\pi f \tau$ with f the HeNe laser frequency of $\sim$474 THz,

Table 3.2. *Summary of the relevant physical properties of a number of large ring lasers. The table lists the area enclosed by the beams, the quality factor Q, the ringdown time τ, the obtained frequency splitting f_{Sagnac} and the inferred sensor resolution S.*

Ring Laser	Area [m^2]	Q -	τ [μs]	f_{Sagnac} [Hz]	S [prad/s/$\sqrt{Hz}$]
C-II	1	5.3×10^{11}	180	79.4	146.2
GEOsensor	2.56	3×10^{12}	1000	102.6	108.1
ER-1	6.25	1.1×10^{12}	375	198.4	79
G-0	12.25	2.5×10^{12}	829	288.6	11.6
G-ring	16	3.5×10^{12}	1200	348.5	12
AR-2	22.6	6.1×10^{9}	≈ 3	≈ 374	316
UG1	367.5	1.2×10^{12}	409	1512.8	17.1
UG2	834.34	1.5×10^{12}	640	2180	7.8
ROMY	62.4	2.6×10^{12}	894	304–554.4	32

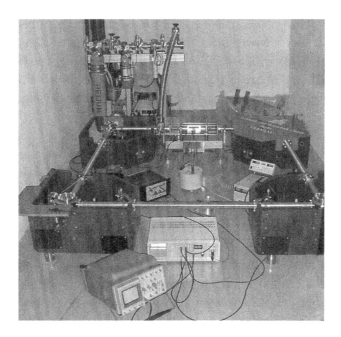

Figure 3.13 The GEOsensor ring laser during its initial construction at the Geodetic Observatory Wettzell. It is now located at the Piñon Flat Seismic Observatory near Anza in Southern California. (Reprinted with permission from [187], ©2022 Springer Nature)

the table lists the ringdown time τ, the actual frequency of the Sagnac beat note and the inferred sensor resolution. It is important to note that all rings are orientated horizontally on the Earth, with the exception of G-0 and ROMY. The former is located vertically along an east/west running wall in Christchurch, and the latter represents an ensemble of four linear independent triangular rings, assembled in the shape of a tetrahedron with the tip pointing downwards at mid-latitude. AR-2 is the latest square open cavity ring, employed by the group in Arkansas. The respective performance parameters were derived from [65]. According to Eq. 3.1, the scale factor of a ring laser gyroscope is the most important figure to look at for improvements when the rate of rotation is a small and nearly constant value, such as in the case of Earth rotation measurements. From that point of view, larger rings inherently provide significantly higher sensor resolution. The other obvious parameters for improvement are reduction of cavity losses and a boost in beam power. However, there is not much room for any gains. The absence of any other intra-cavity components only leaves the accuracy of the beam alignment and the quality of the mirrors available for improvement. Once good and clean super-mirrors have been installed, there is very little room for enhancements in the quality factor, apart from a small difference in the range of a few percent. Higher beam powers are also delicate to achieve, as we quickly leave the regime where a single longitudinal laser mode per sense of propagation exists. For Table 3.2 only one figure of merit, namely the sensitivity inferred from ringdown measurements, has been considered. This does not necessarily mean that such optimal performance is readily obtained at all times.

The enhanced scale factor comes at the expense of decreasing mechanical stability. Larger cavities usually also show a faster degradation of the laser gain medium over time, due to hydrogen outgassing from the walls of the stainless steel beam tubes. The application of getter pumps with sufficient pump power is therefore mandatory. In order to derive global geophysical signals, it is furthermore important to rigidly connect the interferometer to the Earth's crust and to maintain or track their orientation (φ) with extremely high resolution ($\delta\varphi \leq 10$ nrad). Naturally, it is easier to achieve this with larger constructions. However, a heterolithic ring structure inside an artificial cave suffers substantially from Earth strain, thermo-elastic deformations and atmospheric pressure variations, typically known as the cavity effect [86]. The monolithic rings do not show signs of this effect because they are much smaller and stiffer, and the way their supporting platforms are constructed minimizes strain across the entire Zerodur body. Therefore the G-ring remains by far the most stable and highest resolving ring laser structure to date. The larger ring structures nevertheless have the potential for considerable improvements. In order to utilize this, their cavities have to be actively controlled in order to regain the lacking stability through feedback loops, tightly locked to the perimeter.

There are a few other large ring laser structures, which are missing from the table above. The air-filled Arkansas ring laser project (see Section 3.4), employing a Brewster window sealed gain tube, did not report a sensitivity limit or ringdown time for any of their triangular rings. We also note the Italian ring laser project with instruments, dubbed G-Pisa [17, 18] and GINGERINO [21, 54]. In fact, these are slightly modified ring laser versions of the GEOsensor design with geometric dimensions similar to the GEOsensor and G-0. One of these lasers was embedded vertically into the VIRGO gravitational wave antenna as a high sensitivity tilt meter, measuring ambient rotations around one axis of vertical tilt [20]. Another square ring laser called ER-1 based on the GEOsensor design has arms 2.5 m long and is mounted in Christchurch on a thick concrete floor in the new Rutherford Regional Science and Innovation Centre. This means that all these systems are essentially already covered in Table 3.2.

3.5.5 Overview of Existing Constructions

In this section we briefly list the various milestones of the development of large highly sensitive and stable ring lasers in a semi-chronological order. This first phase was governed by two basic paradigms. First, that it was thought to be impossible to build gyros larger than 60 cm perimeter because they would inevitably run on several longitudinal modes. Second, it was thought that zero-expansion materials like Zerodur or ULE were absolutely mandatory in order to obtain a stable interferogram. Both of these design rules were essentially the heritage of aircraft gyro development. Our effort to explore large ring lasers started with the development of the *Canterbury Ring Laser*. A Zerodur baseplate of 1 m^2 in size carried the mirror holders, which in turn were encapsulated by stainless steel corner boxes and pyrex tubes. C-I was not very stable, despite the Zerodur base. The mirror holders, made of steel, were only loosely placed on top of the plate. Due to the many O-ring seals made from Viton, the vacuum was not of sufficient quality. On top of that, it was the early days of the development of IBS super-mirrors. However, the advantage of using mode starvation to avoid multi-mode operation was quickly recognized and exploited. It made C-I a success story. The next generation ring C-II attended to all the lessons learnt from C-I and is shown in Figure 3.14b. The Zerodur plate was massive, and the ultra high vacuum compatible beam path enclosure allowed many months of continuous operation. Furthermore, this ring laser already had much better mirrors available, so that it never locked up. In terms of stability, it gained more than three orders of magnitude over C-I and was the first to demonstrate effects from ocean loading, solid Earth tides. Retrospectively, C-II had already provided the first evidence of diurnal polar motion. C-II turned out to be too small for the goal of detecting variations in Length of Day. This led to the design of G, which

(a)   (b)   (c)

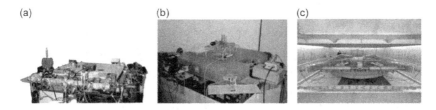

Figure 3.14 The large Zerodur based ring lasers: a) The Canterbury Ring (C-I) developed and operated in Christchurch 1986–1998; b) the second generation Canterbury Ring (C-II), developed in Germany (1994–1996), operated in Christchurch 1997–2011; c) the G ring laser, developed and operated in Germany since 2001. (Photo on right: © A. Heddergott of TU München)

in essence was an upscaled version of C-II and it is made from a massive slab of Zerodur.

In order to provide the necessary scale factor, the design of G seriously violated the prevailing paradigm of the limited permissible cavity length, and it was not obvious if such a large structure would be able to operate on a single longitudinal mode. This led to the construction of another entirely different sensor design, the all-stainless steel vacuum recipient gyros (Figure 3.15a). The first prototype was G-0, named to indicate that it was intended to be the 0th approximation of a *Grossring*. Nobody expected it to show a Sagnac interferogram. The sole purpose of G-0 was to demonstrate that it would allow a single mode of operation. It succeeded and, to the amazement of everyone, it worked very well as a gyro. Based on this reassuring experience, the contract for the 10 million Euro G ring laser could be signed with a fair amount of confidence. In the days before the internet, scientific communication was a lot harder to facilitate. It was not until the end of the century that our research group learnt about the ring laser activity in Arkansas. This firmly installed the concept of over-pressuring the cavity in our operations. The idea of going ever bigger to explore the ultimate design of large ring laser gyros was driving our next goals. The results were the UG-1 and UG-2 ring lasers, eventually filling the entire available space in the Cashmere cavern. While the sensitivity increased and the backscatter reduced as expected, these large structures exhibited serious stability issues. Due to significant cavern deformation in the presence of changing atmospheric loads, the scale factor could no longer be held sufficiently stable. While the two very large UG rings were very interesting from an instrumental point of view, they did not deliver on the driving interest, the extraction of the variations of the Length of Day from an inertial sensor. The sensor drift caused by the outgassing of such a large cavity could not be overcome on the existing budget limitations. The cavern deformations from the variable atmospheric pressure reduced the scale factor stability to several ppm, due to serious beam steering effects over distances up to 40 m.

Figure 3.15 The group of heterolithic rings composed out of stainless steel structures, which are mounted onto massive rock or a large slab of concrete, are shown here: a) the G-0 prototype (on the left wall), b) the UG-2 ring around the perimeter of the cavern, c) the PR-1 structure on the shear wall in the 'West' building and d) the GEOsensor in a vault on the Piñon Flats Geophysical Observatory near San Diego, USA. ((d) reprinted with permission of AIP Publishing from [182], ©2022, American Institute of Physics.)

Figure 3.16 The modified GEOsensor design, dubbed GP2, features additional diagonal tubes, which are used as linear cavities to interferometrically stabilize the area (a). The right panel shows the upscaled Gingerino in the underground Gran Sasso laboratory (b). (Photos adapted from [25])

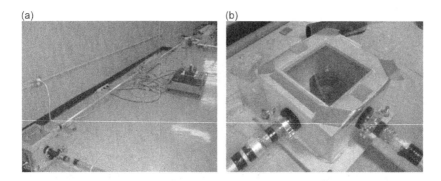

Figure 3.17 View of the testbed operations of an open air active ring laser gyro based on a HeNe Brewster window DC discharge gain cell (a). Laser operation is obtained when the cavity is flushed with dry nitrogen in order to remove air humidity (b). (Photos courtesy of Robert W. Dunn)

Large ring lasers sense rotational oscillations (S waves) from earthquakes on top of the constant rate bias caused by the Earth's rotation. With the inauguration of the International Working Group on Rotational seismology,[1] the need for field observatory-deployable ring lasers led to the construction of moderately sized ($A = 2.56$ m^2), properly engineered ring lasers, which could be installed either on a horizontal pad or a vertical wall. Dubbed the *GEOsensor*, this modular heterolithic design found applications at a vertical wall at the physics department of the University of Canterbury (NZ) (Figure 3.15c) and at the Pinon Flats Geophysical Observatory (Figure 3.15d). A modified version of the GEOsensor design were used by our colleagues from the *Instituto Nazionale de Fisca Nucleare* for prototyping an even more demanding high resolution rotational sensor, aimed at ground based detection of the Lense–Thirring frame dragging effect [34, 221]. This group pioneered the active stabilization of the scale factor by interferometrically tuning the diagonals of a square ring laser [176] to equal length (see Figure 3.16a). They also operate an upscaled version ($A = 12.25$ m^2) of this design in the deep underground facility *Laboratori Nazionali del Gran Sasso* (LNGS) (see Figure 3.16b). Finally, we built the first full scale four-component large ring laser structure, dubbed *ROMY* (ROtational Motion in seismologY). It is designed in the shape of an equilateral tetrahedron with 12 m sides. ROMY is the logical next evolutionary step in high resolution rotation sensing for the geosciences.

The Arkansas ring lasers differ from those of our working group by the fact that the cavities are open to the ambient pressure, and the beam path is gently flushed with dry nitrogen to keep the absorbing humidity out (Figure 3.17). The laser gain

[1] www.rotational-seismology.org

(a)
(b)

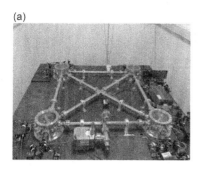
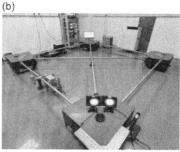

Figure 3.18 Display of the two passive gyros HUST-0 (a) enclosing an area of 1 m^2 and HUST-1 (b) enclosing 9 m^2 operated at the Huazhong University of Science and Technology (Wuhan). (Photos courtesy of Jie Zhang)

medium is confined to a Brewster window-enclosed long gain tube, and the sensor layout is typically in the shape of an equilateral triangle. This open cavity concept and the presence of significant backscatter coupling from mirrors with 99.9% reflectivity makes this type of gyro very sensitive to infra-sound. Although the detected infra-sound frequencies are not caused by ground rotations, this sensor design demonstrates a remarkable sensitivity for it. Details are covered in Section 6.6.

The latest addition to our work is the investigation of the potential of passive gyros for high resolution rotation sensing. Figure 3.18 gives an impression of this setup. The two prototypes HUST-0 (A = 1 m^2) and HUST-1 (A = 9 m^2) mainly differ in size, and their development is progressing quickly.

3.6 Operational Principles and Practice

Proper rotation sensing is only achieved over a fairly narrow parameter set. Here we review the best practices for high resolution rotation sensing.

3.6.1 Mode Structure in Large HeNe Lasers

In order to keep the losses in the cavity as low as possible, there are no components other than the mirrors and the capillary for the laser gain excitation present in the cavity. This leaves only the cross-section of the gain capillary and the collinearity of the gain tube relative to the beam path as variable parameters in order to select the mode structure of the beams in the cavity. In the early days of the C-II ring laser, we operated a 6 mm diameter gain capillary in the system, as depicted in Figure 3.19. Micrometer screws in the x and y directions on each end of the capillary allow for a collinear adjustment of the gain tube relative to the laser beams. From the viewpoint of a laser beam, the diameter of the capillary constitutes a hard aperture, which

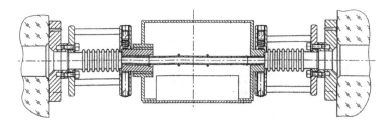

Figure 3.19 A technical drawing of the gain section of the C-II ring laser. Micrometer adjustments on each side allow alignment with respect to the laser beam.

blocks out higher order transversal modes when the required mode volume is larger
than the free aperture of the capillary. A ring laser can be operated on any oscillating
laser mode in principle, but in practice it is essential to use the TEM_{00} for the best
performance. This mode has the lowest divergence angle and also the lowest mode
volume. In addition, this mode delivers the best contrast for the beat note. If the
system were running, say, on the TEM_{10} mode, for example, there would always
be a high probability that a weak TEM_{00} mode would be excited as well, which
would instantly cause mode competition, and the achievable stability of the Sagnac
beat note would be considerably reduced. In the early days of the C-II ring laser,
a gain tube with 6 mm diameter was used. Together with two flat mirrors and two
mirrors with a radius of curvature of 6 m in a diagonal configuration, C-II produced
a spot diameter at the capillary of $d_x = 1.5$ mm in the beam plane and, due to the
astigmatism from the spherical mirrors, $d_y = 1.2$ mm in the orthogonal direction. As
a result, C-II strongly preferred the TEM_{52} mode for lasing. By tilting the gain tube
slightly against the optical axis of the laser beam, a reduction in the effective free
aperture of the capillary could be achieved, so that a variety of different transversal
mode patterns could be excited. Figure 3.20 depicts a selection of them.

However, with the exception of the TEM_{52} mode lasing, the gyro did not produce
useful gyro operation and even tended to lock up when the capillary was skewed too
heavily in the search for working TEM_{00} operation. Although the depiction of the
gain tube as a hard aperture is quite intuitive, it does not fully describe the system.
The laser gain across the capillary is not uniform (see [149], Section 13.8). The
radial gain profile for a DC-excited HeNe laser has the form

$$g(r) = \frac{\alpha I\, J_0(2.405r/R)}{1 - bI\, J_0(2.405r/R)} - \beta I\, J_0(2.405r/R). \tag{3.3}$$

The parameters $\alpha = 2.14$, $b = 0.04$ and $\beta = 0.194$ are numerical values, obtained
from a curve fitting procedure carried out by [233]. $J_0(x)$ is the zeroth order Bessel
function, and I is the current across the plasma. Figure 3.21 provides the radial
gain distribution of a HeNe laser for different currents in the plasma, albeit only

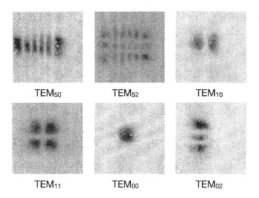

Figure 3.20 With the application of a wide gain tube, there is no proper mode selection achievable. Skewing the gain tube alignment provided an aperture of variable size but increased the losses considerably. All the displayed mode structures were excited in C-II.

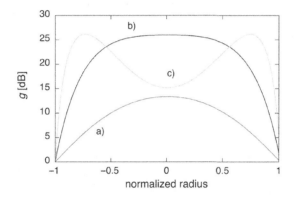

Figure 3.21 The gain profile across the capillary in a DC-excited HeNe laser with the radius normalized. The model calculation has been carried out for three different currents. For low excitation powers (I = 10 mA (a)), the maximum gain is in the center. At higher powers I = 50 mA (b), the gain profile widens. For even high excitation powers I = 160 mA (c), the gain decreases in the center, since the de-excitation by wall collisions becomes progressively less efficient.

strictly valid for a system with DC excitation and for a capillary of rather small cross-section. Due to the fact that the neon atoms have to be de-excited to the ground state by wall collisions, this process is more effective on a narrow gain tube. A wider gain tube requires more excitation power to obtain the same amount of beam power. The calculation in Figure 3.21 captures three scenarios. For low excitation powers, the maximum gain is in the center of the gain tube cross-section (a). For higher currents (b), the gain profile widens progressively because the de-excitation to the ground state by wall collisions becomes less effective in the gain tube center. When the current is raised further, the gain profile caves in around the center (c), pushing the maximum gain further out toward the walls, thus encouraging higher order laser modes to come into existence.

Although our ring lasers in contrast are excited by RF frequencies with beam powers in the range of 2–30 watts, the observations above can be readily transferred to our systems. For the 6 mm gain tube of C-II, higher RF powers are required, which puts the regime of operation closer to scenario (b), so that the modes with a larger spatial cross-section experience more gain and hence are preferred. When the gain tube diameter of C-II was reduced to 4 mm, the power requirements dropped to around 2 watts, and the TEM_{00} mode was always the only excited mode.

3.6.2 Single Longitudinal Mode Laser Operation

The vast majority of all large ring lasers reported here employ a cavity entirely filled with cold Helium–Neon gas. Excitation occurs within a small glass capillary with no Brewster windows and therefore uses the minimum number of optical surfaces within the laser cavity. Typically the cavity is filled to a total pressure anywhere between 2 and 10 hPa of helium and neon in order to increase the homogeneous broadening of the laser transition. It is highly desirable to operate each ring with a 50:50 mixture of ^{20}Ne and ^{22}Ne, which eliminates competition between the bi-directional traveling waves occurring at the center of the gain profile. The ratio of the He and Ne partial pressures in our rings is of the order of 50:1. Excitation occurs within the glass capillary using capacitive coupling of RF power, thereby removing the need for internal electrodes and high voltages. High frequency excitation at approximately 50 MHz and above has the highly desirable, if not essential, effect of avoiding biasing effects such as Langmuir flow.

Typically, ring laser gyroscopes are operated on a single longitudinal laser mode. To achieve this with large ring lasers, a gain starvation approach is used. This involves the servo controlled reduction of the RF excitation power, so that only a counter-propagating pair of longitudinal modes (having the same mode index) at a spectral position close to the line center of the ^{20}Ne and ^{22}Ne transition, is excited above the lasing threshold [171]. As the laser dimensions increase, the separation of neighboring laser modes decreases. Since the combined gain curve is very flat at line center, the differential gain between modes becomes progressively less significant. This led to the very plausible expectation that larger ring lasers without extra mode selection devices would necessarily run on several longitudinal modes and that it would be impossible to achieve single mode operation. Considering that the gain profile is nearly 2 GHz wide, the largest of our lasers support many more than 100 of these longitudinal mode frequencies. However, even for the massive 834 m^2 UG-2 laser, operating with a FSR as small as 2.47 MHz, single mode operation can be readily attained [96]. This is achieved by the homogeneous broadening induced through the use of higher He gas pressures, which has the effect of suppressing mode competition for about ± 100 MHz around the actually excited laser mode, which provides sufficient room to maintain the differential gain as a mode selection mechanism [94].

Although single mode operation is both attainable and remains useful even for very large ring lasers, such as UG-1 and UG-2, the mode starvation process takes some time to establish. At start up, laser oscillation occurs on tens of longitudinal modes. Over a period of around one minute, all but one mode die away, due to the small differences in gain available to adjacent modes. This mode decay process scales approximately as the inverse cube of the laser dimensions [94] and is pressure dependent. Depending on the effective losses in the cavity, it may not be possible to increase the gas pressure to values of 6 hPa and above, because this also reduces the laser gain. Larger cavities have wider beams and make larger gain tube diameters necessary, which in turn reduce the laser gain. All these aspects have to be considered in finding the best sensor configuration.

3.6.3 Phase Locked Multi-mode Operation

We have by now seen that single mode operation, obtained through gain starvation [63], works well even for large cavities, which have a correspondingly small FSR of the order of 2.5–20 MHz. Together with the benefit of enlarged homogeneous line broadening from over-pressuring the cavity [64], an operating laser power range of the order of 20 nW can be readily obtained. The desire to reduce shot noise by increasing the laser power still remains. HeNe lasers offer an alternative approach for the laser to operate on multiple longitudinal modes in a phase-coupled regime and therefore obtain higher output powers [63]. Mode locking in HeNe lasers is very well known [110] and occurs through the intrinsic non-linearity of the gain medium itself. Deliberately running any ring laser gyroscope above its multi-mode threshold yields a regime where multiple longitudinal modes coexist in a generally stable fashion [8, 9, 92]. This occurs without the need for intra-cavity modulation and yields output powers 10–30 times greater than obtained with single mode operation. The observed mode patterns are typified by a central dominant mode surrounded by several minor modes placed symmetrically about the dominant mode. The minor modes tend to be stable over long periods of many days. However, on longer timesscales, small variations in the relative intensities have been observed. These effects are more significant in rings with higher geometric instabilities. In this operating regime, we observe a very clear, strong and stable Sagnac frequency.

In Figure 3.22 we compare the performance of these two operating regimes for the G ring laser by monitoring the very uniform rotation rate of the Earth. All known

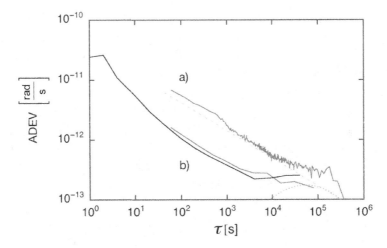

Figure 3.22 ADEV of the G ring laser beat note for a time series, which is corrected for the effect of diurnal and semi-diurnal geophysical signals.The ring was operated both in single mode (a) and self-locked multi-mode (b). The higher beam power in the multi-mode regime reduces the shot noise considerably.

geophysical signals have been removed from the raw measurements. The Allan deviation (ADEV) of a time series of the corrected Sagnac beat note is displayed for single mode (a) and multi-mode (b) operation. While the shot noise is significantly reduced in the case of the multi-mode operation, there are no apparent detrimental effects from the additional co-existing longitudinal modes for a measurement series of several days of duration. At periods of one day ($\approx 10^5$ s), one can see the effect of what appear to be residuals of geophysical signals with periods of around 1 day. These can exist for several reasons. First of all, it could mean that our correction model is incomplete and that it does not remove everything. More likely are effects from Earth crust deformation, caused by variable atmospheric loading. The most likely and most significant contributor, however, is a small orientation error of the ring laser structure, with respect to the Earth body. When the true orientation differs from the assumed orientation, the model corrections are incomplete. We used a geodetic survey of the laser beam positions within the local surveying network of the Geodetic Observatory Wettzell to infer the correct orientation of the ring laser beam plane. Although carefully done, the resolution of the survey is not nearly good enough to produce a reliable metric for the orientation of the ring laser. This is a good example of the scale factor error illustrated in Figure 3.5a. The improvement for heterolithic ring laser structures is not so clear. Due to the reduced mechanical stability, the phase locking effect is less persistent and less stable. Figure 3.23 depicts the Allan deviation ADEV of the vertically orientated G-0 ring laser run on a single longitudinal mode and in a mode-locked regime.

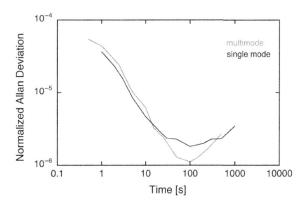

Figure 3.23 ADEV plot of the G-0 ring laser for operation on a single longitudinal mode and multiple modes in a mode-locked regime.

In either case, within a factor of two we obtain a minimum deviation at close to 10^{-6} of the Earth's rotation rate with a roll off in performance after 100 s due to the mechanical instabilities of the system. For mode-locked operation, we see a small advantage near the minimum value as a result of the higher beam powers. Varying the gas pressure in the laser cavity is known to significantly alter the output powers available for which stable mode-locking occurs due to the pressure dependence of the saturation intensity. For any output power above the multi-mode threshold, there are two possibilities: the existence of regions for which stable mode-locking occurs (and hence a stable Sagnac waveform is observed) and other regions in which the laser is free-running with no fixed phase relationship between the different longitudinal modes [9]. We have operated four different heterolithic laser gyroscopes (ranging in size from 2.56 to 367 m^2) in the mode-locked regime. All devices show comparable performance in either mode of operation. While this concept shows some advantages in the short term, we are not using this mode of operation for long term operations. The presence of several additional laser modes in the cavity generates some detrimental side-effects for the backscatter correction process. However, the origin of these effects has not yet been exhaustively investigated, and there may be a way around it.

3.6.4 Operation on Two Different Cavity Modes

The larger the ring cavity, the smaller is the free spectral range. For the G ring laser, the difference is 18.75 MHz. With a sufficiently large gas pressure ($p > 6$ hPa), the ring laser will either switch on with both laser modes on the same longitudinal mode index or with the modes on two neighboring mode indices. Here we have explored the case when the longitudinal mode index differs [195]. In practice this

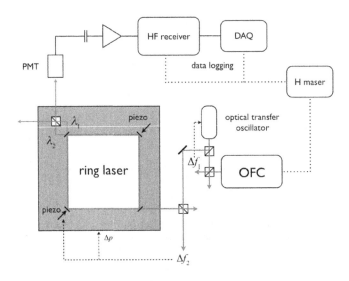

Figure 3.24 Block diagram of the cavity stabilization and beat note recovery setup.

means that the beat frequency from the experienced rotation is biased by 18.75 MHz. Although we only explore the case where the difference in mode index was ±1, comparable results would be obtained for larger differences. In order to use this mode of operation, the optical frequency in the cavity needs to be well stabilized. We have used an optical frequency comb (Menlo Systems, OFC 1500), referenced to the hydrogen maser of the Geodetic Observatory Wettzell. In order to enhance the signal to noise ratio, a HeNe transfer laser has been used. For a reliable lock, a beat signal amplitude of at least 30 dB is required. Here we report on a setup where the transfer laser was locked to the OFC 1500 via the beat note Δf_1, sketched in Figure 3.24. The second feedback loop used the beat note Δf_2, obtained from mixing the transfer laser with the counter-clockwise ring laser optical frequency. Δf_2 was kept constant to within 1 kHz by driving two piezo-actuators simultaneously, which were pressing on the back of the mirror holders of the ring cavity on two diagonally opposite corners. This provided a tuning range of about 600 kHz with a bandwidth of about 2 Hz. In order to cover a much larger tuning range, the air pressure inside the stabilizing tank enclosure could either be slowly raised or lowered.

The experienced FSR biased rotation rate with a frequency of $\Delta f_{\text{Sagnac}} + \Delta f_{\text{FSR}} =$ 348.516 Hz + 18.75 MHz was extracted with a broadband photomultiplier. It was then down-converted to an audio frequency of about 400 Hz utilizing a heterodyne receiver. All subsystems, namely the OFC 1500, the receiver and the data acquisition system, were synchronized with the same hydrogen maser in order to avoid additional frequency drifts and offsets. Cavity length fluctuations in a Sagnac interferometer cancel out when both laser beams are within a few hundred Hz of

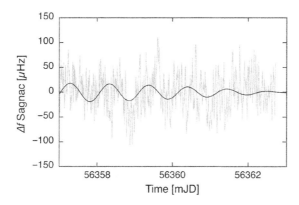

Figure 3.25 The recovered Sagnac beat note from the G ring laser with both laser modes on a different longitudinal mode index and the diurnal polar motion signal superimposed. Since geometrical cavity effects no longer fully cancel out, we loose about one order of magnitude in sensor resolution.

the same optical frequency. When the mode spacing is one or more FSR, these frequency fluctuations can no longer be ignored. For the two wavelengths we obtain

$$\lambda_2 = \frac{I}{I+1}\lambda_1, \tag{3.4}$$

where I is the longitudinal mode index, which in the case of the G-ring is of the order of 25.3 million. A frequency variation of the optical frequency of the order of 1 kHz results in a relative variation of $\lambda_1/\lambda_2 \approx 10^{-8}$. This sets the low end of the noise level to about 40 μHz, which is about a factor of 10 larger than that of the common mode operation. Figure 3.25 depicts the residuals of a longer observation series of six days, carried out in this offset configuration. The rate bias from the Earth's rotation of 348.5 Hz has already been subtracted. No geophysical signals can be recovered at such a high level of the noise floor. For comparison purposes we have also plotted the diurnal polar motion signal on top of this dataset.

When the G-ring is operated on the same longitudinal mode index for both beams, the diurnal polar motion can be readily extracted with good fidelity. An example is shown later, in Figure 3.69. One motivation to operate an optical Sagnac interferometer on two different longitudinal laser modes would be the reduction of backscatter coupling. This has been argued, for example, by Korth et al. [121]. However, this expectation is not observed. Backscatter coupling depends on the mirrors, involves a double scatter process and does not depend on the other beam. We treat this mechanism in detail in Section 3.9.4.

Expressed as a time deviation (Figure 3.26), we see that the operation on two different longitudinal modes is about an order of magnitude worse than for the common mode case. This applies even for a well stabilized cavity. In principle one

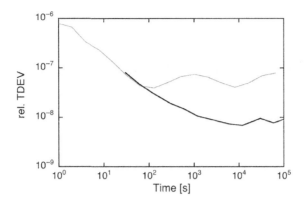

Figure 3.26 The time deviation of a six-day-long observation with the ring laser operated on two different longitudinal modes. Optical frequency fluctuations reduce the sensor resolution by about one order of magnitude.

could most probably improve on these results by producing a tighter lock on the optical frequency of the ring. A major limitation in the obtained results certainly comes from the low regulation bandwidth of no more than 2 Hz. The thickness of the mirror-holding plates of 25 mm and their corresponding weight are responsible for that. In the current installation this cannot easily be changed.

3.6.5 Sensor Resolution

According to [40, 59, 208] and others, the sensitivity limit of a ring laser gyroscope, derived from the irreducible quantum noise for a rotation measurement, is given by

$$\delta\Omega = \frac{cP}{4AQ}\sqrt{\frac{hf}{p_c t}}, \tag{3.5}$$

where P is the perimeter, A is the area enclosed by the light beams of the gyro, $Q = \omega\tau$ is the quality factor of the ring cavity, h is Planck's constant, p_c is the circulating beam power per mode in the cavity and t is the time of integration. For the G ring, the ringdown time was first measured to be $\tau = 1$ ms in 2001. Over the years it reduced to a value of $\tau = 500$ μs in 2007 due to gradual mirror degradation of unknown cause. In 2009 the old mirrors were replaced by a new set of mirrors with an essentially identical coating structure, consisting of a Bragg stack of about 46 alternating quarter-wave layers of SiO_2 and Ta_2O_5. In contrast to the earlier mirrors, which were made from ULE, the new mirrors used fused silica as a substrate. The latter was chosen because a better polish can be achieved. The fused silica material furthermore exhibits a higher mechanical quality factor, which reduces thermally induced substrate noise (see Section 3.10.3) as an effect describable by the fluctuation – dissipation theorem [155]. The new mirrors also showed fewer coating defects when they were initially inspected. However, the effectively achieved overall Q-factors with both mirror sets turned out to be comparable. The obtained ringdown time for the G ring, when turned on after the mirror change, was measured to be 1.2 ms. The total losses of the cavity are inferred from the photon decay time and the free spectral range of the cavity according to

$$\delta = \frac{1}{\tau f_{\text{FSR}}}. \tag{3.6}$$

With the circulating power attenuated by the mirror transmission, the available output power for the detection of the rotational signal is $P_o = P_c \cdot 2 \times 10^{-7}$. For the operational sensitivity of the G ring, when strictly operated on a single longitudinal laser mode with a typical setting for the output beam power P_o of 20 nW leaking through one of the corner mirrors, we obtain a theoretical internal sensor resolution of the G ring laser of 3.2×10^{-14} rad/s/ $\sqrt{\text{Hz}}$. However, in practice this intrinsic sensor sensitivity cannot be exploited, since the internal interferogram cannot be accessed. The light leakage through the mirrors, the quantum efficiency of the light to voltage conversion (η_{qe}) and the actual contrast (K) due to the astigmatism of the laser beams set a practical limit. Equation 3.5 thus becomes

$$\delta\Omega = \frac{cP}{4AQ} \sqrt{\frac{hf}{P_o \, \eta_{\text{qe}} \, K \, t}}, \tag{3.7}$$

with the contrast of the interferogram given as

$$K = \frac{I_{\text{max}} - I_{\text{min}}}{I_{\text{max}} + I_{\text{min}}}. \tag{3.8}$$

This limit can be further reduced by the noise floor of the electronic subsystems, such as the photo-detector, the digitizer, the amplifiers and, very critically, the applied frequency estimator. Effectively, we obtain 1.1×10^{-11} rad/s/ $\sqrt{\text{Hz}}$ and can therefore resolve a rotation rate of 3.5×10^{-13} rad/s after 1000 seconds of integration time [192].

All our ring lasers are very large, compared to an aircraft gyro. The optical path length inside the cavity varies between a few meters and 121.44 meters. With total cavity losses at the level of 44 parts per million, this translates into a very narrow linewidth. The theoretically expected Schawlow–Townes linewidth for the G ring is

$$\Delta v_L = \frac{N_2}{N_2 - N_1} \frac{2\pi h f_0 \Delta v_c^2}{P_L}, \tag{3.9}$$

with a width of the cavity resonance of

$$\Delta \nu_c = \Delta_{\text{FSR}} \frac{1 - R}{\pi \sqrt{R}}. \tag{3.10}$$

Using $\Delta_{\text{FSR}} = c/L = 18.75$ MHz and $R = 0.999956$, $\Delta \nu_c \approx 265$ Hz and hence $\Delta \nu_L \approx 7$ µHz. Since a HeNe laser is a four level system, one may expect to set $N_1 \approx 0$, which assumes full inversion. For an industrial HeNe laser with a capillary diameter of 1–2 mm and a moderate excitation current, this assumption is certainly true. However, our lasers use wider capillaries with 4 mm (C-II), 5 mm (G) and 6 mm (UG-2) diameter. This slows down the wall-collision-induced de-excitation of the Neon atoms from the $1s$ state. As a result the laser gain may reduce and by the process of electron–neon collision pumping may lead to an increase of N_1, a reduction of the inversion [149, 227] and consequently to an increased linewidth. An unperturbed ideal oscillator would operate on a single frequency. Due to the quantum nature of laser light generation, we have to consider that additional white frequency noise broadens the linewidth and generates a Lorentzian shape [208]. The spectral linewidth Γ is then given by

$$\Gamma = \frac{2\pi h f_0^3}{Q^2 P_0}, \tag{3.11}$$

with f_0 the optical frequency, Q the quality factor of the cavity and P_0 the power loss through a mirror. A value of $\Gamma \approx 1.7$ µHz is obtained, which agrees reasonably well with the estimated Shawlow – Townes linewidth from Eq. 3.9 for the G ring laser.

For a large cavity like that of the G ring laser, however, the actual linewidth is defined by the mechanical stability of the ring cavity. Small scale vibrations and the thermal drift of the cavity cause frequency dither to the optical frequency, which reduces the frequency stability of the cavity. When we compare the optical frequency of the ring laser against an optical frequency comb (Menlo Systems: FC1500-250), which is referenced to a hydrogen maser, we observe a frequency stability of the unconstrained ring cavity of 8×10^{-13} at 1 second, corresponding to about 380 Hz. Figure 3.27 depicts the cavity stability as a function of time. A slow drift due to changing temperature is also evident in the measurement. Since these cavity-related effects are in common mode for both senses of propagation, they cancel out for the rotation sensing application.

Actual measurements, each value integrated over 60 s over a total of 8 h on a quiet day show an rms variation of the detected Sagnac frequency estimates of the Earth rate of $\sigma \approx 18$ µHz, as Figure 3.28 illustrates. In the absence of any variations of the scale factor, one obtains a constant lower limit for the gyro resolution according to Eq. 3.5, which only depends upon fluctuations of the laser beam power in the

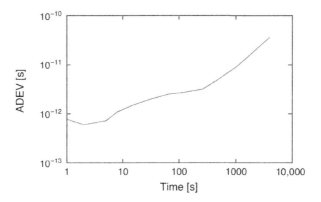

Figure 3.27 The stability of the G ring laser cavity, measured against an optical frequency comb, which in turn is referenced to a hydrogen maser. According to the specifications of the maser, its respective stability is 1×10^{-13} in 1 s.

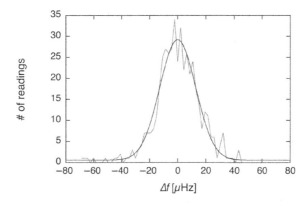

Figure 3.28 The actual resolution of the Sagnac frequency of the G ring laser, established from 1 min. averaged observations over 8 h on a seismically quiet day.

cavity. A typical value for the stability of the beam power due to mode competition of 0.01% over a time period of about 1 hour has been observed from the observation of the intensity in the G ring.

In order to illustrate the mid-term stability of large ring lasers, Figure 3.29 shows an Allan deviation plot of most of the lasers specified in Table 3.2. One can clearly see that a monolithic construction and large size are essential for better performance. This is not unexpected. Compared to UG-1, C-II experiences a larger perturbation from backscatter coupling, due to the much smaller scale factor. The performance of UG-2, on the other hand, falls off despite the enhanced scale factor. This is due to the rectangular sensor layout, which leads to an increased arm length of the instrument with concomitant negative beam steering effects and additional losses

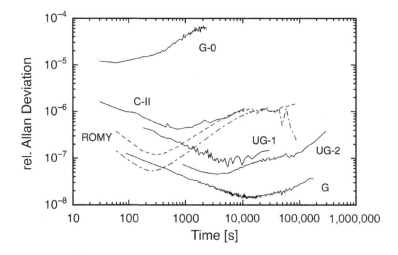

Figure 3.29 Relative Allan deviation of most of the large ring lasers of Table 3.2. The data for the G ring laser was taken before the implementation of a laser cavity stabilization scheme. Two of the four ROMY rings are included. The upper curve is from a slanted cavity; the lower curve is from the horizontal component.

from to the larger beam spot diameters on the cavity mirrors. The G-0 laser, as the only vertically mounted ring laser, does not have the mechanical stability to perform well. This prototype ring laser construction was not built with rotation sensing in mind; it was only intended to test the feasibility of mono-mode operation at this sensor size in order to specify the G ring laser dimensions. The GEOsensor is designed specifically for studies in the field of rotational seismology, where long term stability is not an essential design criterion. We also included one of the angled rings of ROMY in the diagram, together with the horizontal component. The latter exhibits a slightly better stability. Both rings are still unconstrained, so they lack long term stability.

3.6.6 Multi-corner Beam Recombination

In order to study the dynamic behavior of the ring gyro structure of very large rings, it is generally advantageous to operate more than one beam combiner on the instrumentation. This is captured in the sketch shown in Figure 3.30. Each beam combiner produces the same interferogram from the two counter propagating beams in the cavity.

Since the recombination does not take place at the same location, the instantaneous phase angle differs between the two spots. Interfering these interferograms with each other provides the phase offset between the two spots of interference, however with the ambiguity of an unknown integer number of waves. By recording

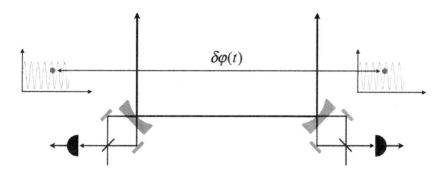

Figure 3.30 An example of the installation of two beam combiners at different corners of the ring laser gyro. Each of these mixing stages shows the Sagnac interferogram, however with a different instantaneous phase angle. Monitoring the beat signal of the two interferograms provides access to the varying separation between the two ends of one arm of the ring structure.

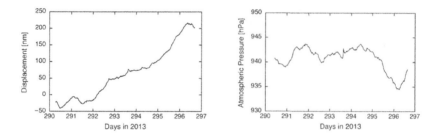

Figure 3.31 The variation of mirror separation over a period of seven days (left). Drivers for the changes are compression from atmospheric pressure and thermal expansion from increasing temperature. The corresponding ambient pressure is shown on the right.

the time evolution of the phase angle, some basic properties of the mechanical stability of the cavity can be inferred. Figure 3.31 shows an example of such a measurement. The phase angle is unwrapped and converted into the variation of the arm length of the G-ring Zerodur structure by using the relationship that $360°$ correspond to $\lambda/2$ in distance. We see a gentle increase in length, which is driven mostly by an almost linear upward trend in temperature of about 1 mK per day on the left-hand side diagram. The ambient pressure over the same period is shown on the right side of Figure 3.31. The two bumps downwards on day 291 and 296 reflect themselves in a change of the mirror separation of the tested arm of the gyroscope in the opposite direction. A lower ambient pressure reduces the compression on the body of the gyroscope, which in turn extends the length of the optical path. The same applies when the temperature increases and expands the cavity. This method is particularly useful to estimate strain effects that act on the arms of a large ring

Figure 3.32 Depiction of the design of the additional beam combiner mounted to G. The structure is made from carbon fiber and is as symmetric as possible.

laser gyro. In the case of the monolithic G-ring, this effect is virtually non-existent, since one of the important design criteria was the elimination of the effects of Earth strain on the laser cavity. However, extracting the variation of the arm length from differential phase changes between beam combiners on different corners comes with a word of caution.

When we compare the results for the G-ring to the calculated variation of the length of one arm, based on the measurement of the instantaneous optical frequency (see discussion in Section 3.7), we find that the numbers obtained here are about one order of magnitude too large. This has two reasons. First, an active ring laser always adjusts the optical frequency when the length of the cavity changes to an integer number of waves around the cavity. The change in optical frequency and the integral cavity length should compensate each other for an ideal cavity, as long as the laser does not flip over to another longitudinal mode index. More important, however, is that we see the recombination of the beams outside the ring cavity. Figure 3.32 shows the installation on G. This means that the respective beam combiners behind each mirror define the endpoints of the inferred distance. So the design of the beam combiners has a significant effect on the measurement quantity of interest. Although the free space interferometer has been set up such that both arms are as symmetrical and as short as possible, we nevertheless see a lot of the imperfections of this extra-cavity signal path. Temperature and pressure effects act in the same way on the free space interferometer as on the cavity. This topic is taken up again in more detail in Section 3.9.8.

Extending this concept to the very large UG type of heterolithic rings with an arm length of 10 m or beyond can nevertheless be very beneficial. From the almost square UG-3 ring laser with an arm length of about 20 m and a beam combiner at each corner, it was possible to reconstruct short term strains for all four arms from the microseismic frequency band around 0.1 Hz with an amplitude

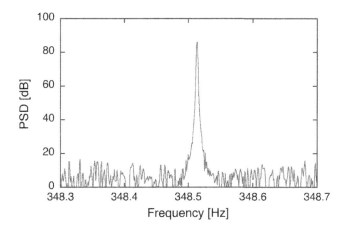

Figure 3.33 A small section from the interferometer spectrum around the Sagnac beat note. The observed peak is caused by the rotation rate of the Earth of 11.34°/h at a latitude of 49.1° north, with a fidelity of almost 90 dB.

of ±5 nanostrains as well as the corresponding strains from solid Earth tides of ±100 nanostrains, which compare well with the values derived from theoretical models [78].

3.6.7 Frequency Demodulation

The prime observable of a large ring laser gyroscope is the beat frequency between the two counter-propagating narrow linewidth CW laser beams in the cavity. For applications in geodesy, this beat note corresponds to the instantaneous rotation rate of the Earth, experienced by a strapped down rigid laser cavity. Apart from some very small variations, caused by geophysical signals at the semi-diurnal and diurnal period, this beat note remains almost constant in frequency. Figure 3.33 shows a typical spectrum generated from a continuous dataset of 15 minutes' duration. The Earth rate of 15°/h sticks out as the only observed signal in this portion of the interferometer spectrum. The measured beat note has a high fidelity of about 90 dB over the noise floor, and there is a difference of 65 dB in signal level between the beat note and the second harmonic frequency (not shown in the figure). This indicates a low level of waveform distortion and therefore only small effects from perturbations caused by backscatter pulling and pushing. While the estimation of the Earth rate from a ring laser is almost a static process, external perturbations to the ring laser body, either by lack of sensor stability or by ground motion, exhibit a different situation. If the ring cavity can slowly shift with respect to the body of the Earth, this will cause a drift of the Earth rate, due to a corresponding change in cavity orientation and by any rotational motion that may have occurred during the slip process. Small oscillatory ground motions around the normal vector on the laser plane give rise to side bands on either side of the dominant Earth rotation rate signal, corresponding to a frequency modulation on the constant part of the Earth rate. Such signals are caused by micro-seismic activity or are induced by surface (Love) waves from earthquakes. We observe the signal voltage $u(t)$ as

$$u(t) = u_0 \cos(\omega_c t + \Delta\phi \sin \omega_{\text{seis}} t), \tag{3.12}$$

where ω_c is the angular frequency caused by Earth rotation, which takes the role as the carrier. The quantity ω_{seis} corresponds to the induced additional oscillatory rotation rate from the seismic process and makes up the modulation signal with $\Delta\phi$ the peak frequency deviation. In the absence of strong motion signals, we typically have $\eta = \Delta\phi/\omega_c \ll 1$, and this allows the approximation

$$\frac{u(t)}{u_0} \approx \cos \omega_c t - \frac{\eta}{2} \cos(\omega_c t - \omega_{\text{seis}})t + \frac{\eta}{2} \cos(\omega_c t + \omega_{\text{seis}})t. \tag{3.13}$$

Figure 3.34 illustrates the result. The depiction of the vector sum of the two counteracting modulation frequencies (here the seismic signals above and below the carrier) illustrates how the modulation frequency is pushed ahead and behind the carrier frequency. While this process alters the carrier frequency, it barely changes the amplitude – in contrast to amplitude modulation, which changes the amplitude but not the carrier. Amplitude modulation can also be observed on large ring lasers on occasion. This is usually a sign that the system is being operated above the multimode threshold, an operational state that has to be avoided.

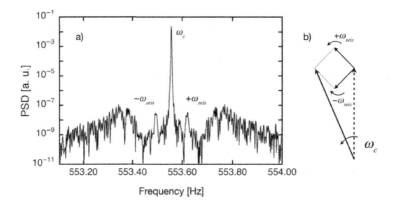

Figure 3.34 A small section from the interferometer spectrum around the Sagnac beat note of the horizontal ring in ROMY. The *carrier* signal induced by the Earth's rotation rate is surrounded by a broader band of frequencies, generated by micro-seismic activity. The origin of the two narrower lines closer to the carrier are caused by an unknown process.

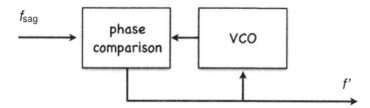

Figure 3.35 Principle of operation of the Sagnac demodulator. A phase comparison unit locks an external VCO to the beat frequency of the ring laser gyro. The error voltage adjusting the VCO corresponds to to the desired FM demodulation signal.

The extraction of the frequency content of earthquake signals can be a demanding task, since this may involve a broad band of different oscillation frequencies, which either co-exist or change in rapid succession. Therefore it is desirable to apply a frequency demodulation approach to separate the seismic signal from the carrier. The overall frequency band of interest reaches from several nHz up to about 20 Hz, covering about eight decades. Applications in seismology evaluate the frequency range between 1 mHz and 20 Hz. Aircraft gyros operate in a different window again, namely in the range between 5 Hz and 4 kHz, in order to respond to aircraft motion with a sufficiently fast update rate for the strapped-down IMU. Frequency counters, auto regressive frequency estimators and FFT based frequency determinations become more accurate as the time of integration expands. Local earthquakes, on the other hand, have high frequency components in excess of 10 Hz, so that faster frequency conversion techniques are required.

Frequency demodulation techniques track the rate of change of the carrier phase. By locking a voltage controlled oscillator (VCO) to the output signal of the gyro, the rate of change can be extracted from the feedback signal of the VCO. Figure 3.35 illustrates this process in a block diagram. The result is a frequency to voltage conversion via a phase-locked loop (PLL). The conversion rate for the G-ring demodulator is shown in Figure 3.36. This diagram also reveals one of the major downsides of any technique that aims to cover a wide frequency range of about five orders of magnitude. While the dynamic range is very suitable for seismic studies, the demodulator does not have the necessary dynamic range to provide a very high frequency resolution. Therefore it will always be necessary to have more than one detection scheme in operation, in order to capture highly dynamic oscillations as well as very slow processes with high frequency resolution. At first glance this seems a little odd, so we have to remind ourselves that the measurement signal is a property of the rigid cavity and that there are no moving mechanical parts involved. In this realization, the cavity properties are probed by light. Figure 3.37 finally shows an example of a mag. 5.4 regional earthquake, which happened in 2003 in the Vosges (France). So far, this has been the closest significant event to the G ring

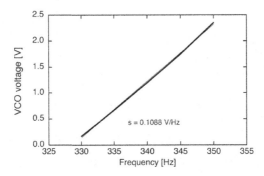

Figure 3.36 The conversion rate of the demodulator, established for the G-ring.

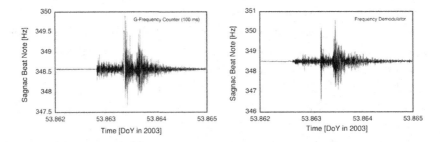

Figure 3.37 Comparison of the mag. 5.4 Feb. 2003 Vosges earthquake, recorded by the G ring laser using a frequency counter and by the FM demodulator technique. The higher bandwidth of the demodulator is evident.

laser. Since the demodulator still resolves frequencies of oscillation higher than 50 Hz, it maintains the full earthquake signature. The frequency counter method in comparison shows strong lowpass filtering effects, due to the 100 ms integration time. The signal excursions have lower peak values, and higher frequencies are clipped by the longer signal integration times.

3.7 Scale Factor

The ring laser equation (see Eq. 3.1) provides a linear relationship between the measured beat note of the interferometer and the rotation rate that the apparatus experiences. The larger the scale factor S, the higher is the sensor resolution. In the simplified expression

$$\delta f = S \cdot \Omega \tag{3.14}$$

with

$$S = \frac{4A}{\lambda P}, \tag{3.15}$$

the area A and the perimeter P provide the dimensions and the optical wavelength λ the nonius of the sensor. Since the P defines the laser cavity, both P and A are tied to the wavelength λ via the lasing condition, which requires that the cavity only supports an integer number I of waves ($P = I \cdot \lambda$). For an ideal square cavity with sides of length a we get:

$$S = \frac{4A}{\lambda P} = \frac{4a^2}{\lambda 4a} = \frac{I}{4}. \tag{3.16}$$

As long as the longitudinal mode index I does not change, we find that the scale factor is independent of the laser wavelength. Small changes in area and perimeter are compensated by a corresponding shift in the optical frequency. In this ideal case, we may also expect that the backscatter coupling will not change. However, in a real system we will always find that the relative distances between the reflection points on the mirrors will not remain the same for all arms of the interferometer, so that the backscatter coupling, and hence the observed beat note δf, will in fact always change. Nevertheless, this is not an effect that is related to the scale factor itself.

The situation changes fundamentally when the optical frequency shifts so far that the ring laser switches to another longitudinal mode, which then experiences more laser gain. Since we overpressure our cavities with gas pressures between 6.5 and 10 hPa, we usually experience mode changes when the optical frequency drifts by more than 100–120 MHz. This corresponds to an increase or decrease of the longitudinal mode index by 6 in the case of G and about 12 for the case of ROMY. This changes the scale factor by $I/(I + 6)$ for G, with I of the order of 25.3×10^6 and by $I/(I + 11)$ for ROMY with $I \approx 57 \times 10^6$. The scale factor for a square ring laser with the optical frequency shifting by more than one FSR can be rewritten as

$$S = \frac{(a + \Delta a) \cdot (f + \Delta f)}{c}. \tag{3.17}$$

For a shift of 112 MHz in optical frequency in the 16 m long cavity of G, we observe a shift of the longitudinal mode index by 6 FSR and a corresponding change in the observed Sagnac frequency by ≈ 165 µHz, which is about 5 parts in 10^7. For the triangular cavity of each of the ROMY rings, the scale factor becomes

$$S = \frac{\sqrt{3}(a + \Delta a) \cdot (f + \Delta f)}{3c}, \tag{3.18}$$

where we observe a shift of about 11 FSR, corresponding to ≈ 215 µHz or 4 parts in 10^7. These examples assume a uniform change over the entire cavity for simplicity.

For the case when the contour of the ring laser is slightly rectangular instead of a perfect square, the geometrical scale factor variation introduced by thermal expansion or pressure induced deformation still remains at a very low level within the regime where no mode change occurs. From a survey during the construction of the G ring laser, it is known that the nominal perimeter is 3 ± 1 mm longer than the design value of 16 m. Therefore we have modeled the cavity as a rectangle with sides $a = 4.002$ m, and $b = 4$ m, and for a worst case scenario we have assumed that the entire variation of the cavity length only affects the longer sides. From the continuous measurement of the optical frequency of the counter-clockwise propagating laser beam, we then computed the length of the resonator as a function of time to infer the effective change of the scale factor. Since the refractive index of the laser gas differs by less than 1 part in 10^6 from the empty cavity, we have neglected this in the calculation. The instantaneous cavity length P is

$$P = \frac{Ic}{f_o + \Delta f}.$$ (3.19)

The scale factor-related effect on the Sagnac frequency from the uniform expansion or contraction of the ring area is then obtained from

$$\delta\omega = \left(\frac{4\left(\frac{P-P_0}{4} + a\right)\left(\frac{P-P_0}{4} + b\right)}{\lambda P} - \frac{4ab}{\lambda_0 P_0} \right) \mathbf{n} \cdot \mathbf{\Omega},$$ (3.20)

with λ_0 and P_0 being initial values. Figure 3.38 represents the expected variation of the Sagnac frequency caused by the uniform variation of the scale factor for a contour very close to an ideal square over the run of a whole week. The effect was computed from the measured optical frequency in the cavity. Variations in Length of Day are causing the observed Sagnac frequency to change at the level of ± 5 µHz. At this level of resolution, the compression or expansion of the ring laser body due

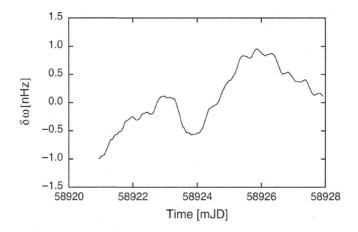

Figure 3.38 Changes to the geometrical scale factor of the 16 m^2 Zerodur structure from thermal expansion and compression induced by varying atmospheric pressure. The variation of the scale factor on the Sagnac beat note is inferred from the optical frequency of the counter-clockwise beam of the G ring laser.

to ambient pressure and temperature changes is entirely irrelevant. However, for the goal of observing the Lense–Thirring frame dragging effect, this variation can no longer be ignored. A solution is the stabilization of the lengths of the diagonals for a near-to-ideal square cavity to a constant value [176, 177]. For a rectangular or triangular cavity such as the UG-2 or ROMY ring laser, this effect is about two orders of magnitude larger.

Apart from a direct variation of the scale factor, there can also be an apparent variation. A changing temperature may also minutely act on the electronic circuitry that is required to run the gyroscope. Amplifiers and photodetectors exhibit a temperature dependence in their signal gain response. As shown in Figure 3.68 in Section 3.9.5, a change in the intensity of the laser radiation inside the cavity gives rise to a bias in the beat note, which for the G ring laser has been found to be as large as -42 μHz per nW of beam power. Stabilizing the beam power to the regime of about 1 pW of variation is required in such a case. None of the ring lasers that we operate exhibits equal intensities for the two beams traveling in opposite directions. While this difference in the beam powers of the G ring is at the level of only 1%, values of even 16% and 20% have been observed in both smaller rings (like the GEOsensor) and in larger rings (like ROMY) on occasion.

As the temperature changes the length of the cavity, the operating lasing frequency adjusts itself accordingly. We have observed temperature-induced shifts of the laser wavelength of more than 20 MHz, even for monolithic sensors made from Zerodur. Compression or expansion from the change of atmospheric pressure acts in a similar way. Under the influence of ambient pressure changes, the length of the cavity either expands or shrinks, depending on whether the pressure goes down or up. This was analyzed in more detail for the monolithic C-II ring laser, before the ring was isolated by an ambient pressure stabilizing vessel. Figure 3.39 displays the result. The mirror holders on the corners of the massive Zerodur block act in a similar fashion to a drum skin. An increasing ambient pressure bends the mirror holders slightly inwards. For all four mirrors of C-II taken together, we observe a shift of the optical frequency of 1.2 MHz per hPa. In comparison, we observe a pressure response of 300 kHz per hPa of ambient pressure in the G laboratory. The effect is smaller for G because the mirror holder plates were chosen to be thicker, in order to make G less sensitive to changes of the environmental parameters. Furthermore, the system can be put under constant pressure underneath a pressure isolating tank, employing a high resolution pressure regulator. With the pressure variation under control and a residual variation of less than 0.1 hPa, only the temperature drift from the annual cycle of the seasons remains as a major perturbing factor.

We illustrate these effects with an example. The G ring laser was operated for about 100 days under an ambient pressure stabilizing vessel, set to a constant value of 968.3 hPa, as shown in Figure 3.40. The temperature in the laboratory follows

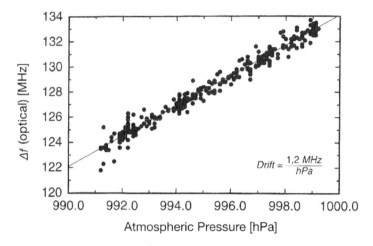

Figure 3.39 The dependence of the optical frequency on ambient pressure variations in the C-II ring laser resonator.

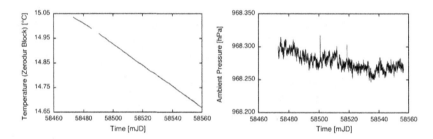

Figure 3.40 The temperature of the ring laser body and the ambient pressure in
the G ring laser laboratory over almost 100 days. While the pressure could be held
constant over this time, the temperature shows a distinct downward trend, following
the annual temperature cycle of the seasons.

the annual cycle of the seasons, which over the length of this measurement series
corresponded to an almost linear decreasing trend. There are no excursions in the
temperature, since the laboratory is well isolated and noone disturbed the measurement series. In addition, we measured the optical frequency of the cavity with an
optical frequency comb several times throughout the measurement sequence. Here
also a strictly linear trend was obtained. The optical frequency drops by 100 kHz
per day, as inferred from the comb measurements. The corresponding effect of the
rotation signal is a similar linear trend, but upwards, as illustrated in Figure 3.41.
There are a few wobbly excursions visible in the residuals of the observed ring laser
beat note. They are evidence of backscatter coupling, and the part with these small
excursions can be ignored for this discussion. A total of 0.4°C variation on the body
temperature of the monolithic ring lasers does not change the geometrical scale

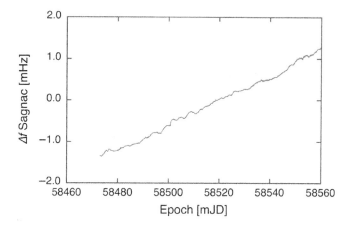

Figure 3.41 The ring laser response – minus a constant value of 348.5156 Hz – of the G ring under a pressure-stabilizing vessel to a linear downward drift in temperature is a corresponding linear upward drift. Superimposed are small excursions caused by backscatter coupling.

factor by a significant amount (see the discussion around Figure 3.38). The cause for such a large drift of the interferogram must therefore be related to a slow and continuously changing backscatter coupling of the laser beams (see Section 3.9.4). Furthermore, we have to take non-reciprocal laser dispersive effects from the influence of the non-linear dynamics of the gain medium into account. This is treated in more detail in Section 3.9.3.

3.7.1 Angular Random Walk

Apart from the desired extremely high sensor resolution, already treated in Section 3.6.5, the most important quantity of a large ring laser structure is the long term sensor stability. Since such instruments are designed to monitor the rotational motion of the Earth, it is not possible to fully separate the sensor behavior from the measurements. Microseismic activity, caused by excitations from wind and ocean waves, varies a lot in strength and can limit the available sensor resolution temporarily. Figure 3.42 illustrates this on the left side panel. Between day 58924 and 58928 in the modified Julian timescale, corresponding to March 15–19 in 2020, one can see a period where the microseismic activity is low and below or at the sensor resolution. When we only use this part of the measurement series, we obtain a sensor noise level of $2.7 \times 10^{-6}{}^{\circ}/\,h/\sqrt{Hz}$. The noise level taken over the entire dataset for comparison is about a factor of two larger, namely $4.6 \times 10^{-6}{}^{\circ}/\,h/\sqrt{Hz}$. Compared with commercial aircraft gyros, which are frequently reset, large ring lasers are required to be extremely stable, ideally over periods of many months.

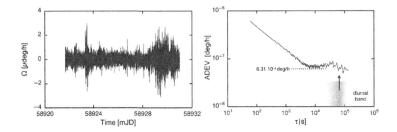

Figure 3.42 The Allan deviation, taken for the geophysically quiet period between day 58924 and 58928 in the modified Julian timescale, reaches a noise level of $2.7 \times 10^{-6 \circ}/\,\mathrm{h}/\sqrt{\mathrm{Hz}}$, while for the angular random walk a value of ARW = $6.31 \times 10^{-8 \circ}/\sqrt{h}$ is obtained.

In the jargon of navigation, the sensor stability is described as an angular random walk. For the comparison between the large ring laser gyros covered in this book and the performance of gyros for navigation we adopt the terminology used in the field of navigation [129].

The angular random walk is usually given in the units "$^\circ/\sqrt{h}$" and is taken from the $-1/2$ slope (white noise contribution) Allan deviation chart at one hour of averaging. This is shown at the right side panel of Figure 3.42. We observe an apparent angular random walk of ARW = $6.31 \times 10^{-8 \circ}/\sqrt{h}$ for the G ring laser. However, this dataset is still contaminated with residual contributions from geophysical signals. Although the orientation-induced contributions of polar motion and solid Earth tides have already been removed from the measurements through the application of theoretical models, there are still some residual signals left in the observations from the diurnal band of excitations. These come most prominently from a small ambiguity in sensor orientation, local deformation effects on the sensor and possibly deficits in the correction models. Since these extra components have a maximum period of around one day, they generate a corresponding systematic bump in the Allan deviation curve, so that the stability floor of the sensor in the $1/\sqrt{t}$ slope cannot be reliably established. Therefore, the obtained value for the angular random walk figure is a little larger than the unperturbed sensor would actually deliver. The correlation time of the gyro output in the absence of such geophysical signals is therefore somewhat longer than the ADEV plot suggests. Since the slope resumes the white noise behavior beyond the diurnal signal band, we can expect that the true correlation time of the sensor may reach well beyond one day.

3.7.2 Beam Wander

The highest sensitivity for a ring laser gyro is reached when the contour has the shape of a square. In this way the area can be maximized, while the losses are still

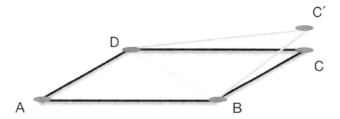

Figure 3.43 A square ring laser, although the most advantages design in terms of sensitivity, has the disadvantage that the area of the contour is not contained in one plane. The square ABCD is the preferred arrangement, while a contour like ABC′D is the likeliest one.

small. This, however, has the downside that the area is not unambiguously determined, because the beam path is not necessarily contained in one plane. Figure 3.43 illustrates this effect. When one of the mirrors steers the beam path out of the desired laser plane, the polarization of the laser beam in the cavity is no longer a clean s-polarization but becomes slightly elliptical, and consequently the losses in the cavity increase. While the criterion of preserving a pure s-polarization is helpful in principle, it turns out that it requires a significant amount of skew in order to reliably establish this change in polarization, which generates extra loss in the cavity due to the higher transmission of the mirrors for the p-polarization. For our heterolithic ring laser structures, a certain undesired amount of non-planarity in the range of up to 180 seconds of arc is not unusual.

For the very large rings like UG-1 and UG-2, this issue of non-planarity is exacerbated by the lack of mechanical stability of the entire ring laser structure. The permanent changing deformation of the Cashmere Cavern floor as a result of the variable atmospheric pressure [86] causes beam steering, both in the horizontal and the vertical plane. However, this effect of ground motion is not restricted to large structures. It even affected the C-II ring laser as a whole by tilting the entire ring laser support [184]. This deformation of an underground cave structure is greatest around the perimeter of a cave, where the corners are. Since our large rings were using as much of the cavern floor space as possible, they were running around the perimeter of the cave. Beam steering by cavern floor deformation causes the orientation of the effective normal vector on the ring laser plane to change, and it affects the enclosed area as well as the perimeter [163]. The repercussions of this are much larger for rectangular structures than they are for squares. Typical beam walk-off dynamics have been observed with three cameras, located behind the mirrors on three different corners, where cameras 1 and 3 were monitoring the shorter arms (21.015 m) and camera 2 one of the longer arms (39.703 m). Examples of the meandering traces of the beams are shown in Figure 3.44.

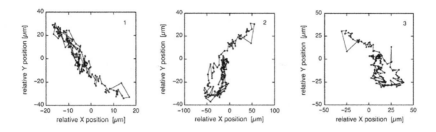

Figure 3.44 The observed beam walk on the two short arms and one of the long sides of the UG-2 ring laser taken at regular time intervals of 20 minutes over a period of 2 days.

Assuming a square ring laser completely planar along the corners $ABCD$, one finds the normal vector **n** representing the entire area. However, if one corner is slightly tilted out of plane, the effective area may be obtained by subdividing the full area into triangles and projecting the normal vector $\mathbf{n_i}$ of each triangle onto the vector of rotation and summing them up. Nearly all the large ring lasers listed in Table 3.2 are orientated horizontally on the Earth. Because Earth rotation is the dominant measurement signal, they show a strong latitude dependence of $\Omega_{\mathrm{eff}} = \sin(\phi + \delta_N)$, with ϕ corresponding to the latitude at the location of the instrument and δ_N representing a tilt toward north. East-west tilts are nearly negligible, since the cosine of an angle representing a small eastward tilt $\delta_E \approx 0$ is so close to 1 that it can be neglected, except for very strong seismic motion.

Due to the long beam trajectories of 39.703 and 21.015 m of the UG-2 laser, small mirror tilts, of the order of a few seconds of arc, cause a noticeable beam displacement on the next mirror. We have computed this variation in ring laser area (a), and it is displayed in Figure 3.45 together with the recorded atmospheric pressure in the cavern (b) during this measurement sequence. There is clearly some correspondence between the variation of pressure loading in the cavern and the variation in area. By using the camera measurements, augmented by the measurement of height variation, and measuring the intensity variations on a photomultiplier over a knife edge, the instantaneous perimeters, areas and true orientations of the two triangular subareas (see Figure 3.43) were inferred by raytracing based on the ABCD matrix formalism. While the area shows a variation at the level of a few parts per billion, the effect on the orientation and the perimeter is much larger. It turns out that the overall variation of the scale factor is in the parts per million regime. This is illustrated in Figure 3.46. We have determined the effective change in the scale factor from the variation of the area, in response to the deformation caused by the atmospheric pressure loading. The overall variation in the scale factor for the period of the experiment is of the order of 6 ppm, which is more than a factor of 1000 worse than required for high resolution rotation sensing in the geosciences. A fuller discussion can be found

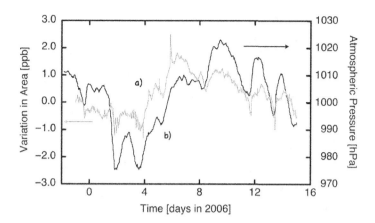

Figure 3.45 The effective area of the UG-2 gyroscope (a) under the influence of cavern deformation is changing at the level of a few parts per billion as the result of the beam wander. Overlaid is the variation of the atmospheric pressure (b) inside the cavern.

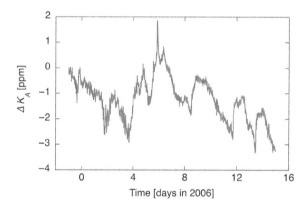

Figure 3.46 Time dependence of the variation of the ring laser scale factor under the effect of pressure loading. (Reprinted with permission of AIP Publishing from [182], ©2022, American Institute of Physics.)

in [163]. The same method had also been applied to the G ring laser; no apparent signs of a beam walk contribution were found in that data. The resolution limit for the measurements on G reached values below 1 μm.

3.7.3 The Effect of Earth Strain on Large Cavities

In Section 3.6.6 we inferred the variation of the arm length of a large active ring laser gyro from multiple beam combiners, which was compromised by the complication that it is not only the ring cavity that contributes to the observed variation in length.

Even small asymmetries in the extra-cavity beam combiner setup cause apparent changes in length. Here we introduce another method, which provides access to the changes in perimeter of very large cavity structures such as the UG-1 and UG-2 setups. Earth strain from the gravitational pull of the moon, with a theoretical value of 20 nanostrain, changes the length of the cavity by 1.5 μm for UG-1 and 2.4 μm for UG-2. On top of this come contributions from cavern deformations, as already discussed in Section 3.7.2. These, however, depend on the atmospheric pressure and are far less predictable. Operating multiple beam combiners at the G ring laser did not provide much useful information because of the superior sensor stability. Due to the massive Zerodur cavity structure, there were no perceptible strain effects on the cavity. Only asymmetries from the extra cavity free space path were found. Here we operate the UG-1 ring laser on two different longitudinal laser modes, in order to establish the area strain experienced by the gyroscope. This approach combines the rate bias (Sagnac signal) from the rotation experienced by the sensor with the geometrical properties of the cavity in the beat frequency. When the two beams in the cavity run on different optical frequencies, cavity length-related effects no longer cancel out. Since the time-varying effect of the perimeter is much larger than variations in the Earth rotation rate, we can directly obtain the combined area strain that the cavity experiences.

Since the FSR of UG-1 is only 3.9 MHz, we operated the gyro at a mode separation of 2 FSR, or 7.8 MHz in optical frequency. A change in cavity length of 2λ corresponds to 7.8 MHz. With $\lambda_1 = I/(I + 2)\lambda_2$ and for a cavity of 77 m, I is of the order of 121.7×10^6. This results in a cavity-related shift for the beat note of 64 mHz/λ (632.8 nm). For the example shown here, we selected a day where pressure effects were almost linear (Figure 3.47 (left)). Therefore, the effects of cavern deformations and tidal strain can be well separated. The 7.8 MHz beat frequency was extracted from a photo-multiplier, amplified and down-converted into the audio frequency range ($\approx$288 Hz) and then digitized. In order to exclude any

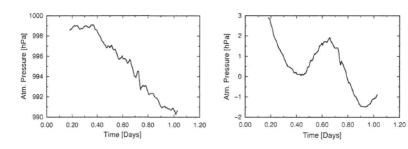

Figure 3.47 For a day of nearly linear pressure variation (left), we obtain the combined perimeter variation of 5 μm caused by pressure induced cavern deformation and tidal area strain on the UG-1 ring laser cavity.

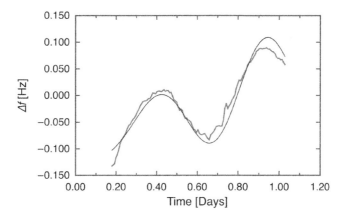

Figure 3.48 The combined effect of area strain and atmospheric pressure variation on a ring laser operating at two different optical frequencies. The observations (thick curve) agree well with a theoretical strain model (smooth curve).

unwanted drift on the measurement signal, both the mixing oscillator and digitizer were synchronized to a GPS controlled reference oscillator. The thick solid curve in Figure 3.48 presents the variation of the length of the cavity expressed as a shift in the measured FSR frequency, and the smooth line shows the expected area strain signal computed from a global Earth strain model. The model values were converted to represent the observed variation in the detector beat note. In order to account for the apparent upward drift due to the atmospheric pressure-induced cavern deformation, a linear drift term was added to best fit the observation. A detailed strain model for the Cashmere Cavern that can predict strain effects from pressure loading does not yet exist. In the beginning of Section 3.7 we investigated the effect of a small variation of the perimeter on the scale factor of a large area gyroscope and found that this is practically negligible for a regular and rigid cavity, as long as the gyro does not change the longitudinal mode index. Changes in the perimeter and area are compensated by the corresponding change of the laser wavelength. The increased homogeneous linewidth from the higher gas pressure in the laser cavity provides a much larger operating range for these very large cavities. From that point of view, these pure strain effects do not really matter. However, cavern deformations usually also cause significant tilt effects on the corner boxes of the gyroscope [184], so that beam steering effects emerge, which then cause considerable degradation to the scale factor stability (see Section 3.7.2). It is worth pointing out that this method provides the actual area strain very reliably – as opposed to the multi-beam combiner solution introduced in Section 3.6.6. Here we are only investigating the cavity behavior, and there are no external components, such as an asymmetric free space beam path, involved. This concept of inferring the strain acting on a long cavity from the beat note of two neighboring longitudinal modes is not limited to traveling wave resonators. It can also be readily applied to linear cavities.

3.7.4 Scale Factor Corrections from Varying Laser Gain

Following the alignment of the UG-2 ring laser to optimize the system for minimum losses and after refilling the laser cavity with a clean supply of Helium and Neon, a measurement sequence of approximately two weeks' duration was started. The Sagnac frequency was recorded with an integration time of 30 minutes. From the experimental setup, one would expect Earth rotation to produce a constant Sagnac frequency, with a variation of up to 1 mHz. However, the measurements reveal an overall downward trend with superimposed systematic excursions of considerable amplitude, where some occur rather sharply, while others follow a much smoother course. Most of these rather erratic departures are related to scale factor variations induced by beam steering effects from pressure loading of the cavern, and these effects have been treated in Section 3.7.2 Other systematic variations arise from the fact that the gyroscopes are HeNe gas lasers, and therefore they suffer from a continuous degradation of the laser gas purity, caused by outgassing from the cavity enclosure. As the laser gain reduces with time, a substantial drift in the measured Sagnac frequency develops. This effect is more pronounced on larger rings, which have considerably more surface area exposed to the laser gas. For the measurement series from the UG-2 ring laser, we applied all the necessary corrections to the scale factor from beam steering, as detailed in Section 3.7.2. This reduced the perturbations of the observed rotation rate noticeably. The remaining trend in the residuals shows a systematic downward drift of the observations, which is shown in Figure 3.49.

The existence of such a distinct systematic feature suggests the presence of an independent bias mechanism in the ring laser cavity. The consistent downward trend is caused by the contamination of the laser gas – primarily with hydrogen but also oxygen, nitrogen and water vapor – via outgassing from the pipes of the cavity enclosure, and this gives rise to an additional concentration-dependent loss factor in the gain medium. The presence of additional hydrogen in the cavity reduces the efficiency of the pumping process by providing additional de-excitation pathways. This effect has been observed in a different context for linear HeNe lasers by [5] and [91]. Since UG-2 has stainless steel tubes enclosing the laser gas along the entire perimeter of 121.435 m, it experiences a substantial amount of outgassing. Measurements of the total gas pressure inside the cavity were made on two occasions, each over a period of 56 days. Overall increases of 0.022 mbar and 0.020 mbar of hydrogen, amounting to $4 \cdot 10^{-4}$ mbar H_2 per day were observed, taking the UG-2 ring laser well into the regime where additional losses from absorption effects

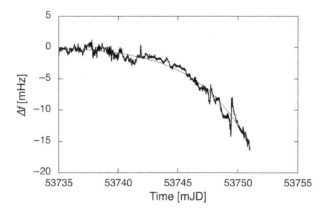

Figure 3.49 Time series of the residuals of the UG-2 ring laser observations with geometrical scale factors corrections applied. There is a systematic exponential trend remaining, which suggests the presence of another detrimental process. (Reprinted with permission of AIP Publishing from [182], ©2022, American Institute of Physics.)

by hydrogen become visible [163]. Because of the need to adjust large ring lasers to single longitudinal mode operation near the laser threshold, a feedback circuit is used to stabilize the gain medium to obtain a constant circulating beam power. Increasing losses in the laser cavity will therefore raise the loop gain accordingly. Following the approach of [12], K_A describes the contribution of the active medium to the scale factor of a large ring laser gyro as

$$\frac{\Delta K_A}{K_A} = \left(\frac{\Delta K}{K} \right)_N - \frac{aG}{1 + xP_o} + NL(\Omega). \tag{3.21}$$

In this equation $\Delta K_A / K_A$ corresponds to the scale factor correction from an arbitrary nominal value due to the active laser medium $(\Delta K / K)_N$ is the constant part of the scale factor correction, independent of dispersion and scatter, from the same nominal value. The second term on the right side allows for laser gain-related contributions, and the last term accounts for non-linear contributions, such as backscatter related coupling, which are neglected in the following discussion. Usually the gain factor G is considered constant with respect to time in ring laser theory. However, for the reasons outlined above one has to account for the progressive compensation of gas impurity-related losses by setting, for example,

$$G = G_0 e^{\alpha t}. \tag{3.22}$$

This choice of G is arbitrary and motivated by the observed behavior of the loss, as shown in [91]. The beam output power of Eq. 3.21 in the required form for large ring lasers with an 1:1 isotope mixture of ^{20}Ne and ^{22}Ne becomes

$$P_o = 2I_s A_b T \left(\frac{G}{\mu} \cdot \frac{\kappa_1 Z_i(\xi_1) + \kappa_2 Z_i(\xi_2)}{Z_i(0)} - 1 \right),$$ (3.23)

with I_s the saturation intensity, A_b the beam cross-section and T the transmission of the laser mirrors. Z_i is the imaginary part of the plasma dispersion function with lasing at a frequency detuning of ξ_n with respect to the corresponding line centers of the two neon isotopes, having partial pressures of κ_1 and κ_2, respectively. The most important part in this equation is the factor $\frac{G}{\mu}$, which represents the gain–loss ratio. This factor is approximately constant over the time of the measurements because of the feedback loop operation. Since there is no drift in the optical frequency involved, the contribution of the plasma dispersion function also remains constant. Therefore, the denominator in Eq. 3.21 can be approximated to $1 + x P_o \cong 1$, so that this equation reduces to

$$\frac{\Delta K_A}{K_A} = \left(\frac{\Delta K}{K} \right)_N - a G_0 e^{\alpha t}.$$ (3.24)

After a non-linear fitting procedure is applied to the dataset of Figure 3.49, one obtains a corrected dataset, as shown in Figure 3.50. This result is the best mid-term performance that was obtained from UG-2. Once the dramatic effect that hydrogen has on the operational stability of our large lasers was fully recognized, this led to a modification of the hardware design of our ring lasers. In the case of the G ring laser, the stainless steel enclosure of the cavity underwent an extensive heat treatment in order to extract as much hydrogen as possible, so that outgassing is considerably reduced. Furthermore, we have introduced a significant amount of porous getter material, corresponding to an active pump area of 36 m². In this way the gas quality has been stabilized, and G allows continuous operation over several years.

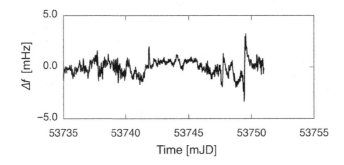

Figure 3.50 Display of the UG-2 observations with the correction for the laser gas contamination applied. (Reprinted with permission of AIP Publishing from [182], ©2022, American Institute of Physics.)

3.8 Long Term Geometric Cavity Stability

The most effective way to stabilize the output of a ring laser gyro is the tight control of the operation frequency of the gyroscope. This is not because the optical frequency has a significant effect on the scale factor; it is because the laser dynamics bias the obtained beat note significantly by small non reciprocal frequency shifts. A small uncompensated offset of, say, 10 μHz of one laser beam with respect to the other corresponds to a fractional frequency shift of 2×10^{-20}. With respect to a Sagnac beat note of 348.5 Hz, such an offset would already be 10 times larger than the effect of the variation in "Length of Day" contribution on the Earth's rotation. As long as we are interested in the correct estimation of variations of the experienced rotation signal, the strategy of tying the optical frequency as tightly as possible to a constant value has empirically been the best approach to date. The absolute value of the measured Earth rotation rate Ω_E for the geosciences is less important than the correct estimation of the variation of Ω_E over time. The low output signal levels of the gyro of 1–20 nW in the single mode regime and the corresponding low SNR for mixing the gyro laser with an external laser frequency reference is the largest obstacle in this process. The situation improves for the mode-locked regime, where beam powers of up to 100 nW are available from the gyro. This has detrimental side effects for the backscatter correction, however. The stabilization methods described in this section have been applied successfully to the various ring lasers of our group over the years.

3.8.1 Fabry–Perot and Iodine Laser Stabilization

Beating of coherent optical frequencies is a well established method in photonics. With a sufficient SNR, one can resolve optical frequencies with an uncertainty of less than 1 Hz. However, to achieve such a spectral resolution, beam powers in the regime of 100 μW up to several mW are required. The available leakage of light from a single mode ring resonator is of the order of $1 \text{ nW} \leq p \leq 20 \text{ nW}$. An iodine-stabilized HeNe laser acting as an optical reference oscillator, on the other hand, has sufficient beam power but smears out most of it, as the system has to sweep the laser mode by ±10 MHz across the selected absorption line of the iodine atom in order to lock the laser. The available intensity of the mixing of the sweeping laser beam with the very weak cavity beam is not nearly enough for a direct beat evaluation. Therefore, we have used a high resolution Fabry–Perot optical spectrum analyzer (Newport Super Cavity SR-130) with a finesse of F = 50,000 to tune the C-II ring laser to a constant optical frequency. Figure 3.51 sketches the applied setup.

The beam coming from the ring laser passes through a Faraday isolator to avoid a perturbation of the ring cavity from reflected light. It is combined with the beam from the Winters 100 iodine-stabilized laser in a beam splitter cube. The light from

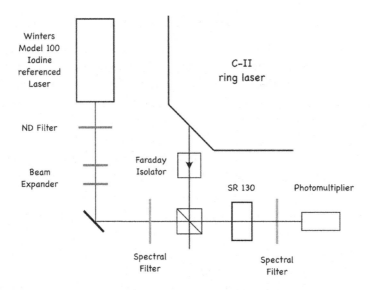

Figure 3.51 Block diagram of the reference laser stabilization setup. The combined light from the ring laser and a Winters 100 iodine-stabilized HeNe laser passes through a high finesse Fabry–Perot Newport SR-130. The light of each beam is detected by a photomultiplier, and the offset in frequency is computed.

the reference laser is collimated and adjusted in beam power to avoid saturation of the photomultiplier. A narrow spectral filter (1 nm bandwidth) reduces the amount of plasma light from the HeNe laser to enhance the SNR. The Fabry–Perot type SR-130 (Newport) has a resolution of 150 kHz, a transmission of 10% and a scanning range of 560 MHz per volt. A high gain photomultiplier and significant flux integration (several seconds) were used in order to locate the ring laser frequency within the Fabry–Perot tuning range. By digitizing the photomultiplier output voltage and directing the piezo of the Fabry–Perot under computer control, it is possible to periodically determine the relative offset frequency of the ring laser beam with respect to the reference laser. By controlling a piezo actuator pushing on one mirror holder of the ring laser cavity, it was possible to close the loop and to keep the C-II ring laser on the same optical frequency over a longer time [183]. Figure 3.52 illustrates this process. Since the scanning and sampling process takes some time, the concept is only applicable for an inherently stable gyro like C-II. The best results were obtained for a step size between 0.2 and 0.4 MHz.

While this method works very well for extreme low light levels, it has a very low bandwidth of 50 mHz or even less, which is only suitable for the compensation of the pressure and temperature drift of the ring laser cavity. The drift of the SR-130 super-cavity is also not negligible and was found to be 0.6 MHz/h, and the scanning and light integration process had to consider this as well. The

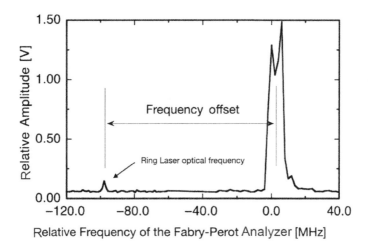

Figure 3.52 Illustration of the frequency stabilization process. The Fabry–Perot analyzer is scanned slowly across a small frequency range and captures the offset between the ring laser and the reference laser in the spectral domain. The ring laser cavity is tuned in a closed loop configuration to keep the spectral offset constant.

spectral resolution eventually obtained has a standard deviation of 300 kHz [183]. It should also be mentioned that this method depends on the condition that the iodine laser remains locked onto the same reference absorption line. While this has been obtained for several weeks on occasion, we have never continuously obtained this condition over more than one month. This method requires a very robust reference laser design. Another drawback of this concept is the small tuning range. Since pushing on a fairly massive mirror holder plate was the only way of tuning the cavity, the tuning range did not exceed 20 MHz. In the absence of a pressure-stabilizing vessel, which came several years later, and good temperature control, this is a serious limitation.

Figure 3.53 shows almost two days of stabilized operations of the C-II ring laser. The excursions on the trace of the Sagnac beat note are evidence to the fact that the regulation bandwidth of this concept is not high enough. The very low frequency trend in the rotation rate results from backscatter coupling. Nevertheless, it is worth pointing out that within the first year of operation, the C-II ring laser has taken the sensitivity of large ring lasers to the Earth's rate from several percent down to $\Delta f / f \approx 2 \times 10^{-5}$. The introduction of the atmospheric pressure-stabilizing vessel eventually achieved another factor of 10, and this marks the best performance level that was achieved with C-II until the devastating Christchurch earthquake in February 2011. Since that time, the ring installation in the Cashmere Cavern underneath the Christchurch Port Hills has not been accessible.

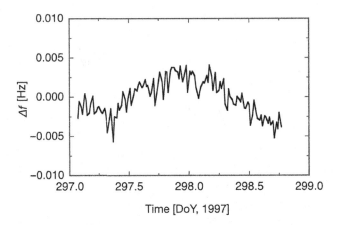

Figure 3.53 With the optical frequency stabilization in operation, the overall stability of the gyroscope improved by a factor of five over the unconstrained ring. However, the sensitivity to pressure induced variations reduced by a factor of ≈100. This left backscatter coupling as the next large error source.

3.8.2 FSR Stabilization

Over-pressuring the ring cavity with total gas pressures of up to 10 mB for the $\lambda = 632.8$ nm transition reduces the overall laser gain. On the other hand, it also increases the homogeneous line broadening of the laser. As a consequence, the gyroscope can be operated on a larger optical power because the regime where mono-mode lasing on the same longitudinal mode index is achieved is then much wider. When the laser power is increased toward the multi-mode threshold, the first additional modes appear at multiples of the FSR, with no additional laser modes excited within the first 100 MHz on both sides of the dominant mode. For the G ring laser with an FSR of 18.75 MHz, this turns out to be 112.4 MHz (6×FSR) or 131.2 MHz (7×FSR). For larger ring lasers, enclosing areas of more 10 m², this becomes an important operational feature because even on rings with a mechanically less stable gyro cavity, the jumps in mode index then become less frequent. The resonant frequency of the over-pressured cavity may shift by much more than just one free spectral range, and the laser mode index still does not change.

When G is operated in the phase-locked multi-mode regime, we find that the dominant laser mode is usually much larger in beam power than the additional modes on either side, each at a distance of 112.4 MHz. Figure 3.54 shows an example obtained from a scanning Fabry–Perot optical spectrum analyzer. Both satellite modes have the same offset from the main mode, and the beat signal between all three laser modes can be readily picked up by a wide bandwidth photomultiplier. All three modes typically operate in a phase synchronized state, so the obtained beat signal

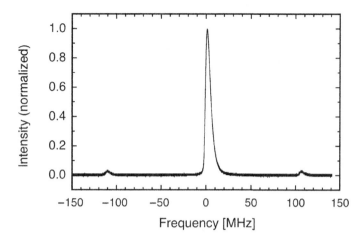

Figure 3.54 Mode structure of the G ring laser under phase-locked conditions, observed with a Fabry–Perot optical spectrum analyzer. The maximum intensity of the secondary modes at a distance of 6 × FSR is 3% of that of the dominant mode.

has a well defined and very sharp frequency, and this can be used for self-referencing laser beam stabilization of the gyro cavity. Since the FSR frequency is inversely proportional to the cavity length, we can stabilize the cavity and hence the optical frequency on this FSR beat note. The FSR is $f_{FSR} = c/P$ with P the perimeter and c the velocity of light. With $P = 16$ m $= I \cdot \lambda_{opt}$, we obtain a ratio of

$$\frac{f_{opt}}{f_{FSR}} = I \approx 25.3 \times 10^6, \tag{3.25}$$

which means that the FSR frequency has to be measured more than seven orders of magnitude more accurately than the desired stability of the optical frequency. In the case of the G ring laser and with an integration time of 30 s, we can establish the FSR frequency to just below 3 mHz. Taking into account that the measurement is based on a quantity of 6 × FSR, the resultant optical frequency stability is within 13 kHz, which appears to be sufficient for the purpose. However, due to the required long integration time of the FSR frequency, the bandwidth of this feedback loop is only 0.03 Hz wide, and it therefore can only compensate for a slow drift. On the other hand, the compression or expansion of the entire cavity length from temperature or atmospheric pressure changes remains the dominant effect that has to be compensated, and these are very slow processes.

Figure 3.55 depicts the block diagram of the experimental setup. The FSR signal is extracted from the counter-clockwise-propagating optical beam by a broadband photomultiplier. The beat frequency of 112.4 MHz is detected with a narrowband radio receiver, amplified, and compared to the local high-fidelity hydrogen maser.

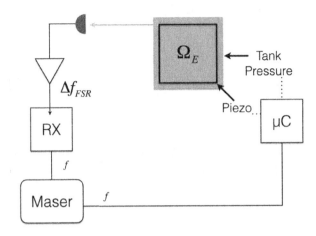

Figure 3.55 Block diagram of the two possible stabilization schemes. One can either lock the optical frequency of the interferometer to a H-maser or measure the free spectral range, compare it to the H-maser and hold in constant.

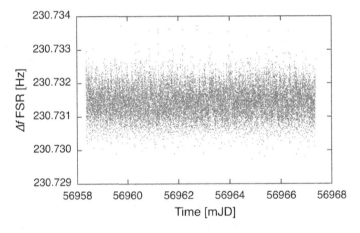

Figure 3.56 Time series of the radio-stabilized FSR beat note, which is held constant to within 2 mHz. This corresponds to variations of the optical frequency of the Sagnac interferometer of less than 10 kHz over almost 10 days.

A microcontroller determines an error signal from this beat note in order to keep this FSR beat frequency constant and feeds back to the laser cavity by driving a combination of a piezo actuator and the ambient air pressure inside the pressure tank. Driving the pressure inside the vessel in addition to the piezo stage essentially extends the range of the adjustment. A typical time-series of the controlled FSR frequency is shown in Figure 3.56.

The advantage of this control system is that it is straightforward and comparatively simple to set up. Another advantage is the fact that the exploited beat note has a high fidelity. The inherent passive stability of the G ring laser is beneficial, but not a necessary prerequisite. The G cavity changes this beat frequency very slowly and over a small range only. We have stabilized a number of large stainless steel gyros in this way and so extended the time between mode changes considerably. However, in all cases we have reached the end of the adjustable range for the cavity length controller sooner or later, as there was not a large enough stroke in the mechanical part of the system. In order to make the range larger, there are two piezo actuators operated on the G ring laser in parallel, each bending a 25 mm thick mirror holder plate very slightly, at the level of a few nm. This gives a full adjustment range of about 3 MHz for the optical frequency in the cavity. Increasing or reducing the ambient pressure inside the pressure vessel provides an additional adjustment range of 330 kHz per hPa. Taken together, the two tuning methods provide enough range for the G ring laser. However, any of our heterolithic stainless steel cavities require a much larger tuning range. This can be achieved by shifting an entire corner assembly in and out in a radial direction through a combination of a piezo and a stepping motor translation stage.

For Figure 3.57 we set up the G ring laser to stabilize the cavity of the FSR frequency in closed loop. A Menlo Systems FC1500-250 optical frequency comb (OFC) was used to monitor the optical frequency of the ring laser cavity, as shown in Figure 3.58. Both the radio receiver and the optical frequency comb are referenced to the h-maser, EFOS 18, of the Geodetic Observatory Wettzell. The middle panel in Figure 3.57 represents the FSR beat note. We note the vertical gap in the signal distribution, which was caused by a numerical artifact of the applied *Buneman Frequency Estimator* (see Section 4.3.2 for explanation) but has no detrimental effect on this type of measurement. The controller managed to keep the FSR frequency within the desired 3 mHz of the FSR signal. As a consequence, the optical frequency remained within a ±10 kHz wide window, as displayed in the top panel of the Figure 3.57. There is one larger outlier in the comb measurements and two small data gaps. These are not effects from the ring laser but are caused by the marginal contrast in the beat signal between the comb and the ring cavity. Although a 100 μW mono-mode transfer laser was part of the setup, the lock between transfer laser and comb remained marginal. The lower panel of Figure 3.57 shows the improved stability of the Sagnac beat note from the controlled ring perimeter. This stabilization scheme does not hold the unwanted backscatter pulling entirely constant, as it does not move the cavity length in a fully symmetric fashion, which would maintain the phase of the backscatter-induced shift of the beat note. However, it significantly reduces the variability of the backscatter coupling as compared to the entirely uncontrolled cavity. The signal contributions of all known geophysical signals have already been removed from the measured Sagnac frequency in Figure 3.57. Nevertheless, one can still observe a small amount of drift.

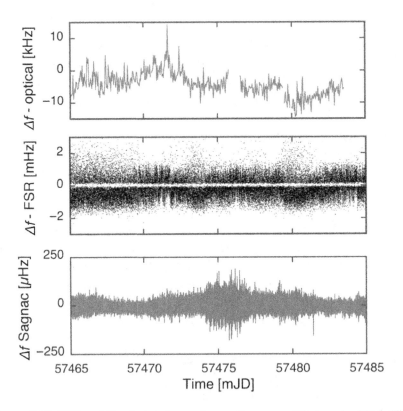

Figure 3.57 FSR-stabilized ring laser observations over 20 days in 2015. The top panel shows the stability of the observed optical frequency in the ring laser cavity. The middle plot presents the FSR frequency, which was used to stabilize the interferometer observations. The lower diagram depicts the observed Sagnac frequency. Stabilizing the optical frequency (to a constant value) improves the available sensor stability significantly.

There is also some variability in the noise level evident in the observations. Such increased noise levels are induced by a stormy weather pattern, either on the Atlantic ocean or over the observatory. As the ambient microseismic noise floor comes up, there are sufficient Love wave-type excitations contained in the ground motion to cause the ring laser structure to dither around the vertical axis and hence show the response from tiny rotational excitations. For this method of operation, it is important to note that the gyro has to be operated in the multi-mode phase-locked regime. This requires a larger value for the ratio of the intensity relative to the saturation intensity. In the presence of a small beam power difference between the two counter-propagating laser modes, one can observe that this difference becomes larger at higher beam powers. The result is a non-reciprocal frequency offset, caused by dispersion, which depends on the circulating laser powers. For the 32 m long

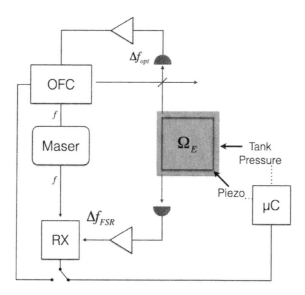

Figure 3.58 Depiction of the setup used for the measurement presented in Figure 3.57. The FSR measurement setup has been extended to use an optical frequency comb and a transfer laser (not shown) to monitor the FSR stabilization.

cavities of the ROMY ring laser with an FSR of 8.3 MHz, we find that it is very difficult to obtain phase locking on three modes, as there is no readily accessible phase-locked laser mode regime. The ring laser chose to either run on a single longitudinal mode; or when set to higher intensities, it broke out into seven or more laser modes of comparable intensity, which also showed significant mode competition.

3.8.3 Optical Frequency Comb Stabilization

Summarizing the results obtained so far from the various approaches to the stabilization of the optical frequency inside a large ring laser cavity, we come to the following conclusions:

- The geometrical scale factor does not change when temperature and/or pressure change the cavity length by less than ± 100 MHz and the longitudinal index of the laser mode does not change.
- When the optical frequency drifts in the cavity, it changes the phase angle of the backscattered laser intensity and hence the amount of backscatter coupling. Therefore, it is desirable to stabilize the optical frequency as well as possible.
- Running the ring laser cavity at higher intensities is not desirable because this increases the effects from dispersive frequency pulling, due to the observed

difference in the optical beam power between the two counter-propagating laser modes in the cavity.

- The advantage of a lower shot noise threshold is negligible in comparison to a reduction in the null-shift error.
- Practically, it is better to operate the cavity frequency stabilized as well as possible and then apply model corrections, rather than aiming for a moderate stabilization and a more refined model correction.
- While optical frequency stabilization at a low bandwidth already helps a lot on stable large monolithic gyros, it is expected that high bandwidth stabilization will still provide a significant improvement.

The result is a stabilization concept that references the ring laser cavity straight to a hydrogen maser via a Menlo Systems FC1500-250 optical frequency comb. Since the ring laser runs on a single longitudinal laser mode, it has a very low beam power of about 20 nW; we have to use half of the 100 μW transfer laser beam power to lock the transfer laser to the counter-clockwise beam of the gyro by keeping the beat note Δf_1 constant with a feedback loop. Then we use the other half of the transfer laser beam power to beat this signal with the optical frequency comb and then stabilize the beat note Δf_2 by feeding an error signal back to the ring laser. Since the comb uses an erbium-doped fiber for the laser radiation amplification, the setup required a frequency doubler stage to allow beating with the HeNe wavelength. This conversion caused extra losses and reduced the available beam power. When the beat notes had a sufficient SNR of more than 30 dB, it was possible to stabilize the ring laser to about 1 kHz of the optical frequency. Figure 3.59 provides a block diagram of the experimental setup.

Over a period of several days, we maintained the optical frequency in the ring laser cavity to better than 1 kHz by amplifying the beat note Δf_2 and using a frequency to voltage converter. As a consequence of this rigid locking scheme, the signature of the systematic errors in the Sagnac beat note could be drastically reduced [194]. Figure 3.60 illustrates the result obtained after the numerical removal of the geophysical signals. There are almost no deviations from random noise left. The signal excursion toward the end of the dataset is a measurement signal (local perturbation) and not noise. Figure 3.61 shows the corresponding Allan deviation for this measurement series (b) compared to the performance of the same system without stabilization (a). The controlled system predominantly exhibits white noise behavior. With a lower limit of 3×10^{-9} of $\Delta\Omega/\Omega_E$, it presents the best resolution that we have obtained from the G ring laser so far, albeit only for a short time. Toward the end of the dataset (b), the curve appears to begin to roll off from the dashed line, as it enters the regime of periods around 1 day. However, this is not necessarily an indication that the system is approaching its resolution limit (flicker

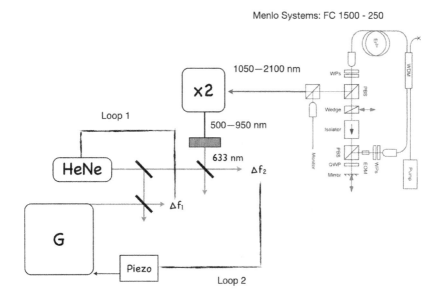

Figure 3.59 Illustration of the ring laser stabilization scheme based on a frequency comb and the application of an HeNe offset laser. The offset laser is locked to one of the laser modes in the gyro, while a piezo actuator is used to keep the beat note between the offset laser and the frequency comb constant.

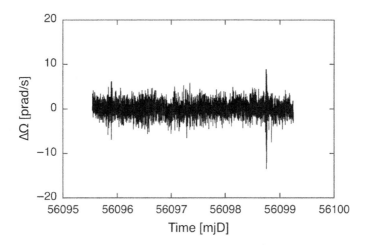

Figure 3.60 Time series of the residuals of the observed rotation rate Ω_E when the optical frequency in the ring cavity was locked to a hydrogen maser via an optical frequency comb. For better visibility, the known geophysical signals have been removed numerically from the observation data.

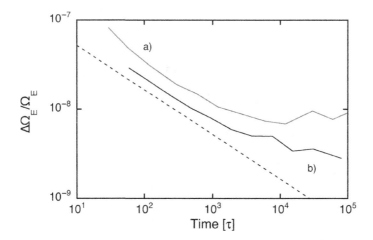

Figure 3.61 Time deviation of the observed rotation rate for the G ring laser. When the ring was locked to a hydrogen maser (b), a sensitivity of 3 parts in 10^9 of the Earth's rotation rate was achieved after one day. The unconstrained system (a) typically rolls off at about 1 part in 10^8.

floor). The measurements are reaching the diurnal window of the geophysical signals of the Earth, and we must still expect some uncorrected contributions, which result from model deficits of the applied corrections. For example, an error in the scale factor or an error in the established orientation of the sensor normal vector relative to the instantaneous Earth rotation axis (see Figure 3.22 in Section 3.6.3) would become visible here. So far we have not been successful in maintaining the closed loop operation with a bandwidth of about 10 Hz for more than several days, due to the low light level in the ring cavity and insufficient beam power for the link between the offset laser and the frequency comb. Nevertheless, this mode of operation is currently the most promising approach for stable high resolution Earth rotation sensing.

3.9 Ring Laser Error Corrections

3.9.1 Toward Absolute Scale Factor Determinations

Ring laser measurements have two key dimensions, the precision at which the Earth rotation rate can be estimated and the accuracy of the obtained values. In terms of precision we operate the G ring typically at the level of 1 part in 10^8 or slightly below. In terms of accuracy, however, the error limit at the time of writing is at the level of approximately 2×10^{-6}, and this has several reasons. G is a single axis gyro, and although the position of the laser plane on the platform has been carefully surveyed, we still have some uncertainties with respect to the orientation of the plane of the laser beams relative to the instantaneous Earth rotation axis. The resolution of the survey with respect to the orientation of the laser plane in space is optimistically judged to be maybe as good as 0.025 mrad, which corresponds to a bias error of 13 ppm. By making use of the known Earth rotation value and the introduction of small orientation angle adjustments in a process of minimizing the visibility of the diurnal polar motion signal (see Section 3.6.3), the effective orientation error can be reduced by, say, one order of magnitude. Only a full three axis gyro will improve this orientation-related error further and in a reliable way (see Section 3.12). Although the orientation provides a significant contribution to the systematic measurement error of a single axis gyro, it is not the only geometric error source to consider. Dispersion in the gain medium and the mirror coatings, lateral displacement of the beam caused by the effective thickness of the mirror coatings (Goos–Hänchen displacement), and finally the refractive index of the laser gas in the cavity require small corrections for the scale factor, approximately at the level of 1×10^{-8}. Due to the importance of this matter, we reproduce here the main results from [98], where these corrections were originally described.

Dispersion from Mirrors and the Plasma

The G cavity is filled with a mix of neon and helium gas, consisting of 0.100 hPa ^{20}Ne, 0.098 hPa ^{22}Ne and 9.682 hPa ^{4}He. The pressures have been measured at a temperature of 285.5 K, using a MKS Baratron 626AX11MCD pressure gauge with a nominal accuracy of 0.25%. The refractive index of helium was calculated to be 0.3183×10^{-6} using data and algorithms given in [212] and that of neon as 0.01254×10^{-6} using data from [31]. Assuming that we can add the refractive indices at the low pressures of this application, we obtain $n = 1 + 0.3308 \times 10^{-6}$. Any dispersion for n is too small to matter, since a change in wavelength of 1% will cause a change in refractivity by as little as 0.001%. In order to establish the optical frequencies in the ring laser, we assume the instrument is non-rotating and the Sagnac effect is zero as the two frequencies degenerate into the mean frequency of both. Resonance requires the total phase variation around the cavity to be an integer multiple of 2π rad. The total phase consists of several contributors, which are:

(i) **Cavity perimeter phase:** for perimeter P,

$$\varphi_P = 2\pi P / \lambda = 2\pi P n / \lambda_{\text{vac}}, \tag{3.26}$$

   where λ is the wavelength in the gas medium inside the cavity, n the refractive index of the laser gas medium and λ_{vac} the vacuum wavelength.
(ii) **Gouy phase:** The Gouy phase shift occurs in the propagation of laser beams and causes the phase velocity within a narrow laser beam to be slightly larger than that of a plane wave of the same frequency [204]. For a square cavity of

side length L and identical spherical mirrors of radius of curvature R, the Gouy phase around the cavity for the TEM$_{00}$ mode is

$$\varphi_G = 2 \left[\arccos \left(1 - \frac{L\sqrt{2}}{R} \right) + \arccos \left(1 - \frac{L}{R\sqrt{2}} \right) \right]. \qquad (3.27)$$

(This is derived from equation 19.21 in [204], making use of symmetry in a square cavity with spherical mirrors and 45° angles of incidence). The Gouy phase associated with a laser cavity is independent of both the wavelength and the refractive index of the medium in the cavity.

(iii) **Mirror reflection phase:** For an ideal Bragg stack mirror, at the design wavelength the reflected amplitude has a phase of π relative to the incident amplitude, and after four reflections round a square cavity there should be a zero net reflection phase change. A non-ideal layer thickness of the mirror coatings may cause phase anomalies at the design frequency, and furthermore, the phase varies nearly linearly with the optical frequency at frequencies near the design value [57]. We define φ_M as the mean *departure* from a phase change of π in reflection for the four mirrors in the cavity,

$$\varphi_M = \varphi_{M0} + \Delta f \frac{\partial \varphi_M}{\partial f}, \qquad (3.28)$$

with Δf the deviation in frequency from the central design value.

(iv) **Phase changes in the laser gain medium:** The spectral gain of the laser gain medium is accompanied by a frequency-dependent phase shift φ_K. This varies approximately linearly with the few hundred megahertz near the center of the laser gain curve [96]. Therefore φ_K can be written as

$$\varphi_K = \varphi_{K0} + \Delta f \frac{\partial \varphi_K}{\partial f}, \qquad (3.29)$$

with Δf the frequency deviation from the gain curve peak. φ_K is proportional to the single-pass fractional gain, which at equilibrium is equal to the round-trip cavity loss, and this in turn is inversely proportional to the ringdown time. In practice, for a HeNe laser system with approximately equal quantities of ^{20}Ne and ^{22}Ne, the gain curve is closely symmetrical about the frequency of maximum gain. The laser operating frequency is always close to this peak gain, and φ_K has a zero-crossing very close to peak gain, so for present purposes it is accurate enough to take the term φ_{K0} as zero within 0.0001 rad.

For the longitudinal mode number N, the lasing condition for the square ring cavity is:

$$\varphi_P - \varphi_G - 4\varphi_M - \varphi_K = 2\pi N. \qquad (3.30)$$

We express the path-length phase as $\varphi_P = 2\pi P n f_{\mathrm{opt}}/c$, where f_{opt} is the optical frequency in the cavity and c the vacuum light velocity. Rearranging this expression for the optical frequency yields

$$f_{\mathrm{opt}} = \frac{c}{nP}(N + (\varphi_G + 4\varphi_M + \varphi_K)/2\pi). \qquad (3.31)$$

It is useful to define a *reduced optical frequency* f^*_{opt} as $cN/(nP)$, related to f_{opt} by

$$f^*_{\mathrm{opt}} = f_{\mathrm{opt}} - \frac{c}{nP}(\varphi_G + 4\varphi_M + \varphi_K)/2\pi, \qquad (3.32)$$

which is equivalent to an observed optical frequency "corrected" for phase-changing effects related to details of the cavity, but not the refractive index. In order to determine the exact number N of waves around the cavity, we make use of the free spectral range F of the cavity. Furthermore we combine the two frequency-dependent phase terms from the mirrors and the gain medium into a single expression $\varphi_D = 4\varphi_M + \varphi_K$. By putting values $N + 1$ and N into Eq. 3.31 and taking the difference, we get

$$F = \frac{c}{nP}\left[1 + F\frac{\partial \varphi_D}{\partial f}/2\pi\right]. \qquad (3.33)$$

The beat note can be readily observed in a gyro by operating it in the multi-mode regime (see Sections 3.6.3 and 3.8.2). The mode spacing is *independent* of N, provided that we remain within the region of linear dispersion for φ_D. It is useful to define the *restored mode spacing* F^* by

$$F^* = \frac{c}{nP} = F\left[1 - F\frac{\partial \varphi_D}{\partial f}/2\pi\right]. \qquad (3.34)$$

Starting from the observable mode spacing F, we apply a correction for the dispersion induced frequency pulling and obtain:

$$N = f^*_{\mathrm{opt}}/F^*. \qquad (3.35)$$

To obtain the optical frequency from the counter-clockwise beam from the G cavity, we used an optical frequency comb (Menlo FC1500-250-WG) together with a transfer laser to compensate for the low beam power ($p \approx 20$ nW) from the ring cavity. Figure 3.62 shows an illustration of the setup. The optical frequency comb is referenced to a hydrogen maser. The power of a transfer laser is split into two parts. One of the beams is used to lock the transfer laser to a tooth of the comb by keeping the beat signal Δf_1 constant. The other beam from the transfer laser is mixed with the laser mode from the ring laser in order to obtain the beat note Δf_2. By operating the comb on several slightly different repetition rates, the optical frequency of the ring laser could be determined as $473,612,720.56 \pm 0.15$ MHz. Then we operated

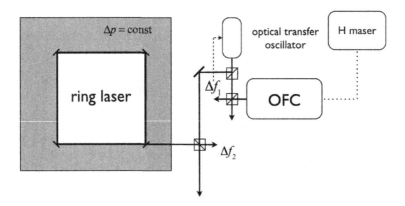

Figure 3.62 Illustration of the setup for determination of the optical frequency in the cavity. The optical frequency comb (OFC) is referenced to a hydrogen maser and locked to a transfer laser, by keeping Δf_1 constant. The other part of the transfer laser signal is mixed with the ring laser mode to establish the value of the optical frequency in the ring.

the laser in the same setup, that was used for the stabilization of the cavity length on the free spectral range, as detailed in Section 3.8.2. In this way the mode spacing was measured as $F = 18.734380691$ MHz. This value allows a preliminary calculation of the perimeter as $P = c/F$ and provides 16.0023 m. Assuming a square, the mean side length then is 4.0006 m. Using the measured radius of curvature of the mirrors $(4.05 \pm 0.04$ m), Eq. 3.27 gives the total Gouy phase as $\varphi_G = 6.487 \pm 0.045$ rad. For the phase shift associated with the mirror coatings, our best estimate of φ_{D0} is 0.045 ± 0.012 rad. With these values applied to Eq. 3.32 we get:

$$f_{\text{opt}}^* = 473,612,701.08 \pm 0.15 \text{ MHz} \tag{3.36}$$

The uncertainty is principally from the optical frequency measurements, with a nearly insignificant contribution from the terms of Eq. 3.32. Our best initial estimate for the factor $\partial \varphi_D / \partial f$ in Eq. 3.34 is the sum of the dispersion due to the four mirrors $(-5.595 \times 10^{-8}$ rad/MHz) and to the gain medium with 66.7 ppm single-pass gain $(-2.575 \times 10^{-8}$ rad/MHz), giving a total of -8.170×10^{-8} rad/MHz. Applied to Eq. 3.34 this value provides the restored mode spacing:

$$F^* = 18,734,385.425 \text{ Hz} \tag{3.37}$$

Considering that $F^* = c/nP$, we obtain a more accurate value for P as follows:

$$P = 16.00225212 \text{ m} \tag{3.38}$$

Finally, the square-bracketed dispersion correction factor on the right-hand side of Eq. 3.34 recurs in later expressions so it is useful to denote it as D:

$$D = 1 - F\frac{\partial \varphi_D}{\partial f}/2\pi. \qquad (3.39)$$

We can evaluate D accurately from $D = F^*/F$ as $1 + 2.5266 \times 10^{-7}$.

The Goos–Hänchen area correction

Reflected beams from structured dielectric surfaces, such as the super-mirrors in the G cavity, experience a lateral shift [47]. This is caused by the beam entering the coating, so that the effective surface of reflection sits below the top of the coating. The magnitude γ of the shift, measured perpendicular to the beam direction, is given by

$$\gamma = \frac{-\lambda d\varphi}{2\pi d\theta}, \qquad (3.40)$$

where θ is the angle of incidence and φ is the plane-wave reflection phase [47], which is polarization dependent. Furthermore, we experience a small change in the direction of the beam by an angle α, caused by the curvature of the mirror. For a square cavity, the Goos–Hänchen effect causes displacements of the beam positions on the mirror. Figure 3.63 shows a typical situation with a greatly enhanced Goos–Hänchen shift. The dashed line shows the ideal beam path in the absence of the Goos–Hänchen shift. The solid line illustrates the beam steering for a beam in the clockwise direction. The symbols a and b represent the displacements at arrival and departure, respectively, where a positive sign indicates that the resteered beam is outside the dashed line. Detailed analysis shows that the value of α controls the orientation of the truncated square defined by the dashed line in the figure. It is clear that $a + b = \gamma$ in every case, to first order in γ, irrespective of α. The area enclosed

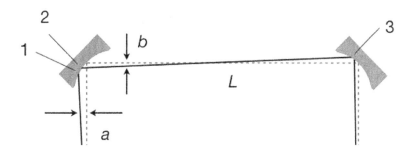

Figure 3.63 Visualization of the Goos–Hänchen displacement on one arm of a square ring laser. (Adapted with permission from [97], ©2022, The Optical Society.)

by the beam of the ring laser is determined by the average outward displacement of the beams relative to the dashed line in the figure. This is just the average of the quantities a and b and equals $\gamma/2$, so the total area enclosed by the beam is:

$$A = L^2 + 4L\gamma = L^2(1 + 2\gamma/L). \tag{3.41}$$

Since $L = P/4$ for a square cavity, in terms of the perimeter we get

$$A = \frac{P^2}{16}(1 + 8\gamma/P). \tag{3.42}$$

An important point is that the phase length of each side of the gyro is not changed by the GH effect. This may be seen by considering the sub-reflections from the layers of the mirror. The layers are designed such that the reflections are phase-coherent. The center of the departing beam in Figure 3.63 is at the point marked 2 on the left-hand mirror, but it has the same phase as if the center were at the point marked 1. The path from 1 to the point marked 3 on the right-hand mirror has the same length to first order in γ as the path for zero GH shift. Given the structure of the mirror dielectric layers, the mirror reflection phase derivative with angle can be evaluated numerically, and the calculated GH shift for the G is $\gamma = 0.1759$ μm. This leads to an area correction factor of $(1 + 8\gamma/P)$ in Eq. 3.42 to give the value $1 + 8.78 \times 10^{-8}$. However, this requires that the contour of the beams form a sufficiently accurate square. The planarity errors of G are below 0.05 mm, and the cosines of the associated dihedral angles are within 2 parts in 10^{10} of unity. Alignment procedures based on auto-collimation ensure that the corners of G are all within 10 seconds of arc of their required angle and the length of all arms are equal to within ± 0.15 mm. The associated perturbations to both area and perimeter are first order in this error, but the first-order perturbations cancel in Eq. 3.42, leaving second-order perturbations of around 3 ppb, verified by Monte Carlo modeling.

Scalefactor Corrections to the Sagnac Frequency

For the corrections derived in this section so far, we have treated a non-rotating ring cavity. In order to derive the Sagnac splitting, it is common to look at the difference in propagation time of two counter-traveling beams in the cavity [95] (see also Section 2.2) in a rotating frame. For one edge of a rotating polygon defined by the vertices V_1, V_2, the rotation causes a transit time delay of

$$\Delta\tau_{12} = \frac{2A_{12}\Omega}{c^2}, \tag{3.43}$$

where A_{12} is the sub-area defined by the perimeter segment V_1V_2 and the center of rotation. Ω is the experienced rotation rate of the cavity relative to an inertial frame. This delay is positive for a co-rotating beam and negative for the anti-rotating direction. In the absence of Fresnel drag from the intra-cavity laser gas, there is no extra effect on the experienced propagation delay, since the propagation in helium and neon is isotropic [162]. By summing over all sub-areas of the contour, the entire enclosed area A of the light beams is eventually obtained. These time delays generate extra phase terms, which according to Eq. 3.31 will generate different optical frequencies f_{opt1} and f_{opt2} in the two directions, and we get the phase terms

$$\varphi_{S1} = \frac{4\pi A \Omega f_{opt1}}{c^2} \quad \text{and} \quad \varphi_{S2} = \frac{-4\pi A \Omega f_{opt2}}{c^2}. \tag{3.44}$$

Inserting these terms into Eq. 3.31 for f_{opt1} and f_{opt2} at the same mode number N and forming the difference gives

$$f_{opt1} - f_{opt2} = \frac{c}{2\pi n P} \left[(f_{opt1} - f_{opt2}) \frac{\partial \varphi_D}{\partial f} + \frac{8\pi A \Omega f_{opt}}{c^2} \right]. \tag{3.45}$$

We relabel $f_{opt1} - f_{opt2}$ to f_S to denote the Sagnac frequency, and f_{opt} is the mean optical frequency as before. Rearranged for Ω we obtain

$$\Omega = f_S \frac{cnPD}{4Af_{opt}}, \tag{3.46}$$

with the dispersion factor D as defined in Eq. 3.39. Similarly expressed, for the Sagnac frequency we get

$$f_S = \frac{4A\Omega f_{opt}}{cnPD} = \frac{4A\Omega}{n\lambda_{vac}PD}. \tag{3.47}$$

Although Eq. 3.46 and Eq. 3.47 are highly accurate Sagnac equations, they do not obviously relate to the mode number in the cavity. With the application of Eq. 3.31 to replace the optical frequency, we finally get:

$$f_S = \frac{N\Omega}{4} \frac{(1 + 8\gamma/P)}{n^2 D} \left[1 + \frac{\varphi_G + \varphi_D}{2\pi N} \right] \tag{3.48}$$

Equation 3.48 has the same general form as introduced in the beginning of Section 3.7. However, there are now four correction factors included. We have derived them explicitly for the G ring laser, and Table 3.3 lists their values.

Table 3.3. *Correction factors for the scale factor of the G ring laser.*

No.	Quantity	Value for G	Estim. Uncertainty (ppb)
I	$(1 + 8\gamma/P)$	$1 + 8.78 \times 10^{-8}$	1.6
II	$1/n^2$	$1 - 6.616 \times 10^{-7}$	1.6
III	$1/D$	$1 - 2.527 \times 10^{-7}$	0.2
IV	$1 + \frac{(\varphi_G + \varphi_D)}{2\pi N}$	$1 + 4.11 \times 10^{-8}$	0.3

The last factor (IV) is the only one that depends weakly on N. Even when the ring laser is operated on a neighboring mode, at the current level of performance, this will not alter the value of the correction. Furthermore, the modification obtained here to the scale factor of the G ring laser is mostly a property of the cavity itself, which barely changes over time. Therefore, we can combine the values of Table 3.3 and obtain:

$$f_S = \frac{N}{4}\Omega(1 - 7.854 \times 10^{-7}). \tag{3.49}$$

Although this correction is apparently small, it represents a significant reduction of the systematic bias error in ring laser gyroscope applications in geodesy.

3.9.2 Stabilization of Diagonally Opposite Corners

So far, we have concentrated on the aspect of *stability* of large ring laser gyroscopes, since this is the main aspect that rotation sensing in the geosciences requires. The demand for *accuracy* on the recovery of the exact rotation rate of the Earth by a high resolution rotation sensing device is required in particular in the field of fundamental physics [34]. Apart from operational stability, this goal requires the exact knowledge of the scale factor, the exact orientation of the effective beam plane within the gyro and complete control over the laser dynamics in the cavity, as well as a perfect modeling of the backscatter coupling. A measurement device for the highly resolved recovery of the Lense–Thirring frame dragging on the surface of the Earth will always consist of a group of rotation sensors tied to each other, because this promises to be the best option to fulfill the requirement to establish the orientation accurately to ≈ 1 nrad from any of the gyro components to the instantaneous Earth rotation axis vector. Only when a local sensor array with a clear (interferometric) reference from each beam plane to the respective others is set up can it locally establish the Earth rotation vector accurately enough. Then it will be possible to deduce the tiny offset from the Earth rotation rate, which is caused by the relativistic frame dragging effect at the level of $\Delta\omega/\Omega_E \approx 10^{-9}$.

Other effects, such as geodetic precession, are of similar magnitude. In order to maximize the scale factor, most large ring lasers are designed in a square shape (for an explanation, see Section 3.5.1). Unlike for triangular structures, the effective gyroscope area of a square cavity is not accurately defined, even if the perimeter is tightly stabilized to a constant value. There are two degrees of freedom remaining. The ring structure can be sheared off to one side, or one of the corners can be shifted out of plane. Controlling the two diagonals of a square structure provides the necessary control over the geometrical part of the scale factor, and this played an important role in the discussion of the scale factor corrections in Section 3.9.1,

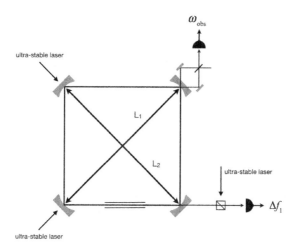

Figure 3.64 Schematic of the active stabilization process of a square ring laser scale factor. An ultra-stable reference laser is used to lock the diagonals L_1 and L_2 to the same optical frequency. The mirrors of the diagonals are shifted in position to a point that both diagonals also exhibit the same FSR. Finally, the beat note between the reference laser and the optical frequency from the gyro Δf_1 is exploited for perimeter adjustment. (Schematic adapted from [176]).

where the radial distance of the mirrors from the center and the flatness of the laser beam plane were significant contributors to the error budget.

This requirement has been studied in Pisa (Italy) and is documented in [176]. From this work the design goal emerged, that the scale factor can be sufficiently controlled when the absolute length of the diagonals are made equal and the perimeter is controlled in order to obtain a regular square shape. When the cavity shape is that of an exact square, the length of the perimeter has a minimum. The full stabilization concept of a square ring laser treats diagonally opposite mirrors as independent linear passive Fabry–Perot cavities. Each of these cavities is stabilized by a Pound–Drever–Hall locking scheme to an ultra-stable reference laser independently but on the same optical frequency, by tuning the length. Figure 3.64 illustrates the concept.

In order to adjust the two diagonals to the same geometrical length, the FSR of the diagonal cavities is excited by an additional modulation signal on the reference laser. The diagonal mirror spacing is then tuned to the same FSR value for both resonators. A length stabilization better than 1 part in 10^{11} has been demonstrated for two independent linear Fabry–Perot test cavities of the same length as the GP2 ring laser diagonals. The determination of the two FSR frequencies has been achieved to a few parts in 10^7, which corresponds to a difference in cavity length of 0.5 μm. At the time of writing, the stabilization concept is in the process of being implemented on the GP2 ring laser.

3.9.3 Error Contributions from the Active Cavity

Several mechanisms in a real world ring laser cause departure of the experimentally measured Sagnac frequency from the idealized value. In a generalized form, we can rewrite Eq. 3.1 as

$$\Delta f = K_R(1 + K_A)\mathbf{n} \cdot \mathbf{\Omega} + \Delta f_0 + \Delta f_{bs}, \qquad (3.50)$$

where $K_R = 4A/\lambda L$ is the geometrical scaling factor of an empty ring laser cavity. K_A accounts for additional contributions due to the presence of an amplifying laser medium, while Δf_0 allows for mode pulling and pushing because of dispersion and Δf_{bs} takes the coupling of the two laser beams in the presence of backscatter [11] into account. These last two effects are well established in the ring laser literature (e.g. [12, 236]) and are usually both very small and almost constant for the ring lasers discussed here. Ring laser applications in geodesy and geophysics require ultra-stable sensors with a relative resolution of $\Delta\Omega/\Omega_E < 10^{-8}$. Therefore, it is important to understand the nature and variability of these error contributions. We begin with the most significant error contribution, which is the backscatter coupling.

3.9.4 Backscatter Coupling

Due to unavoidable reflections, the laser light, scattered off an intra-cavity surface, is coupled into the beam traveling into the opposite direction. In our case the scattering occurs at the cavity mirrors themselves. According to [11] the beat frequency disappears entirely if the rotation rate falls below a threshold value of

$$\Omega_L = \frac{c\lambda^2 r_s}{32\pi Ad}. \qquad (3.51)$$

A is the area enclosed by the cavity, d is the diameter of the beam and r_s is the magnitude of the backscatter contribution from the mirror. Values for r_s are experimentally taken from the ringdown time measurements. For ring lasers of the size of C-II or smaller, Ω_L can reach a significant amount, but this is not enough to cause frequency lock-in by the Earth rate, as the primary source for the gyro excitation. The lock-in threshold of the major instrumentation of our working group ranges from 0.2 to 0.01 Hz and is several orders of magnitude away from lock-in, even for the smaller of our rings. This means that backscatter coupling always provides a serious error source, in particular when long term observations are required. To deal with this important problem adequately, we reproduce this error mechanism from the publication [97] of our working group here in this section in a very detailed treatment.

One can relate the backscatter coupling to the common and differential mode effects of the superimposed instantaneous intensities and the phases of the two counter-propagating laser beams and the corresponding back scattered light [11,

141, 226]. From the nonlinear coupled differential equations already introduced in Section 2.6, the frequency offset due to backscatter can be derived:

$$\dot{I}_1 = \frac{c}{P}\left[\alpha_1 I_1 - \beta I_1^2 - \theta_{12} I_1 I_2 + 2r_2 \sqrt{I_2 I_1}\cos\psi\right], \tag{3.52}$$

$$\dot{I}_2 = \frac{c}{P}\left[\alpha I_2 - \beta I_2^2 - \theta_{21} I_1 I_2 + 2r_1 \sqrt{I_1 I_2}\cos(\psi + \varsigma)\right]. \tag{3.53}$$

Equation 2.34 is used here in a slightly compacted representation,

$$\dot{\psi} = \omega_S - f_L\left[r_1\sqrt{\frac{I_1}{I_2}}\sin(\psi + \varsigma) + r_2\sqrt{\frac{I_2}{I_1}}\sin\psi\right] + \kappa(I_2 - I_1)\tau, \tag{3.54}$$

where $f_L = c/P$ is the frequency spacing of the cavity longitudinal modes, ω_S is the unperturbed Sagnac angular frequency; the parameters α, β and θ are (single-pass) gain, self- and cross-saturation coefficients, respectively; κ is a coefficient describing the differential frequency shift due to differential intensity-related dispersion in the gain medium (cross-dispersion); and τ is the cold cavity ringdown time. The parameters r_1 and r_2 are the fractional backscatter-coupled amplitudes and, ς is the backscatter phase. This formulation based on the intensities I_1 and I_2, including the laser dynamics by incorporating the laser gain medium with self- and cross-saturation effects, leads to a set of coupled non linear differential equations, which have to be solved. However, the effects of backscatter are not related to the laser process itself. Scattering happens at the mirrors by diffuse reflections, and it is therefore a property of the *passive* ring cavity [97, 206]. This means that a *linear* analysis approach can be chosen in order to obtain the required corrections for the perturbed counter-propagating optical frequencies in the ring cavity.

Passive Cavity Analysis

We assume a rotating ring cavity, which in the absence of backscatter would show a beat note ω_S, corresponding to Eq. 3.1. Furthermore it is assumed that ω_S varies sufficiently slowly that the circulating beams inside the cavity can be regarded as being in a steady state. For a large gyro, rigidly tied to the body of the Earth, this is always the case. To show the generality of this treatment, we may assume that a laser beam is injected at resonance into the cavity from an external source, similar to the experiment of Ezekiel and Balsamo [71, 175] and later [121] and [137]. We call this beam 1. Backscattering will now cause some leakage into the other sense of propagation. This gives rise to the counter-propagating beam 2, which is very slightly off resonance but still well within the empty cavity bandwidth. Nevertheless, due to the low losses of the cavity, the amplitude of beam 2 builds up to quite significant levels. Beam 2 will scatter back into beam 1 by the same

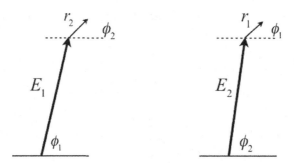

Figure 3.65 Illustration of the backscatter process, where a small portion of light is scattered from one sense of propagation into the respective other sense, with r_1 and r_2 the backscatter amplitudes, thus altering the phase angle of each beam by vector addition.

mechanism, causing a phase change by phasor addition and a shift of the original resonant frequency. This is illustrated in Figure 3.65.

One immediate consequence of this *double-backscatter* process is that it does not help to run the interferometer at two different longitudinal modes spaced apart by one or more FSR, since each beam will perturb itself; this is on top of the additional complication that the cavity related variations are no longer common mode effects, so that they do not cancel out entirely. This latter effect produces a relative resolution limit of the optical frequencies of $\lambda_1/\lambda_2 \approx 10^{-8}$ in an effort to explore this mode of operation, which as a result of the limited regulation bandwidth is about 1 order of magnitude worse than in common mode operation (see details of Section 3.6.4). Evidence for the double backscatter process can be taken from the following observation. When the G-ring laser is operated at lower gas pressures ($p < 4$ hPa), it can be started either on the same or split on a different longitudinal mode index for the two beams. When it starts split, due to the reduced homogeneous line broadening, the backscattered light of the stronger beam seeds a new laser oscillation in the opposite direction. For G it takes about four seconds, and the weaker mode then jumps to the same longitudinal laser mode as the stronger beam. Mode competition from the newly seeded beam quickly starves the original beam one FSR away.

We choose a reference point where we specify all amplitudes and we omit spatial phase variations. We characterize the scattering processes that injects beam 1 amplitude into beam 2 by the complex amplitude fraction r_1 and similarly r_2 for the scattering of beam 2 into beam 1. The reference point has been arbitrarily chosen, and if it were shifted along the beam, the arguments of both r_1 and r_2 would change in opposite senses. The complex product $r_1 r_2$, however, does not depend on the choice of the reference point. The argument of $r_1 r_2$ is a phase angle that is identified with

ς, the backscatter phase in Eqs. 3.52–3.54. With these preliminaries, the oscillation of beam 1 (at the perturbed optical resonance frequency) is then $A_1 e^{j\omega_1 t}$, where $\omega_1 = \omega_{10} + \Delta\omega_1$, with the perturbation $\Delta\omega_1$ still to be determined. The circulating complex amplitude a_{21} of beam 2 resulting from scattering from beam 1 is given by

$$a_{21} = \frac{A_1 r_1}{1 - \mu_2 e^{j\frac{\omega_S - \Delta\omega_1}{f_L}}}, \tag{3.55}$$

where μ_2 is the (real) amplitude propagation factor ($\mu_2 < 1$) for one circuit of beam 2 in the cavity. The amplitude reinjected back into beam 1 is then

$$\Delta A_1 = r_2 a_{21} = \frac{A_1 r_1 r_2}{(1 - \mu_2) - j\mu_2 \frac{\omega_S - \Delta\omega_1}{f_L}}, \tag{3.56}$$

where we have expanded the complex exponential to first order, and $(1 - \mu_2)$ now represents the fractional round-trip loss of amplitude of beam 2. The reinjected amplitude in beam 1 induces a phase change of

$$\begin{aligned} \Delta\varphi &= \mathrm{Im}\frac{\Delta A_1}{A_1} \\ &= \mathrm{Im}\frac{r_1 r_2}{(1 - \mu_2) - j\mu_2 \frac{\omega_S - \Delta\omega_1}{f_L}} \end{aligned} \tag{3.57}$$

by phasor addition. The corresponding shift of the resonant angular frequency for beam 1 is

$$\begin{aligned} \Delta\omega_1 &= f_L \Delta\varphi \\ &= f_L \mathrm{Im}\frac{r_1 r_2}{(1 - \mu_2) - j\mu_2 \frac{\omega_S - \Delta\omega_1}{f_L}} \\ &= \frac{f_L(1 - \mu_2)\mathrm{Im}(r_1 r_2) + \mu_2(\omega_S - \Delta\omega_1)\mathrm{Re}(r_1 r_2)}{(1 - \mu_2)^2 + \mu_2^2 \frac{(\omega_S - \Delta\omega_1)^2}{f_L^2}}. \end{aligned} \tag{3.58}$$

Repeating the derivation for beam 2, we get an expression similar to Eq. 3.58, namely

$$\Delta\omega_2 = \frac{f_L(1 - \mu_1)\mathrm{Im}(r_1 r_2) - \mu_1(\omega_S + \Delta\omega_2)\mathrm{Re}(r_1 r_2)}{(1 - \mu_1)^2 + \mu_1^2 \frac{(\omega_S + \Delta\omega_2)^2}{f_L^2}}. \tag{3.59}$$

To get a simpler expression for $\Delta\omega_2 - \Delta\omega_1$, the perturbation to the Sagnac frequency ω_S, we assume the losses for the two beams to be nearly the same and replace μ_1 and μ_2 by a single value μ. Assuming small perturbations, terms like $(\omega_S \pm \Delta\omega_{1,2})$ are well approximated by ω_S. Subtracting Eqs. 3.58 and 3.59 then leads to an expression for the error in the estimated Sagnac frequency:

$$\Delta\omega_S = \Delta(\omega_2 - \omega_1) \approx \frac{-2\mu\omega_S \text{Re}(r_1 r_2)}{(1-\mu)^2 + \mu^2(\frac{\omega_S}{f_L})^2}. \tag{3.60}$$

Equation 3.60 indicates that a passive ring cavity used to measure rotation rates suffers from backscatter-induced systematic errors, and these do not require the inclusion of any nonlinear effects from the laser gain medium. The expression also indicates that for given values of r_1 and r_2, a larger round-trip loss $(1-\mu)$ gives a smaller perturbation. The expression derived in 3.60 is for very small perturbations, when $(1-\mu)$ is almost zero. It is numerically almost identical to the expression obtained from numerical integration of Eqs. 3.52–3.54. So it provides an alternative description and must not be treated as an additional newly introduced mechanism.

Inclusion of a Gain Medium

So far the discussion has assumed a passive cavity with external laser injection. By introducing intra-cavity gain, which is just enough to overcome the cavity losses, we convert the system to an active ring laser gyro. This, however, gives rise to dispersive effects, which cause an additional perturbation to the resonant frequencies. The pulling of the active cavity frequencies toward the line center [204] reduces the observed Sagnac frequency, typically at the level of a few parts in 10^7. This, however, is not related to backscatter and is not treated here. Backscatter and other effects may cause unequal beam intensities, and this may result in a *cross-dispersion* perturbation of the Sagnac frequency. It is taken into account by the parameter κ in Eq. 3.54. This effect is usually very small and will be ignored here. For steady state lasing, the intensities of beams 1 and 2 adjust so that with partial gain saturation, each beam has just sufficient gain to overcome the round trip losses and to be self-sustaining. In the double-backscatter process, the amplitude reinjected into, for example, beam 1 not only changes the phase of A_1 but also its amplitude. The fractional amplitude increase is $\text{Re}(\Delta A_1/A_1)$, and the condition for a self-sustaining beam is $(1 - \mu_{\text{eff1}}) + \text{Re}(\Delta A_1/A_1) = 0$. Making use of Eq. 3.56 and its equivalent for beam 2, we get the expressions

$$(1 - \mu_{\text{eff1}}) = -\text{Re}\left(\frac{\Delta A_1}{A_1}\right) = f_L \frac{\text{Im}(r_1 r_2)}{(\omega_S - \Delta\omega_1)} \tag{3.61}$$

and

$$(1 - \mu_{\text{eff2}}) = -\text{Re}\left(\frac{\Delta A_2}{A_2}\right) = f_L \frac{\text{Im}(r_1 r_2)}{(\omega_S + \Delta\omega_2)} \tag{3.62}$$

for substitution into Eqs. 3.58 and 3.59. For simplicity several terms and factors have been omitted from Eqs. 3.61 and 3.62 because they are insignificant. With typical parameters for large ring lasers, the left-hand sides of Eqs. 3.61 and 3.62 are small with typical values around 10^{-8}. Thus, when substituted into Eqs. 3.58 and 3.59 the terms $(1 - \mu_{\text{eff}})^2$ in the denominators become negligible and the factor μ_{eff} in the numerator is close enough to unity to be ignored. After those simplifications we find that $\Delta\omega_1$ and $-\Delta\omega_2$ satisfy the same equation. We conclude that for an active ring laser gyro, unlike for the passive case, $\Delta\omega_2 = -\Delta\omega_1$ is a very close approximation and we replace $\Delta\omega_1$ by $-\Delta\omega_2$. With these simplifications Eq. 3.59 gives the expression

$$\Delta\omega_2 = -\frac{f_L^2 \text{Re}(r_1 r_2)}{\omega_S + \Delta\omega_2} + \frac{f_L^4 (\text{Im}(r_1 r_2))^2}{(\omega_S + \Delta\omega_2)^3} \tag{3.63}$$

and the full perturbation to ω_S is $2\Delta\omega_2$.

Analysis of Beam Modulations

In order to use Eq. 3.63 for the correction of measured ring laser data, r_1 and r_2 must be known. In practice we may deduce them from the intensity modulations of beam 1 and beam 2. Each of the two beams contains a small admixture of the respective other beam from the backscatter coupling and the difference in the resonant frequencies results in observable beat frequencies. The circulating amplitude of beams 1 and 2 are $A_1 e^{j\omega_1 t} + a_{12} e^{j\omega_2 t}$ and $A_2 e^{j\omega_2 t} + a_{21} e^{j\omega_1 t}$ respectively, with

$$a_{12} = \frac{-j A_2 r_2 f_L S}{(\omega_S + \Delta\omega_2)} \tag{3.64}$$

and

$$a_{21} = \frac{-j A_1 r_1 f_L S}{(\omega_S + \Delta\omega_2)} \tag{3.65}$$

with

$$S = 1 + \frac{j \text{Im}(r_1 r_2) f_L^2}{(\omega_S + \Delta\omega_2)^2} \tag{3.66}$$

The effect of S is mainly a very small phase change with a magnitude of less than 10 mrad.

The complex amplitudes b_1 and b_2 of the intensity beats, obtained by inclusion of the complex conjugates to the signals and squaring, are

$$b_1 = \frac{-2j A_1^* A_2 r_2 f_L S}{(\omega_S + \Delta\omega_2)} \tag{3.67}$$

and

$$b_2 = \frac{-2j A_1^* A_2 r_1 f_L S^*}{(\omega_S + \Delta\omega_2)}, \tag{3.68}$$

where asterisks denote complex conjugation. We simplify slightly by defining m_1 and m_2 as the *fractional intensity modulations* of the beams, $b_1/A_1A_1^*$ and $b_2/A_2A_2^*$, respectively. Equations 3.67 and 3.68 are then combined to give

$$r_1 r_2 = \frac{m_1 m_2^* S^2 (\omega_S + \Delta\omega_2)^2}{4 f_L^2}. \tag{3.69}$$

The argument of $r_1 r_2$ in Eq. 3.69 must equal the argument of the product $m_1 m_2^* S^2$ as all other factors are real. Consequently, the backscatter phase ς appearing first in the Eqs. 3.52–3.54 finally show up as the phase difference between the beat modulations of beam 1 and 2, modified by a small phase shift caused by the factor S^2. If we write M for $\mathrm{Re}(m_1 m_2^* S^2) - [\mathrm{Im}(m_1 m_2^* S^2)]^2/4$, then Eqs. 3.69 and 3.63 can be combined to end up with a linear equation in $\Delta\omega_2$:

$$\Delta\omega_2 = \frac{-(\omega_S + \Delta\omega_2)}{4} \tag{3.70}$$

Noting that the *observed* Sagnac frequency is $\omega_{\mathrm{obs}} = \omega_S + 2\Delta\omega_2$, this may be transformed to the more convenient form

$$\omega_S = \omega_{\mathrm{obs}} \frac{(4+M)}{(4-M)}. \tag{3.71}$$

Equation 3.71 cannot yet be checked against integrations of Eqs. 3.52–3.54 except for special cases, because we have not considered saturation in the laser gain medium. Without gain saturation, when β and θ in the rate equations are set to zero, we have an unstable situation whenever $\mathrm{Im}(m_1 m_2^*) \neq 0$. During the integration, amplitude injections cause the beam intensities to change in opposite directions until the weaker beam is extinguished. However in the case that can be checked, the Sagnac frequency corrections from Eq. 3.71 agree with the numerical integrations at a few parts per billion of the Sagnac frequency.

Inclusion of Gain Saturation

Gain saturation in the laser active gain medium in general may affect the amplitudes of the observed beam modulations m_1 and m_2 and also the Sagnac frequency perturbation. It is advantageous to transform the observed beam intensity modulations to a common-mode (CM) and differential-mode (DM) representation, via the relationships for the complex amplitudes:

$$m_C = \frac{m_1 + m_2}{2} \qquad m_D = \frac{m_1 - m_2}{2} \tag{3.72}$$

From Eq. 3.72 it can be shown, that the factor $\mathrm{Re}(m_1 m_2^*)$, occurring in the principal term of Eq. 3.70, is equal to $|m_C|^2 - |m_D|^2$, which implies that the Sagnac frequency perturbation can, to a good approximation, be expressed as the sum of two components, calculable from the CM and DM beam modulations separately. We then investigate pure CM and DM solutions to the Eqs. 3.52–3.54. A pure CM solution has both I_1 and I_2 equal to $I_0 + i_C e^{j\omega_S t}$. This can be realized by putting $r_1 = r_2$ and $\varsigma = 0$. Equations 3.52 and 3.53 then become identical:

$$\dot{I} = f_L I[\alpha - (\beta + \theta)I + 2r \cos \psi] \tag{3.73}$$

We take I_0 to be a solution of Eq. 3.73 for $r = 0$. Then assuming $i_C \ll I_0$ so that fluctuations in ψ can be ignored, Eq. 3.73 becomes

$$j\omega_S i_C = f_L I_0 [-(\beta + \theta)i_C + 2r], \tag{3.74}$$

from which we obtain

$$i_C = \frac{-2j I_0 r f_L}{\omega_S} \frac{1}{1 - \frac{j(\beta+\theta)f_L I_0}{\omega_S}}. \tag{3.75}$$

By comparison with the earlier Eqs. 3.67 and 3.68, we see that for CM beam intensity modulation, gain saturation has modified the beat amplitude by the factor

$$F_C = \frac{1}{1 - \frac{j(\beta+\theta)f_L I_0}{\omega_S}}. \tag{3.76}$$

Similarly we look for a pure DM solution to Eqs. 3.52 and 3.53, namely for $I_1 = I_0 + i_D e^{j\omega t}$, $I_2 = I_0 - i_D e^{j\omega t}$. This can be realized by putting $r_1 = r_2$ and $\varsigma = \pi$. Equations 3.52 and 3.53 are now the same, except for a change in sign. We solve Eq. 3.52 to get

$$i_D = \frac{-2j I_0 r f_L}{\omega_S} \frac{1}{1 - \frac{j(\beta-\theta)f_L I_0}{\omega_S}}. \tag{3.77}$$

The beat amplitude has been modified again. For the DM case we have obtained the factor

$$F_D = \frac{1}{1 - \frac{j(\beta-\theta)f_L I_0}{\omega_S}}. \tag{3.78}$$

To estimate the effect on the Sagnac frequency perturbations, we use Eq. 3.54. In the pure CM case, when $I_1 = I_2$, Eq. 3.54 takes a simple form that allows an analytical solution. The result is

$$\omega_{obs}^2 = \omega_{S0}^2 - 4f_L^2 r^2, \tag{3.79}$$

which exactly satisfies Eqs. 3.61 and 3.62. This means that the CM Sagnac frequency perturbation expressed in terms of r is unchanged, despite the filtering of the beam modulation expressed by Eq. 3.75. As a result, when we use the measured beam modulation to estimate the perturbation, we must apply a correction to restore the effect of the filtering. For the pure DM case, we can substitute the beam modulation solution Eq. 3.77 into Eq. 3.54 and after manipulation we obtain

$$\overline{\Delta\dot\psi} = \frac{2f_L^2 r^2 \mathrm{Re}(F_D)}{\omega_S}, \tag{3.80}$$

where on the left-hand side, the overbar denotes averaging over one cycle of the Sagnac frequency. When compared with Eqs. 3.61 and 3.62, we find that the gain saturation has modified the Sagnac frequency perturbation by the factor $\mathrm{Re}(F_D)$. On the other hand, $|m_D|^2$ has been modified by the factor $F_D F_D^*$. Remarkably these are the same: both are equal to $[1 + (\beta - \theta)^2 f_L^2 I_0^2 / \omega_S^2]^{-1}$. That is, both the DM beam modulation and the associated Sagnac frequency perturbation are changed by gain saturation, but the relationship between them is preserved. Consequently, the measured DM beam modulation does not require correction for the filtering shown in Eq. 3.77.

Finally we consider the general case of arbitrary modulations m_1 and m_2 of the beams, giving a mix of CM and DM modulations. We assume that the CM and DM components independently perturb the Sagnac frequency in accordance with Eqs. 3.71 and 3.72. This is done by comparison with integrations of Eqs. 3.52–3.54. The above derivations assumed very small beam modulations, $i_C, i_D \ll I_0$, and beams of equal power. In practice it is found that they produce satisfactorily accurate results for fractional beam modulations less than $\approx 2\%$.

Cross-Dispersion

So far, cross-dispersion has been disregarded. Although the process enters dynamically via the coefficient κ in the Eqs. 3.52–3.54, the effect is accurately quantified simply by the difference in *mean* intensities of beams 1 and 2. Any processes that cause such imbalances will generate Sagnac frequency perturbations if κ is nonzero. Double backscatter can cause such beam imbalances. In beam 1, $\mathrm{Re}(\Delta A_1/A_1)$ generates an amplitude gain or loss depending on sign. The change in single-pass fractional *intensity* gain is $2\mathrm{Re}(\Delta A_1/A_1)$. From Eqs. 3.61, 3.62 and 3.69 we get

$$2\mathrm{Re}\left(\frac{\Delta A_1}{A_1}\right) = \frac{2f_L \mathrm{Im}(r_1 r_2)}{\omega_S} = \frac{\omega_S \mathrm{Im}(m_1 m_2^*)}{2f_L}. \tag{3.81}$$

This is for beam 1. For beam 2, the change in gain is in the opposite sense. The resulting changes in beam intensity are equal in magnitude and in opposite directions. Their magnitude can be calculated from Eq. 3.52, which implies at steady state $\alpha - \beta(I_0 + \Delta I_1) - \theta(I_0 + \Delta I_2) + \mathrm{Re}(\Delta A_1/A_1) = 0$ and also $\alpha = (\beta + \theta)I_0$ and $\Delta I_2 = -\Delta I_1$. We obtain

$$I_2 - I_1 = \frac{\omega_S \mathrm{Im}(m_1 m_2^*)}{f_L(\beta - \theta)}. \tag{3.82}$$

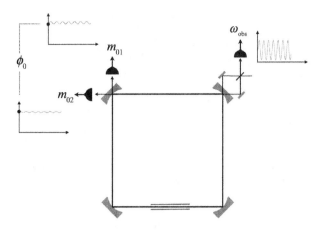

Figure 3.66 For the backscatter correction we have to read three channels.

The difference in beam intensities *increases* for longer ringdown time τ, which is not apparent from Eq. 3.82 but arises because of the way that β and θ depend on cavity loss. For a HeNe laser with an equal mix of isotopes ^{20}Ne and ^{22}Ne, backscatter-induced cross-dispersion Sagnac perturbations are typically more than an order of magnitude smaller than the usual backscatter perturbations.

Experimental Implementation of Backscatter Corrections

Here we look at a practical implementation of backscatter corrections to the observations of the interferogram made with a large ring laser gyroscope.

Figure 3.66 illustrates the process. The prime observable is the sampled oscillation of the interferogram ω_{obs} from the combined beam of the ring laser. Two additional photo-detectors sample each beam individually. From the recorded data of these detectors, we obtain the intensity of each beam (direct or DC component of the recorded signal) and the alternating component for each beam, from which we obtain the fractional beam modulations m_{01}, m_{02} and the difference in phase ϕ_0 between them. For the necessary corrections, several steps have to be carried out:

(i) Correct these first modulation estimates for bias: $m_1 = m_{01}(1 + m_{01}^2/4)$ and $m_2 = m_{02}(1 + m_{02}^2/4)$. This corrects for the measured mean beam power being proportional to $A_1 A_1^* + a_{21} a_{21}^*$ rather than the required $A_1 A_1^*$.

(ii) Apply an angular correction to ϕ for the S^2 factor in Eq. 3.69. $\phi = \phi_0 + \frac{1}{2} m_1 m_2 \sin \phi_0$.

(iii) Calculate the squares of the common-mode (CM) and differential-mode (DM) modulations: $m_C^2 = (m_1^2 + m_2^2 + 2 m_1 m_2 \cos \phi)/4$ and $m_D^2 = (m_1^2 + m_2^2 - 2 m_1 m_2 \cos \phi)/4$.

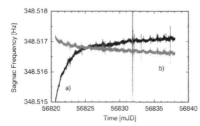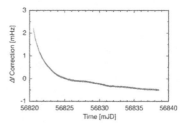

Figure 3.67 Example of a measurement series with the backscatter corrections applied. The left panel shows the raw data (a) and the corrected observation (b). The computed corrections from the auxiliary detectors are shown on the right.

(iv) Correct m_C^2 for gain saturation by multiplying: $1 + (\beta + \theta)^2 f_L^2 I_0^2 / \omega_S^2$. (The assignment of the values of β and θ are outlined in appendix A of [97]). For systems like the G or the G-0 ring laser, a value of 1.033 is used for the CM filter factor.

(v) Form the term M in Eq. 3.70: $M = m_C^2 - m_D^2 + \frac{1}{4}m_1^2 m_2^2 \sin^2 \phi$.

(vi) Use Eq. 3.71 to estimate the unperturbed Sagnac frequency: $\omega_S = \omega_{\text{obs}}(4 + M)/(4 - M)$.

Experimental Results from Backscatter Corrections

With the backscatter correction process integrated into the data logging program, the performance of the G ring laser can be significantly improved. Figure 3.67 shows one example. The curve (a) on the left panel shows the raw observations. Prior to the observation, the laboratory was prepared, which disturbed the temperature stability due to human interaction. It took almost two days to get back to a temperature equilibrium. Therefore, the observation drifts significantly at the beginning of the measurement series. However, this effect was well captured by the backscatter correction process. The panel on the right side shows the correction values obtained from the auxiliary two detectors for m_{01} and m_{02}. The result is the corrected time series (b). While we would expect a horizontally flat curve for a perfect correction, we still observe an almost linear downward trend. This indicates that the applied corrections have not captured all effects. The most likely cause of the remaining drift is an over-estimation of the DC intensities in the cavity. Apart from an error in the estimation of the correct numerical value representing the intensity, there is also the possibility of an additional contribution of residual plasma light, leaking through to the detector, despite the application of a narrow spectral filter. The other components involved, namely the AC amplitudes from the backscattered light and the relative phase between the backscatter contributions on the two beams in contrast, can be estimated much more accurately, due to their periodic oscillatory behavior. When we compare the percentage of the backscattered light for the G ring

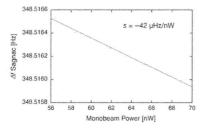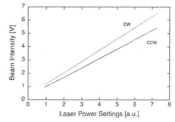

Figure 3.68 Observed dependence of the obtained Sagnac beat note on the intensity inside the ring resonator (left). Over the small range where gyroscope operation is feasible, this effect was linear. The intensity in the cavity over the entire working range does not increase at the same rate (right).

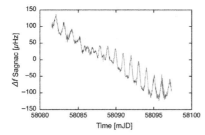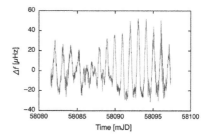

Figure 3.69 A time series of backscatter-corrected measurements of Earth rotation is shown on the left. With the drift of the optical frequency in the cavity inferred from a comparison with a later time series, a correction value of -13.7 µHz/MHz was found. With this correction applied, the dataset displayed on the right improved significantly.

laser and the horizontal component of the ROMY ring laser array, where both rings exhibit a comparable ringdown time of 0.9–1 ms, we find that the peak value for the necessary backscatter correction is about four times larger for G. The calculated spot size of the respective laser beams on the mirrors cover an area of 8.14 mm^2 for ROMY and 2.72 mm^2 for G. Extending these observations to include the C-II and the 6.5 m^2 ER-1 ring laser, we obtain a linear relationship of the amplitude of the backscattered light with respect to the enclosed area of 0.012% per m^2 for square gyros.

3.9.5 Non-reciprocal Cavity Effects

For an ideal ring laser, the two counter-propagating laser beams would be identical with respect to beam size and intensity. However, in practice there is always a noticeable difference in the beam intensities for all of the large ring lasers without exception. Differences in the intensities of more than 10% have been found for C-II and PR-1, while the G-ring shows a beam power difference of only about 1%. The underlying reasons are not fully understood, and it is currently believed that the super-mirror coatings exhibit some minute anisotropy in the form of birefringence, causing the effective cavity Q to be slightly different for the two senses of propagation. The observed effect is also fairly variable for different power settings and appears to also be sensitive to optical frequency detuning from line center. Using a model accounting for dispersive frequency detuning and hole burning [226], the resultant bias, suspected of giving rise to a contribution to the null-shift offset, becomes

$$\Delta f_0 = \frac{c}{2P} \cdot \left(\frac{\xi}{\eta} L(\xi) \frac{Z_i(\xi)}{Z_i(0)} G_0 \right) \cdot \Delta I, \tag{3.83}$$

with ξ the cavity detuning from line center, $Z_i(\xi)$ the imaginary part of the dispersion function, G_0 the gain and ΔI the observed difference in intensity. $L(\xi) = [1 + (\xi/\eta)^2]^{-1}$ is the Lorentzian, with η the relative value of the homogeneous broadening to the Doppler broadening. Equation 3.83 also shows that it is very important to keep the gain G_0 constant, which can be very difficult for a gas laser in the presence of gain medium degradation through hydrogen outgassing from the large surface of the stainless steel enclosed cavities. An example from the 16 m long G ring laser cavity illustrates this effect. In a measurement sequence we have slowly increased the beam power and observed the corresponding effect on the Earth rotation-induced Sagnac beat note. Figure 3.68 (left side) shows the result. The obtained beat note dependence in this power range was linear and showed a variation of the interferogram with a slope of $s = -42$ µHz/nW. In order to resolve the rate of rotation of the Earth with a stability of 1 part in 10^9 and assuming that this is the only non-reciprocal effect to consider, we will require a control over the optical beam power of the G cavity to be more stable than 10 pW. Alternatively, the reciprocity between the two beams in the laser cavity has to be increased. However, as Eq. 3.83 also suggests, the ring laser beat note not only strongly depends on the laser gain, where a beam power difference between the two laser beams generates a significant measurement offset (see Figure 3.68 right side). The interferogram also suffers from dispersion due to the detuning of the optical frequency. In Figure 3.69 we revisit our previous example, where backscatter corrections (detailed in the previous section) are already applied. The bumpy line represents the Sagnac interferogram with the polar motion signal on it (left panel). The gyro was operated under an ambient pressure stabilizing vessel, and temperature was the only remaining variable parameter. After an initial perturbation, the beat note was decreasing almost linearly, following the slope of the annual temperature curve in the underground laboratory. At the time of this measurement we did not have the possibility to measure the optical frequency in the cavity on a continuous basis. However, from a later similar measurement

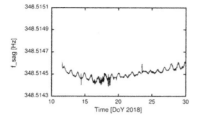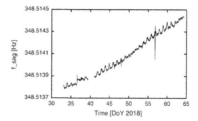

Figure 3.70 Raw observations from a longer series from the G ring laser, with a change in the RF excitation power of 6% as the only difference between the operational parameters of the two observation series. There is a significant difference in the observed drift rate.

series, where the optical frequency was available from an optical frequency comb, a drift rate of -200 MHz/K for the optical frequency could be established for the Zerodur structure of G. This obviously induced a trend of -13.7 µHz of the Sagnac beat note per MHz of optical frequency shift in the cavity. As Figure 3.68 additionally suggests, the obtained drift rate is not a given fixed value but is also highly dependent on the circulating beam power during the measurement. This has been experimentally tested by operating the G ring for two long observation series under the same conditions but at different power settings, which causes the beam power ratio between the two beams to change as indicated in Figure 3.68. Figure 3.70 illustrates the result. For the panel on the left side, G runs at an intensity 6% lower than for the graph on the right side, with a corresponding change in the ratio of the powers of the two counter-propagating beams of about 1%. Apart from a significant offset in the observed Sagnac beat note value, there is also a considerable difference in the observed residual drift rate of the interferometer. Since this situation does not change when the backscatter corrections are applied to the observations, as presented in Figure 3.71, it has to have a different cause.

At the time of these measurements, the optical frequency comb was not available. In order to analyze the dispersive bias more conclusively, we evaluated another data set, where the optical frequency could be tracked as well. The observed rotation rate over the entire measurement period is displayed in Figure 3.72. Day 58460 in the modified Julian date corresponds to August 12, 2018.

The raw measurements as recorded by the data logger are displayed in curve (a). They exhibit a steep linear upward drift. After applying the measured backscatter correction, detailed in Section 3.9.4, the dataset (b) is obtained. The drift is significantly reduced, along with some of the structure in the line of the raw measurements. This curve is a little shorter than that of the raw measurements because the backscatter correction procedure failed for the trailing end of the curve as a result of insufficient fidelity of the phase offset determination due to the small backscatter

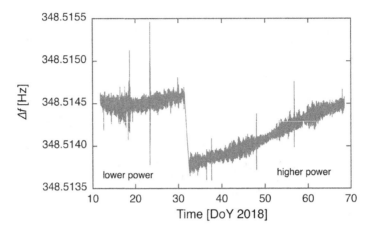

Figure 3.71 The two data series shown in Figure 3.70 with the backscatter corrections applied. The initially observed high drift rate remains, which indicates that there is an additional mechanism involved.

amplitude. As the stability of the scale factor is not a concern in this experiment, the remaining upward trend of the observed rotation rate must be caused by the drift of the optical frequency in the cavity from another biasing mechanism. Unequal beam powers are caused by a different cavity Q-factor for each sense of propagation. When we allow for this observed fact in the calculation of the dispersion related frequency offsets of the cavity frequencies, detailed in Section 2.5, we obtain a shift of the beat frequency of 8 μHz per 100 kHz of optical frequency shift. When this is applied to our backscatter-corrected observations, we obtain the curve (c) in Figure 3.72. While the temperature-induced drift could be entirely removed, there is still some structure in the observations remaining, exhibiting a peak to peak variation of ±100 μHz over the entire length of this particular measurement series. It is believed that most of this variation is due to shortcomings of the auxiliary data extraction.

3.9.6 Subtle Cavity Effects

The Effect of the Optical Frequency on the Plasma Brightness

When the G ring is operating under stable conditions – that is, in the vicinity of the turning point of the annual temperature curve and under stable atmospheric pressure inside the pressure vessel – the system experiences a low drift of the optical frequency in the cavity of about 8 MHz in two months. While the optical frequency gently drops, in this particular case we find a similar trend in the brightness of the

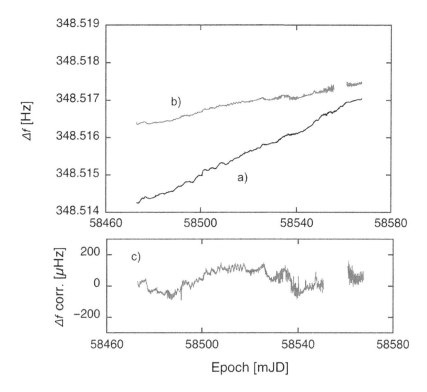

Figure 3.72 The observed raw Sagnac frequency in the presence of temperature drift (a), and the same dataset with the measured backscatter correction applied (b). A further reduction in the remaining drift is obtained when the effect of dispersive pulling as a result of unequal beam powers is included (c). Day 58460 in the modified Julian date corresponds to the August 12, 2018.

laser gain medium (plasma), although the feedback loop for the beam power of the regulated beam remained constant. Figure 3.73 illustrates this effect.

Over a period of more than 60 days, the optical frequency slowly and consistently drifted over a range of 8 MHz (left side scale of Figure 3.73). Apart from two incidents when the system was briefly disturbed by human intervention, this frequency drift was very consistent and without significant excursions, as one can see from the curve of the optical frequency. (The measurements during the period of perturbation have been removed). The brightness of the plasma glow of the laser gain medium was monitored simultaneously by a photo diode with completely independent instrumentation. A spectral filter with a 1 nm passband around the laser wavelength of 633 nm was used to extract the spontaneous emission from the plasma glow (right side scale of Figure 3.73). The observed brightness voltage dropped continuously, developing a similar trend to the optical frequency. On day 59128 in the modified Julian date, we started to lock the gyro cavity to the hydrogen maser of

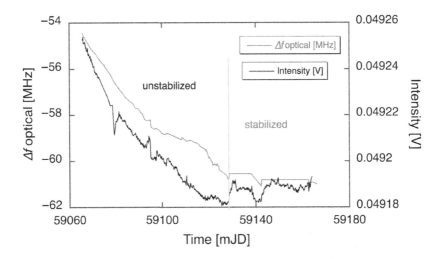

Figure 3.73 Under pressure and temperature stabilized conditions, we observe that the optical frequency drift and plasma brightness trend shows the same correlated trend. The plasma brightness remains constant in this example when the optical frequency is locked to a reference.

the observatory. This removed the trend from both quantities – the optical frequency and the plasma glow.

The Effect of Variable Beam Power

We observe a comparable drift for the ratio of the beam powers in the cavity and the observed Sagnac beat note. The most plausible explanation for this correspondence would be that the unconstrained optical frequency was drifting toward the line center, which would make the laser excitation process more efficient. Since the feedback loop adjusts the brightness of the regulated laser beam to remain constant, more efficient laser radiation excitation results in a reduction of injected RF-power, and that causes a reduction in plasma brightness. Does this have any effect on the observed Sagnac frequency? There is an effect, but there is no definitive and simple answer to the underlying mechanism, as Figure 3.74 illustrates.

Apart from the effect of the optical frequency on the plasma brightness, there is also a variation in relative intensity between the two beams. While the intensity of the clockwise beam does not change, since it is part of the feedback loop circuit, the counter-clockwise beam intensity dropped almost linearly by a little more than 4% over the entire measurement period of more than 100 days, namely by 40 mV,

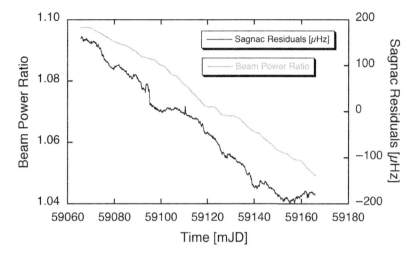

Figure 3.74 The trend in beam power ratio correlates reasonably well with the trend of the residuals of the Sagnac frequency. The lower curve displays the Sagnac frequency residuals, and the upper curve reflects the beam power ratio of the two cavity beams.

starting at 0.97 V. This effect is not temperature induced, since the temperature showed a strong non-linear behavior in this time period. Neither does it depend on the optical frequency, since the slope does not show any change in trend when the optical frequency is stabilized. Unequal relative beam powers cause a non-reciprocal frequency bias in the Sagnac beat frequency, which appears to be the most obvious effect that we observe here. While we can clearly observe this effect, so far we have no convincing mechanism identified that drives it.

Null-shift from the Fresnel Fizeau Effect

Here we refer to another non-reciprocal bias creating a potential issue. Although we expect this effect to be present in our systems, it has not yet been unambiguously identified in our observations (see next section below). This null-shift bias occurs when the laser gain medium is moving along the center of the gain tube in a convective flow, driven either by temperature gradients in the plasma or an electric field, like in the case for DC voltage excitation. The resultant frequency bias δv is [39]

$$\delta v = \frac{2(n^2 - 1)vl}{\lambda P}, \tag{3.84}$$

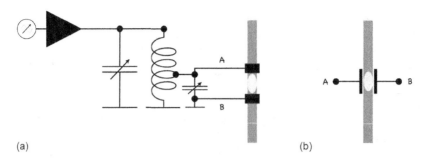

(a)   (b)

Figure 3.75 Illustration of the plasma excitation setup. The 50 ohm asymmetric radio frequency, RF is transformed into a high impedance nearly symmetric frequency. With the electrode arrangement parallel to the laser beam (a), some residual plasma flow might be generated. This is not the case for the arrangement in the orthogonal direction (b).

where n is the refractive index of the low pressure gain medium, v the velocity of the convection and l the extension of the flow region. Finally, λ and P are the wavelength of the laser oscillation and the total length of the cavity. Applied to the G ring laser hardware, we can assume $l \approx 20$ mm and $P = 16$ m. For a rather large value of $v = 4$ mm/s, this would then produce a null-shift bias of $\delta v = 47$ μHz, a value that would hurt our measurements considerably, in particular if v turned out to be variable. However, the scenario presented here assumes the existence of a convection current, with the main flow in the center of the gain tube going into one direction only. This is very unlikely, and we have not yet clearly identified this effect. For the detection of variations in Length of Day, any perturbation coming from this effect is required to contribute by less than 1 μHz.

Non-reciprocal Effects from the Gain Medium Excitation

In Section 3.9.5 we reported an intensity-related bias effect of -42 μHz/nW. A non-reciprocal frequency shift, which increases with the available beam power, suggests the presence of some plasma interaction. So it is necessary to review the laser excitation scheme of our gyroscopes. Figure 3.75a sketches the applied concept. The RF-frequency (40–90 MHz) coming from an oscillator is amplified to a power of 0.1–20 watts into an asymmetrical load of 50 ohms. Since a low impedance signal would neither strike nor maintain the plasma, an impedance transformer assembly similar to that shown in Figure 3.75a is inserted, to obtain the high impedance symmetric RF-signal that can strike and maintain the plasma and therefore excite the laser beams of the gyroscope. The figure also indicates the arrangement of the electrodes around the capillary of the ring laser. The size of the plasma is of the order of about 2 cm, and the plasma is located mostly between the two electrodes. The electric field vector in this configuration is oriented in parallel to the laser beams.

Since the chosen RF-matching may not be truly symmetrical, the plasma is then also not entirely symmetrical with respect to the two electrodes. This probably causes a very small power-dependent local gas flow as discussed in the previous section above, which leads to the observed non-reciprocal frequency shift. By re-arranging the electrodes, such that the electric field vector is oriented orthogonal to the two laser beams (see Figure 3.75b), this bias could be considerably reduced. In this new arrangement we obtain for a range of the laser beam power between 5 and 15 nW a residual power-related bias of -2.5 µHz/nW.

3.9.7 Null-shift Corrections from Non-linear Plasma Dynamics

The presence of an active gain medium inside the cavity of a traveling wave Sagnac interferometer gives rise to frequency pulling and pushing effects on the counter-propagating laser beams. The most obvious evidence for an asymmetry is the difference in observed beam power in any large ring laser that we have built or which we have seen in operation. Unequal beam powers indicate a difference in loss between the *clockwise* and the *counter-clockwise* cavities. We recall the slightly rearranged differential equations 3.85–3.54 set up by F. Aronowitz [11] that govern the operations of a laser gyro. The non-linear plasma dynamics have been studied extensively by our colleagues in Pisa (Italy). Here we summarize their results and follow the interpretation as provided in the references [16, 54, 55]. It is

$$\dot{I}_1 = \frac{c}{P}\left[\alpha I_1 - \beta I_1^2 - \theta I_1 I_2 + 2r_2\sqrt{I_2 I_1}\cos(\psi+\epsilon)\right], \quad (3.85)$$

$$\dot{I}_2 = \frac{c}{P}\left[\alpha I_2 - \beta I_2^2 - \theta I_1 I_2 + 2r_1\sqrt{I_1 I_2}\cos(\psi-\epsilon)\right], \quad (3.86)$$

$$\dot{\psi} = \omega_s + \sigma_2 - \sigma_1 + \tau_{21}I_1 - \tau_{12}I_2 - $$
$$- \frac{c}{P}\left[r_1\sqrt{\frac{I_1}{I_2}}\sin(\psi-\epsilon) + r_2\sqrt{\frac{I_2}{I_1}}\sin(\psi+\epsilon)\right], \quad (3.87)$$

where I_1 and I_2 are the respective intensities of the two beams, ψ the instantaneous phase difference and ω_s the angular frequency. The quantities $\alpha_{1,2}$ and $\sigma_{1,2}$ represent values for amplification minus losses, $\beta_{1,2}$ the self-saturation and $\theta_{12,21}$ and $\tau_{12,21}$ the cross-saturation between the two beams. The backscatter amplitudes and phase are given by $r_{1,2}$ and ϵ. As studied in a Monte Carlo simulation in [16], the random fluctuations of intensity ratio I_1/I_2, the backscatter amplitudes $r_{1,2}$ and the backscatter phase ϵ are polluting the observed Sagnac beat note as the relative noise in operational laser parameters is converted to frequency noise by the free spectral range (FSR). As a consequence, large ring lasers are less affected than smaller ones. For a ring laser such as G, values for the operational parameters for the Lamb theory of a laser, $\alpha_{1,2}, \sigma_{1,2}, \beta_{1,2}, \theta_{12,21}, \tau_{12,21}, r_{1,2}$ and ϵ, can be extracted by observing the Sagnac beat note in a stationary HeNe ring laser at low power together with the respective intensities of the individual laser beams as a function of time. The mixture of the laser gas contains 0.1 hPa ^{20}Ne, 0.1 hPa ^{22}Ne and 9.8 hPa of ^{4}He. The laser intensity at the outside of the cavity is usually as low as 13 nW, and the clockwise beam is held constant by a low bandwidth digital feedback loop, which servos the clockwise beam to a constant reference voltage. The ringdown time τ has been measured to $\tau = 0.001 \pm 0.00005$ s, which translates into a total loss of the cavity of 54 ppm. The observed beam intensities I_1 and I_2 are converted to the dimensionless Lamb units by

$$I_{1,2} = \frac{|\mu_{ab}|^2(\gamma_a + \gamma_b)}{4\hbar^2 \gamma_a \gamma_b \gamma_{ab}} \cdot \frac{P_{\text{out}1,2}}{2c\epsilon_0 A_w T}, \tag{3.88}$$

where

$$\mu_{ab} = \sqrt{\pi\epsilon_0 \frac{\lambda^3}{(2\pi)^3}\hbar A_{ik}} \tag{3.89}$$

is the electric-dipole matrix element between the upper ($a = 3s^2$) and the lower ($b = 2s^4$) laser state. The decay rates[2] for a total gas pressure of $p = 10/1.33$ Torr (conversion hPa to Torr) are $\gamma_a = (8.35 + p) \cdot 10^6$ Hz $= 16$ MHz, $\gamma_b = (9.75 + 40p) \cdot 10^6$ Hz $= 311$ MHz and $\gamma_{ab} = (\gamma_a + \gamma_b)/2 = 164$ MHz. The quantity $\eta = \gamma_{ab}/\Gamma_D$ corresponds to the ratio of the homogeneous and the Doppler broadening of the laser transition. For G we observe a value of $\eta \approx 0.193$. Here $\Gamma_D = \sqrt{2k_B T_p/m_{Ne}}/\lambda$ with $k_B = 1.3806488 \times 10^{-23}$ J/K – the Boltzmann constant, $T_p \approx 360$ K – the plasma temperature. The decay rate of the laser radiation is $A_{ik} = 3.39 \times 10^6$ Hz. With $P_{\text{out}1,2} \approx 13$ nW, the output powers of the two beams, the cross-section of the beam waist in the plasma is calculated as $A_w = 19 \times 10^{-6}$ m^2, and the transmission of the mirrors is given as $T = 0.23$ ppm. Finally, Eq. 3.88 contains the dielectric constant of the vaccum ϵ_0 and the reduced Planck constant $\hbar$.

The coefficients of Eqs. 3.85–3.87 can be calculated by making use of the plasma dispersion function, which describes the broadening of the profile of the laser transition:

$$Z(\xi_{1,2}) = 2i \int_0^\infty e^{-x^2 - 2\eta x - 2i\xi_{1,2}x} \, dx, \tag{3.90}$$

---

[2] Paschen notation

where $\xi_{1,2} = (\omega_{1,2} - \omega_0)/\Gamma_D$ is the detuning from the line center, normalized to the Doppler width for the beams 1 and 2. The frequency detuning $\xi_{1,2}$ depends on temperature of the plasma and the gas pressure inside the cavity. Since our lasing medium is composed of two different isotopes, the variables ξ, η and Γ_D have to be expressed for each isotope separately. For $\eta \ll 1$, which is a good approximation for large HeNe ring lasers, $Z(\xi)$ can be written as

$$Z_I(\xi) \simeq \sqrt{\pi}e^{-\xi^2 - 2\eta} \tag{3.91}$$

and

$$Z_R(\xi) \simeq -2\xi e^{-\xi^2}. \tag{3.92}$$

The indices I and R denote the imaginary and real parts of the plasma dispersion function. With these approximations in place, we can compute the Lamb coefficients in the following way:

$$\alpha_{1,2} = \frac{G_0}{Z_I(0)}[kZI(\xi_{1,2}) + k'ZI(\xi'_{1,2})] - \mu_{1,2} \tag{3.93}$$

$$\beta_{1,2} = \alpha_{1,2} + \mu_{1,2}$$

$$\sigma_{1,2} = \frac{f_{FSR}}{2}\frac{G_0}{Z_I(0)}[kZ_R(\xi_{1,2}) + k'Z_R(\xi'_{1,2})]$$

$$\theta_{12} = \frac{\Gamma G_0}{Z_I(0)}\left[k\frac{Z_I(\xi_{1,2})}{1 + (\xi_m/\eta)^2} + k'\frac{Z_I(\xi'_{1,2})}{1 + (\xi'_m/\eta)^2}\right]$$

$$\tau_{12} = \frac{\Gamma f_{FSR}}{2}\frac{G_0}{Z_I(0)}\left[k\frac{Z_I(\xi_{1,2})\xi_m/\eta}{1 + (\xi_m/\eta)^2} + k'\frac{Z_I(\xi'_{1,2})\xi'_m/\eta)}{1 + (\xi'_m/\eta)^2}\right].$$

Here G_0 denotes the laser gain, $\Gamma = (\gamma_a + \gamma_b/(2\gamma_{ab})$ and $\xi_m = (\xi_1 + \xi_2)/2$ and k is the fractional amount of the respective isotope. The primed parameters refer to ^{22}Ne and the unprimed to ^{20}Ne. For the case of the ring laser G, we have $k = k' = 1/2$. For a measurement series of 50 days duration (between December 16, 2020 and February 5, 2021) the gain for the clockwise beam in the Eq. 3.93 is estimated as 69.5×10^{-6}, and for the unregulated counter-clockwise beam we obtained 72.8×10^{-6}. This corresponds to a difference in loss of 3 ppm.

A solution for the ring laser equation at steady state has been given by [16] and was elaborated further in [54] and [55]. It applies for the case of a large HeNe gyro, operating near the laser threshold at steady state in the mono-mode regime and takes the form:

$$I_1(t) \simeq \frac{\alpha_1}{\beta} + \frac{2\sqrt{\alpha_1\alpha_2}\, r_2 \left(\frac{L\omega_s \sin(t\omega_s + \epsilon)}{c} + \alpha_1 \cos(t\omega_s + \epsilon)\right)}{\beta\left(\alpha_1^2 + \frac{L^2\omega_s^2}{c^2}\right)}$$

$$- \frac{2cr_1 r_2 \sin(2\epsilon)}{\beta L\omega_s} \tag{3.94}$$

$$I_2(t) \simeq \frac{\alpha_2}{\beta} + \frac{2\sqrt{\alpha_1\alpha_2}\, r_1 \left(\alpha_2 \cos(\epsilon - t\omega_s) - \frac{L\omega_s \sin(\epsilon - t\omega_s)}{c}\right)}{\beta\left(\alpha_1^2 + \frac{L^2\omega_s^2}{c^2}\right)}$$

$$+ \frac{2cr_2 r_1 \sin(2\epsilon)}{\beta L\omega_s} \tag{3.95}$$

$$\psi_0(t) \simeq \frac{c\left(\sqrt{\frac{\alpha_1}{\alpha_2}}\, r_1 \cos(\epsilon - t\omega_s) + \sqrt{\frac{\alpha_2}{\alpha_1}}\, r_2 \cos(t\omega_s + \epsilon)\right)}{L\omega_s}$$

$$+ t\left(\omega_s - \frac{2r_1 r_2 (\frac{c}{L})^2 \cos(2\epsilon)}{\omega_s}\right). \tag{3.96}$$

The unperturbed interferometer beat note f_s is finally obtained after some approximations, such that a substitution term K, with

$$K(t) = \sqrt{\frac{\alpha_1}{\alpha_2}}\, r_1 \sin(\epsilon - t\omega_s) - \sqrt{\frac{\alpha_2}{\alpha_1}}\, r_2 \sin(t\omega_s + \epsilon), \tag{3.97}$$

only has a very small average contribution for frequencies much below the actually measured Sagnac angular frequency ω_m. With $|K| \ll L\omega_m$, the null-shift contribution $\delta_{ns} \ll \omega_m$ and the first and second order expansion in K and δ_{ns}, one finally arrives at an expression for the corrected Sagnac frequency

$$2\pi f_s = \omega_s \simeq \omega_{s0} + \omega_{ns1} + \omega_{ns2} + \omega_{K1} + \omega_{K2} + \omega_{nsK}, \tag{3.98}$$

where ω_s is the corrected value for the observed angular frequency of the interferogram caused by the experienced rate of rotation. The quantity ω_{s0} corresponds to the contribution of the backscatter coupling and in a comparison with the approach presented in Section 3.9.4 is in good agreement. It is by far the largest contributor to the required corrections. The terms ω_{ns1} and ω_{ns2} correspond to the first and second order expansions of the null-shift contribution, and the same applies for ω_{K1} and ω_{K2} with respect to K. The last element is ω_{nsK}, which represents the mixed term. In detail these terms are given as [54, 55]:

$$\omega_{s0} = \left(\frac{1}{2} \sqrt{\frac{8c^2 r_1 r_2 \cos{(2\epsilon)}}{L^2} + \omega_m^2} + \frac{\omega_m}{2} \right) \tag{3.99}$$

$$\omega_{ns1} = -\delta_{ns} \times \left(\frac{\omega_m}{2L \sqrt{\frac{8c^2 r_1 r_2 \cos{2\epsilon}}{L^2} + \omega_m^2}} + \frac{1}{2} \right) \tag{3.100}$$

$$\omega_{ns2} = \delta_{ns}^2 \times \frac{2c^2 r_1 r_2 \cos{(2\epsilon)}}{(8c^2 r_1 r_2 \cos{(2\epsilon)} + L^2 \omega_m^2) \sqrt{\frac{8c^2 r_1 r_2 \cos{2\epsilon}}{L^2} + \omega_m^2}} \tag{3.101}$$

$$\omega_{K1} = K \times \left(-\frac{\omega_m}{2L \sqrt{\frac{8c^2 r_1 r_2 \cos{(2\epsilon)}}{L^2} + \omega_m^2}} - \frac{1}{2L} \right) \tag{3.102}$$

$$\omega_{K2} = K^2 \times \frac{2c^2 r_1 r_2 \cos{(2\epsilon)} \sqrt{\frac{8c^2 r_1 r_2 \cos{2\epsilon}}{L^2} + \omega_m^2}}{(8c^2 r_1 r_2 \cos{(2\epsilon)} + L^2 \omega_m^2)^2} \tag{3.103}$$

$$\omega_{nsK} = \frac{\delta_{ns} K}{2 \sqrt{8c^2 r_1 r_2 \cos{(2\epsilon)} + L^2 \omega_m^2}}. \tag{3.104}$$

For practical purposes it is usually possible to neglect the second order terms ω_{ns2} and ω_{K2}, as well as the mixed term ω_{nsK}, because they are too small to matter. For the measurement series mentioned above, we have calculated the respective corrections from the backscatter amplitudes r_1 and r_2 and difference in phase between the modulation on each beam (2ϵ). The Lamb coefficients were obtained from the amount of the measured backscattered light and the intensities of the two beams by making use of the relationship

$$\alpha_1 = \beta \left(DC_{cw} + \frac{AC_{cw}^2}{4DC_{cw}} \right) + \frac{AC_{cw} AC_{ccw} \omega_m \sin{(2\epsilon)} L}{DC_{ccw} 4c} \tag{3.105}$$

$$\alpha_2 = \beta \left(DC_{ccw} + \frac{AC_{ccw}^2}{4DC_{cw}} \right) - \frac{AC_{cw} AC_{ccw} \omega_m \sin{(2\epsilon)} L}{DC_{cw} 4c} \tag{3.106}$$

and

$$r_1 = \frac{AC_{ccw} \omega_m}{\frac{2c \sqrt{DC_{cw} DC_{ccw}}}{L}} \tag{3.107}$$

$$r_2 = \frac{AC_{cw} \omega_m}{\frac{2c \sqrt{DC_{cw} DC_{ccw}}}{L}} \tag{3.108}$$

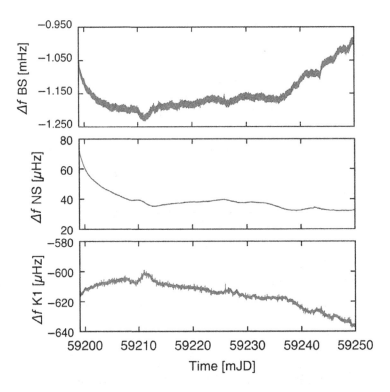

Figure 3.76 The corrections for backscatter (top), null-shift (middle) and the K term (bottom) for a continuous observation of Earth rotation of a series of 50 days.

together with the operational parameters of the G ring laser. These were the total laser gas pressure in the cavity of 10 hPa, the cross-section of the laser mode in the plasma capillary of 19×10^{-6} m^2, the mirror transmission of 0.23 ppm and the cavity length of 16 m. Figure 3.76 illustrates the result for the corrections, which change only very slowly since G is a stable construction. The largest contribution for the correction comes from the backscatter coupling. The null-shift correction contributes 40–80 µHz to the correction, which is still significant for Earth observation. Furthermore, it is interesting to note that the K_1 term predominately corrects for a significant offset between the measured and the true rotation rate of the Earth. This is an important step for an improved accuracy of Sagnac interferometers. Since the G ring laser is very stable, neither the null-shift nor the offset term vary by a large amount. The overall contribution of the plasma dynamics to the error corrections for G is of the order of 3%.

3.9.8 *Effects on the Interferometer from Extra-cavity Components*

Every ring laser concept requires some extra-cavity components in order to superimpose the two counter-propagating laser beams to obtain the interferogram. As

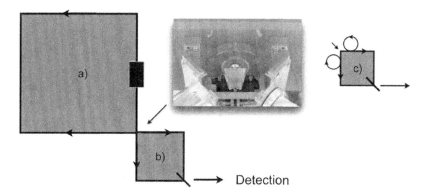

Figure 3.77 A general ring laser layout constitutes a combination of a Sagnac and a Mach–Zehnder interferometer, illustrated as a) and b). Minute back-reflections coupling back on each light beam individually can cause a significant apparent drift on the Sagnac signal from the ring laser gyroscope c).

a minimum, one requires two mirrors and a non-polarizing beam splitter with a splitting ratio of 50%. Alternatively, one can apply a Kösters prism. This offers the advantage of minimizing the free space path of the laser beam, which is important to keep the phase of the interferogram stable. Figure 3.77 illustrates this setup. A very small portion of the two counter-propagating laser beams from the Sagnac cavity (a) passes through one of the mirrors of the ring laser cavity and enters the beam combiner section, which essentially constitutes a Mach–Zehnder interferometer (b). The best way of minimizing the free space path of the laser beam in a phase sensitive interferometer is given by the application of a Kösters prism, as shown in Figure 3.77. The beam splitting coating is located on the interface of the two half-prisms in the middle of the glass body. The shape of the prism provides total reflection and perfect superposition of the laser beams at the beam splitting coating. However, since the prism needs three degrees of freedom for proper alignment (translation parallel to the mirror surface, rotation around the vertical axis and tilt), there need to be a small gap between the mirror holder of the G ring laser and the Kösters prism. This causes a small amount of back-reflection at the ring laser–prism interface of each laser beam onto itself as indicated in (c). As a result of temperature- and atmospheric pressure-induced variations of the effective length of the free space path, the phase of the back-coupled laser frequency changes slowly, and we observe a systematic shift of the phase of the interferogram, showing up as an apparent drift of the ring laser beat note.

In order to recognize and safeguard against frequency drifts induced by extra cavity components, we use the unavoidable backscatter coupling as an indicator of proper gyroscope operations. Intra-cavity backscattering is entirely independent of any external optical component and therefore indicates the correct beat frequency,

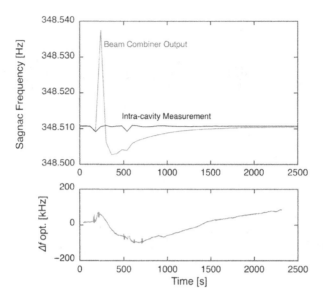

Figure 3.78 Simultaneous observation of the combined beam of the ring laser (including extra cavity components) and the intra-cavity beat notes from the backscattered signals (with no extra cavity components). As the beam combiner is warmed up by the airflow from a hairdryer, the obtained interferogram deviates from the true interferometer signal, while the intra-cavity frequency estimates are not affected. The small amount of perturbation to the interferometer itself is shown by the variation of the optical frequency in the lower part of the figure. It barely exceeds 200 kHz.

albeit with backscatter coupling still in effect. To verify this observation, we have used a hairdryer to gently blow warm air onto the Kösters prism while the ring laser was in operation. As expected, the Sagnac frequency derived from the beam combiner shoots off by more than 25 mHz during the warm up of the prism (see Figure 3.78). After the warm airflow stops, it takes almost one hour before the extra- and intra-cavity frequency estimates start to agree with each other again. The two small dips in the intra-cavity Sagnac frequency estimates indicate when the person with the hairdryer has entered and exited the otherwise isolated laboratory. The lower part of Figure 3.78 shows the behavior of the optical frequency of the counter-clockwise-propagating laser beam during this experiment, as measured with an optical frequency comb against the reference maser of the observatory. The optical frequency changed by about 200 kHz, indicating that the interferometer cavity itself was mostly unperturbed by this experiment.

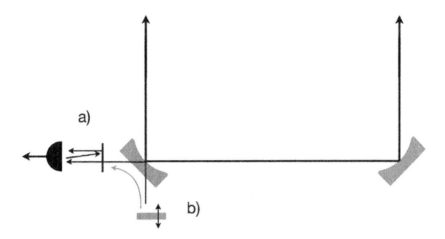

a)

b)

Figure 3.79 Illustration of the effect of extra cavity reflections on the backscatter estimates. Double reflections from the detector and the spectral filter cause slowly changing signal levels of the backscatter amplitude (a). Diffuse external back-reflections from the other beam act in a very similar way (b).

The second interference problem is indicated in Figure 3.79 and concerns the backscatter correction process (see Section 3.9.4). In order to determine the amount of coupling between the two counter-propagating beams, it is necessary to establish the ratio between the amount of backscattered light from the other beam and the genuine DC brightness of each laser beam itself. To distinguish between the laser radiation and the broadband background light from the plasma, a narrow spectral filter (1 nm passband) is employed (see Figure 3.79a). Multiple back reflections between the sensitive area of the photodiode (silicon substrate) and the backside of the spectral filter caused significant multiple reflection interferences. This effect, once recognized, was very hard to avoid, due to the very limited space at the ring laser corners. Figure 3.80 shows the quantitative effect that these back reflections generated on the G ring. The observed raw Sagnac frequency, corrected for the geophysical signals (a), does not show any significant perturbation other than the usual noise level and some low frequency variations of the backscatter coupling. Once the compromised backscatter correction is applied, we introduce roughly periodic perturbations of the order of 10 μHz in amplitude (b). These are caused by the changing phase relationship in the multiple reflections between the detector and the spectral filter, as illustrated in Figure 3.79a. The problem was exacerbated by the fact that the photo-sensitive surface of the photodiode detector is partially reflecting, as well as the spectral filter. The gap between the reflecting surfaces was

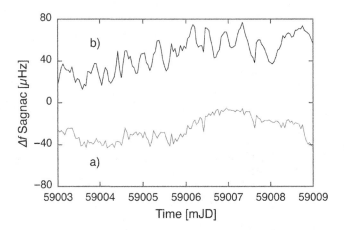

Figure 3.80 Example of a time series taken by the G ring laser with the unperturbed raw Sagnac frequency (a) and the compromised backscatter correction applied (b). When the refractive index varies in response to atmospheric pressure changes, it generates variable levels of interference.

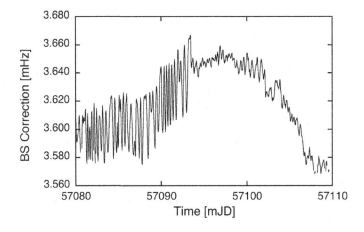

Figure 3.81 Illustration of the amount of extra cavity interference of one beam onto the other. Due to the variable atmospheric pressure, the interference pattern goes through maxima and minima as the refractive index changes the effective path length.

a free space of about 1 cm. Changing ambient atmospheric pressure could therefore modify the effective path length via the refractive index, causing either constructive or destructive interference, with amplitude variations in the range of about 1%.

Another even less obvious extra-cavity perturbation is illustrated in Figure 3.79b. This process is light coupling from one beam onto the other, but from outside of the cavity. Light reflecting from any surface that the other beam hits, is back-reflected onto itself. It returns to the cavity mirror but bounces off the rear side and superimposes with the opposite beam as it is propagating to the photodiode, giving rise to an additional beat note, indistinguishable from the intra-cavity backscatter signal. Again, this is an external reflection passing through a free space distance, which is subjected to the variable refractive index of the air. Figure 3.81 illustrates this effect. In the first half of this time series, the reflection was coming from a low reflecting beam dump. In the second half, the effect was considerably reduced by utilizing some absorbers, but it did not disappear. When the absorber shown in Figure 3.79b was replaced by a small absorber attached to a speaker membrane, the situation improved a lot when the membrane was vibrated at a rate of 136 Hz with amplitudes much larger than a wavelength of the optical beam. The vibration scrambled the phase relationship and thus made the two signals incoherent with respect to each other. One of the consequences was that the intensity readings of each laser beam have to be taken at different corners in order to avoid extra-cavity mixing.

Eliminating the Effects of Extra-cavity Components

In order to avoid bias effects from the Kösters prism, the gap between the cavity mirror holder and the prism was increased, so that the back reflection is placed outside the useful aperture. On top of this, we added a diagnostics feature into our data logging system. Apart from the beam recombination alone, we also determine the intra-cavity beat note on each of the two laser beams. This effect has been exploited already in Section 3.9.4. In the presence of a sufficient SNR, we obtain the Sagnac beat note three times. If all three agree, there is no issue from the extra-cavity components, and we can trust the measured Sagnac frequency obtained from the beam combiner. When they do not agree, the measurement series is compromised, and the observations cannot be used.

Variations in the backscatter coupling can cause significant variations in amplitude of these intra-cavity beat notes, so that sometimes the clockwise beam has better contrast, and sometimes the signal is recovered with better resolution from the counter-clockwise beam. The beat note from the extra-cavity Mach–Zehnder configuration always has an excellent SNR well in excess of 80 dB. Figure 3.82 shows an example taken over more than 50 days. We have subtracted the Sagnac beat obtained with the Mach–Zehnder beam combiner from each of the intra cavity beat notes. A valid observation is obtained when at least two of the three frequency estimates agree to within the SNR of the respective intra-cavity Sagnac frequency estimate. The counter-clockwise beam beat note (a) in Figure 3.82 was considerably weaker than the one from the clockwise beam (b). In order to show this more clearly, we have displaced the clockwise beam by a constant offset value of 100 μHz. The fact that both graphs have a zero mean indicates that all three beat note estimates agree and that there is no difference between the intra-cavity and the Mach–Zehnder Sagnac estimate.

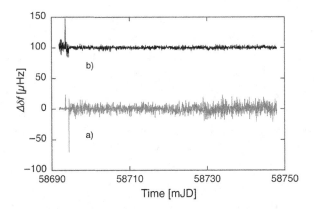

Figure 3.82 In order to guard against an extra cavity-induced drift effect, we subtract the beat note obtained from the beam combiner from the two intra-cavity beat notes. In this example the beat note from the counter clockwise beam (a) is much weaker than the clockwise beat note (b) due to a much lower SNR. Curve (a) has been offset by 100 µHz for better visibility.

For the two detectors, logging the DC intensities of each laser beam and the backscatter amplitude, there is no easy solution available for the avoidance of multi-path signal perturbations. Spacing the detector and the filter sufficiently far apart for a proper separation of the reflection from the signal is about the only practical solution. The scrambling of the phase of the externally reflected opposite laser beam makes the phase incoherent but still adds up to 4% to the DC intensity, while the AC component is believed to be correct. Furthermore, we are forced to operate the intensity detectors on two different corners of the interferometer, which is not desirable. A general problem for the correct processing of backscatter correction is the difference in gain between the detectors and the phase offset that is generated by the electronics of these detectors. Luckily the correction only requires the ratio between the AC and DC quantities and not their absolute values.

3.10 Mirrors

Ring laser mirrors need to be highly reflective. This is achieved by a stack of quarter wave-thick alternating layers of high and low refractivity material, thus forming a Bragg grating. The high refractive index material is typically either tantalum pentoxide (Ta_2O_5) or titanium dioxide (TiO_2), with a refractive index in the range of $n = 2.275$, and the low refractive index material silicon dioxide (SiO_2), with a refractive index in the range of $1.40 \leq n \leq 1.55$ [87]. In order to make the coating resistant and match the top surface to the open air or vacuum, respectively, a half wave overcoat of silicon dioxide is applied at the end of the coating process. In order to reduce losses from either absorption or scattering, the coating has to be dense, amorphous and free of coating defects. Transmission losses can be adjusted by the number of layers in the Bragg stack. The reflectivity of the mirrors ultimately achieved depends on the density of the deposition, the homogeneity of the layer thickness and on the exactness of approximating a thickness of $\lambda/4$ of the desired wavelength range of the mirror application. Today, the best result for a high reflectivity coating in the visible domain has been obtained from an ion beam sputtering process (IBS) [232]. In contrast to electron beam vaporization, where the target material is heated by an electron beam, this process dislodges target material by bombardment with an Ar^+ ion beam in a cold process. It is not required to heat the substrate in order to control the layer composition. For low loss mirrors, it is important to achieve a high packing density in order to obtain a homogeneous coating layer. The amount of oxygen present in the coating chamber is carefully controlled for each layer of the stack, so that the deposited material acquires the right stoichiometry for the required index of refraction, thus minimizing the absorption losses in the thin film coating. After the argon ions are accelerated, they pass an area where electrons are emitted in order to neutralize the ions before the argon beam hits the target. Highly reflective low loss mirrors require exquisitely polished mirror substrates with a residual rms roughness of the order of 0.01 nm, because any remaining surface structure prints through all of the coating layers, thus increasing the scatter loss. Therefore, the IBS coating process begins with an Ar^+ bombardment of the substrates in order to knock adhesive molecules and other contaminants off the carefully cleaned super-polished substrate surface. It is also necessary that the substrates are free of bubbles and inclusions and of the right curvature, again for the sake of reducing the losses and to avoid aberrations of the reflected wave fronts. Furthermore, the substrates must be made from a low thermal expansion material, which ensures that the thicknesses of the layers stay within the boundaries of the specifications. Finally it is desirable to use substrates with low mechanical loss, fused silica in our case, in order to minimize limitations from thermal noise (see Section 3.10.3).

3.10.1 Modern Dielectric Super-Mirrors

The resolution of a ring laser improves when the cavity losses are reduced by as much as possible. Most of the losses of our ring laser gyroscopes occur at the mirrors. There are three basic processes that generate mirror loss, namely transmission, scattering and absorption. For typical gyro mirrors, used in our installations, the transmission loss is specified to be $T = 0.2$ ppm. While the transmission loss can be well controlled by varying the stack size of the coatings, scattering and absorption losses are not so easily accessible, partly because some of the scattering, in particular when diffuse and incoherent, behaves like absorption, and partly because the total integrated scatter is very hard to measure. While the mirrors are responsible for the bulk of the losses, they are not the only source. Another contributor for loss is the gain capillary, which facilitates the laser gain on one hand but also acts as an aperture to select the TEM_{00} mode as the preferred mode and discourages higher order transverse laser modes to be excited.

In practice it is a very common experience that mirrors, even when they come from the same coating run, differ greatly with respect to their losses. This has several causes. The effective residual surface roughness of the substrate plays an important role for the scattering properties of a mirror. Although an rms surface roughness of 0.01 nm is a common specification for a modern super-mirror, this requirement is not necessarily met, despite the application of a careful polishing process. Secondly, there may be tiny zones of contamination from dust, either on the mirror at the time of the coating process or from the handling during the installation process. Most often, however, the variability of the mirror quality is caused by the integration of defects into the thin film deposition in the IBS coating process. Defects are irregularities, which can either be holes or lumps in the thin film, and the density of these defects can vary significantly from one mirror to the next in a batch of the same coating run.

The sensitivity of a Sagnac interferometer to the experienced rotation rate $\delta\Omega$ depends on the linewidth of the laser (see Section 3.6.5) and therefore the quality factor ($\delta\Omega \propto 1/Q$). Once installed in the cavity, the overall losses, caused by all mirrors and the capillary, can be established by a ringdown measurement. Figure 3.83 provides an example of such a ringdown time measurement, where an exponential decay term is fitted to the voltage obtained from a photomultiplier of sufficient bandwidth. From Eq. 3.6 we compute the overall cavity losses $\delta = (\tau f_{fsr})^{-1}$ as $\delta = 66$ ppm. What proportion of these losses is attributable to the capillary is not known, but it is certainly not insignificant. The capillary acts as a mode selection device and has to find a fine balance between additional loss and the undesirable excitation of a TEM_{01} laser mode. The best ringdown time ever achieved with a particular mirror set in G is 1.1 ms, which is equivalent to $\delta = 48$ ppm. Since the mirrors are fixed and cannot be moved, it is only the gain tube that can be positioned. The diameter of the capillary is 5 mm, and when it is aligned to provide the best ringdown time of 1.1 ms, it is at the verge of permitting the TEM_{01} mode to lase as well. In principle it does not matter which of the various transverse laser modes oscillates in the cavity. However, since the TEM_{00} mode has a smaller mode volume (cross-section), it is almost certain that both modes will coexist in the cavity when

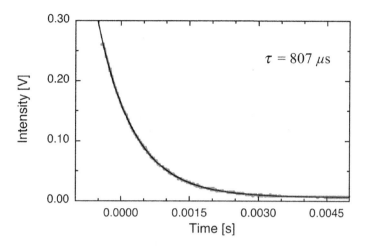

Figure 3.83 Example of a typical ringdown measurement on the G ring laser. From the exponential decay fit, a value for ringdown time of 807 µs is obtained.

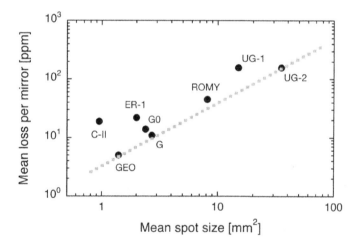

Figure 3.84 The mirror losses scale as a function of the area taken up by the laser spots on the mirror [78].

TEM_{01} is permitted to lase. In that case, mode competition would be unavoidable, and no stable rotation rate could be observed. So, for the example described here, we can attribute at least 18 ppm of the losses to the capillary and on average 12 ppm on each mirror. Due to a very inhomogeneous distribution of coating defects over all mirrors, it is very likely that some of the mirrors will have higher losses than others.

From discussions with the manufacturers of super-mirrors, it emerged that the absorption loss[3] is considered to be of the order of 8 ppm for the wavelength $\lambda = 632.8$ nm. This will then leave $\approx$4 ppm of scatter losses for each mirror, since the transmission on each mirror is only 0.2 ppm. The absorption losses are still much higher than the predicted limit of 1 ppb [26], which indicates that the deposition homogeneity is still a technical challenge. When we look at common mirror behavior across all the various instruments that our group has built and operated, a general pattern emerges, which is consistent with the considerations above. As the illuminated area on the mirrors in the cavity increases, the losses increase in proportion [78]. The earlier mirrors in the older systems, C-II and G-0, perform worse than the new generation mirrors in the rest of the systems. For all other rings we find that they perform at about the best possible level. From this observation we may conclude that the limitation in performance is inherently due to the coating itself, is either caused by absorption or scatter (opaqueness) and that these effects are fairly uniform across all mirrors, so that Figure 3.84 suggests a global coating loss of 3 ppm/mm^2 for the current IBS mirror coating technique. This general observation is also reflected in Section 3.9.4, where the backscatter amplitude is found to scale linearly with the gyro area. These general coating properties are independent of additional effects from mirror degradation. An inferior substrate polish, which is printing through all layers of the coating, thus causes a deterioration in performance for an independent reason. Defects in the coatings or contaminations on the surface push the gyro performance further away from the ideal. It will require some progress in the coating technique to improve the gyro performance for a given scale factor. It is therefore generally beneficial to make the beam spots as small as possible. It also means that making ring lasers ever larger will not continue to improve their sensitivity indefinitely.

3.10.2 The Effect of Spoiled Transmission

The best way to change the transmission of a Bragg stack mirror is to vary the number of coating layers, which corresponds to a significant modification in the production procedure of the thin film coatings. Unless specified specifically in this way, the fabrication process of mirrors with higher transmission involves an additional overcoat, which acts as a quality factor spoiling bandpass. This extra coating layer shifts the wavelength of highest reflection slightly, which increases the transmission at the operational wavelength. We explored mirrors with a corresponding modification from 0.2 ppm to about 2 ppm. Although not obvious, this quality factor

---

[3] R. Lalezari (FiveNine Optics) and L. Pinard (LMA), private communications (2012).

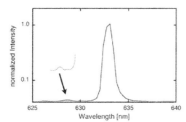 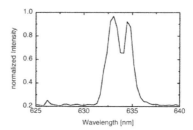

Figure 3.85 Low loss mirror with a spoiling overcoat bandpass to enhance trans-mission. This encouraged additional oscillation on neighboring transitions, such as 629.4 nm (left) and 635.4 nm (right).

of a spoiling overcoat has a very detrimental effect on the operations of a large laser gyroscope, which may vary from one specific set of mirrors to another. Once a set of such mirrors were put into the G ring laser, we noticed a significantly degraded gyro performance, which turned out to be caused by mode competition, as radiation from additional laser transitions started to coexist. In one case we observed a very small amount of additional radiation from the $3s_2 \longrightarrow 2p_5$ transition at 629.4 nm. The intensity ratio was of the order of 335 : 1. In another case we obtained a significant amount of additional $3s_2 \longrightarrow 2p_1$ radiation at 635.2 nm, which is illustrated in Figure 3.85. With mode competition acting, the observations of the G ring laser became very unstable, both in the short term and in the long term. Figure 3.86 shows a small section of the observation of Earth rotation signals. While the typical scatter averaged over 30 s of observation at the time of the measurement for low loss mirrors did not exceed 100 μHz, we now observed an increase of the noise level of a factor of approximately 3. On top of this, the signal variation over several hours almost reached a level of 1 mHz, a factor of about 100 worse than normal. In this particular case only two out of four mirrors had this additional overcoat.

3.10.3 The Effect of Thermal Noise on the Cavity Mirrors

The mirrors in the ring laser cavity are the only parts of the instrument that directly interact with the laser beam, and they define the cavity properties. Low total loss increases the Q-factor of the cavity, which in turn increases the instrument's sensi-tivity for rotation sensing. Low scatter loss is also critical for the reduction of the backscatter coupling of the two independent optical oscillators. Cavity ringdown measurements reveal the combined intra-cavity losses, including mirror transmis-sion, absorption and scattering. For the G ring laser, for example, we currently obtain a ringdown time of $\tau = 1$ ms. With a free spectral range of 18.75 MHz, the losses μ_L are then computed from $\mu_L = 1/(\tau \cdot FSR) = 53 \times 10^{-6}$. This

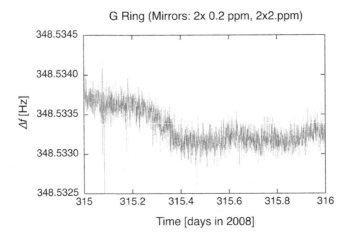

Figure 3.86 Ring laser observations in the presence of mode-competition, due to an additional overcoat on the mirrors, intended to increase transmission. The measurements are highly unstable.

means, on average, each mirror contributes 13 ppm to the loss budget. From the specifications of the mirrors, the transmission loss is 0.2 ppm, and the absorption is reported[4] to be of the order of 6–8 ppm. However, the inferred sensitivity for rotations of $\delta\Omega = 3.2 \times 10^{-14} \text{rad/s}/\sqrt{\text{Hz}}$ as introduced in Section 3.6.5 is not readily observed. In fact there is a discrepancy of about three orders of magnitude between the actual measurement and the theoretical intra-cavity sensitivity. Most of this numerical discrepancy is explained by the difference in the photon flux between inside and outside the cavity.

Several instrumental limitations reduce the sensor resolution. Apart from detector noise, the quantization noise in the digitizer, the limited resolution of the beat frequency estimator and noise processes in the feedback loop for the laser intensity stabilization reduce the effective sensitivity. The quantization noise from the Kinemetrics *Basalt* data logger has been established to be as low as 0.172 μV, and this is far below the sensor resolution. The noise level of the intensity stabilizing feedback loop also presents no concern, since it is more than an order of magnitude smaller than the noise limit presented by the frequency estimator (see Figure 3.88). Thermal noise from the mirror substrates and coatings, however, may contribute significantly to the error budget. This type of noise does not average out; hence it presents a serious concern. It is not intuitively clear why this kind of noise source can be a complication in a ring laser gyro, since the frequency fluctuations should cancel in a Sagnac interferometer as common mode effects.

---

[4] R. Lalezari (*Five Nine Optics*) and L. Pinard (*LMA*): private communications.

However, despite all efforts to avoid it, there is a small difference in beam power between the two counter-propagating laser beams, so frequency fluctuations from coating noise do not cancel entirely. The effect of thermally induced coating noise has been intensely studied for the application of high-Q optical reference cavities in precision metrology [154, 155]. It was found that the mechanical loss ϕ of the materials used to set up the cavity, namely the mirror substrates, the coatings and the spacers, are the important quantities for the evaluation of thermal noise. From the fluctuation-dissipation theorem [134], it follows that the mechanical quality factor and the Young's modulus of the materials are critical parameters. Applied to our monolithic ring laser gyroscopes, we at first restrict the discussion to the mirrors and the coatings, as the cavity bodies typically contribute to a much lesser extent (1% in [155]). In particular for the G ring laser, where the entire 4×4 m cavity is formed by massive Zerodur bars resting on a 4.25 m diameter Zerodur disc with a weight well in excess of 9 t, we ignore the effect of the *spacers*. However, we need to keep in mind that this may be an oversimplification. The power spectral density of the thermal noise displacement $G_m(f)$ of the mirrors including the loss contribution from the coatings is

$$G_m(f) = \frac{4k_B T}{\omega} \frac{1 - \sigma^2}{\sqrt{\pi} E w_0} \phi_{\text{sub}} \left(1 + \frac{2}{\sqrt{\pi}} \frac{1 - 2\sigma}{1 - \sigma} \frac{\phi_{\text{coat}}}{\phi_{\text{sub}}} \frac{d}{w_0} \right). \tag{3.109}$$

In accordance with [154, 155] we use $T = 300$ K, and the Poisson ratio $\sigma = 0.17$ both for ULE and fused Silica. The Young's modulus is $E = 6.8 \times 10^{10}$ Pa, the loss $\phi_{\text{sub}} = 1/(6 \times 10^4)$ and the radius of the beam $w_0 = 900$ μm. The expression in brackets in Eq. 3.109 provides the necessary correction for the coating thermal noise contribution, where $d = 5$ μm is the thickness of the coating, and $\phi_{\text{coat}} = 4 \times 10^{-4}$ – the coating loss. Substituting these numbers into Eq. 3.109 and considering 4 intra-cavity mirrors under an angle of incidence of 45° provides a combined thermal length fluctuation of $\sqrt{G_L} = 3.4 \times 10^{-17}$ m/$\sqrt{Hz}$ at 1 Hz. This calculated length fluctuation translates into optical frequency noise of 1 mHz by using the relationship $\sqrt{G_L}/L = \sqrt{G_v}/v$, where v is the optical frequency of 473.612 THz and L = 16 m – the length of the cavity.

According to [114] we cannot neglect the consideration of the spacer from this discussion, since its contribution scales linearly with the size of the cavity. For the G ring laser this contribution will therefore dominate the Brownian motion noise budget. The authors use the following expression to account for the spacer between the mirrors:

$$G_m(f) = \frac{4k_B T}{\omega} \frac{L}{\pi E (R_{\text{sp}}^2 - r_{\text{sp}}^2)} \phi_{\text{sp}}, \tag{3.110}$$

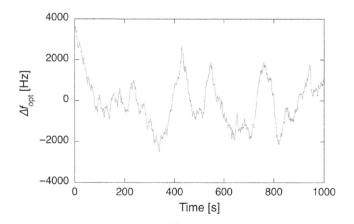

Figure 3.87 Observed fluctuations in the optical frequency in the ring cavity of G. An optical fs-frequency comb was locked to the ring laser to down-convert the optical frequency into the microwave regime. The reference oscillator is the ULISS cryogenic sapphire oscillator, providing an instability two orders of magnitude below the recorded signal.

where R_{sp} is the outer radius of the spacer, r_{sp} the radius of the borehole and all other terms are as defined above. Since the body of G is made from Zerodur, we have to use $E = 9.03 \times 10^{10}$ Pa and $\phi_{sp} = 1/(3 \times 10^3)$. With the substantial simplification of approximating G with the design of a reference cavity, we obtain $\sqrt{G_L} = 5.5 \times 10^{-17}$ m/$\sqrt{Hz}$ at 1 Hz, which is about a factor of two larger than the contribution from the substrates and the coating, leading to an overall thermal noise contribution of 2.6 mHz.

From direct measurements of the fluctuation of the optical frequency in the ring cavity, we obtain random walk variations close to ± 10 kHz over nearly 10 hours, corresponding to a frequency stability of 2×10^{-13} at 1 second, when the optical fs-frequency comb is locked to the ring laser, and we compare the down-converted optical frequency to a cryogenic sapphire oscillator with an instability two orders of magnitude lower than the optical ring cavity. The spectrum of this beat note exhibits a clear $1/f$ characteristic, with no other distinct features that would suggest mechanical vibrations or effects from seismic perturbations. Figure 3.87 illustrates this with a small section from this measurement series. The concomitant frequency drift from temperature effects on the cavity has been removed by a linear regression process and is not discussed here.

The Allan deviation of a $1/f$ signal trend eventually becomes constant and does not improve with longer averaging times. With the length fluctuations converted to fractional frequency fluctuations and the observed PSD denoted as $S_y(f)$, the limit of the Allan deviation σ_y of the fractional frequency fluctuations y is given by the

the expression [114]

$$\sigma_{\delta f} = \sqrt{2 \ln(2) S_y(f) f}.$$ (3.111)

A traveling wave resonator usually has two optical frequencies y_1 and y_2, which are almost identical and differ only by the Sagnac frequency offset. Their PSD is then given by $S_{y1(f)}$ and $S_{y2(f)}$. Since their cross-power spectral density is almost the same, it can be represented by

$$S_{y1y2}(f) \approx (1 - \epsilon) S_{y1}(f).$$ (3.112)

The Sagnac frequency δf and their PSD $S_f(f)$ then is:

$$\delta f \approx y_1 - y_2 \rightarrow S_f(f) = S_{y1}(f) + S_{y1}(f) - 2S_{y1y2}(f) = 2\epsilon S_{y1}(f).$$ (3.113)

Applied to Eq. 3.111 the expression becomes

$$\sigma_{\delta f} = \sqrt{2 \ln(2) S_{\delta f}(f) f} = 2\sqrt{\ln(2) \epsilon S_{y1}(f) f}.$$ (3.114)

The value for ϵ is obtained from the fluctuations of the optical frequency in the cavity and the fluctuations of the beat note in the interferometer. These are

$$\delta f_{\mathrm{rms}}^2 = \int S_{\delta f}(f) df = 2\epsilon \int S_{y_1}(f) df$$ (3.115)

and

$$y_{1\,\mathrm{rms}}^2 = \int S_{y_1}(f) df.$$ (3.116)

Inserting the experimentally established values from the G ring ring laser, we can determine the common mode rejection value:

$$\epsilon = \frac{1}{2} \frac{\delta f_{\mathrm{rms}}^2}{y_{1\,\mathrm{rms}}^2} = \frac{1}{2} \left(\frac{150\ \mu\mathrm{Hz}}{1.2\ \mathrm{kHz}} \right)^2 = 7.8 \times 10^{-15}.$$ (3.117)

Applied to Eq. 3.114, this yields $\sigma_{\delta f} = 3.8 \times 10^{-9}$ for the G ring laser and corresponds to a lower limit of 1.3 µHz for the resolution of the Sagnac frequency due to the Brownian motion noise experienced by the G ring.

Figure 3.88 puts our best observation and the various contributions of several important noise sources into perspective. It reveals that our estimate for the limit is slightly higher than the actually measured boundary. We believe that a more realistic model of the resonator structure will correct that. Thermal mirror substrate and coating frequency noise has turned out to be the single most important remaining error source, since it follows a $1/f$ dependence. The numerical frequency estimator and the feedback loop that stabilizes the intensity of the laser radiation in the cavity also have detrimental effects, but these are less significant than the thermal frequency

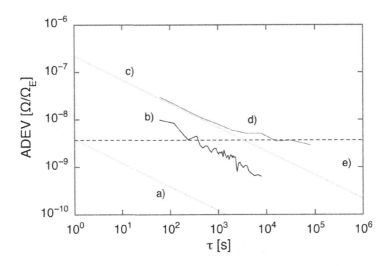

Figure 3.88 Allan deviation plot of important noise sources of the G ring laser. The fluctuations of the intensity stabilization loop mark the lower end (a). The noise limit from the frequency estimator is given by (b). The inferred quantum limit (c) corresponds well with the measured Sagnac frequency (d). The computed Brownian motion e) is in reasonable agreement, despite a significant simplification of the cavity model.

noise caused by the mirrors. They average down as white noise, and the noise level is significantly below the sensor resolution.

Although the calculations for the noise sources are consistent with the observations, they are based on model computations. The models of [155] and [154] were well tested and found to be in good agreement with a large number of optical reference cavities. In a Sagnac interferometer, one might expect a higher degree of common mode rejection on the mirror and coating thermal frequency noise. However, since our beam combiner represents an inverted interferometer configuration, the common mode noise rejection is reduced, as non-correspondent parts of the beam cross-section are superposed at the beam splitter. Nevertheless this leaves some uncertainty and therefore requires some experimental evidence. Therefore, we have taken a continuous dataset of one year of ring laser observations and applied corrections for backscatter, Chandler motion and tilt, as well as the model subtraction for the diurnal polar motion and semi-diurnal Earth tides. The remaining power spectrum therefore only contains the noise floor of the G ring laser in the frequency range between 0.01 Hz and 50 nHz. The observations in Figure 3.89 are in reasonably good agreement with the calculated expectation from mirror and coating noise.

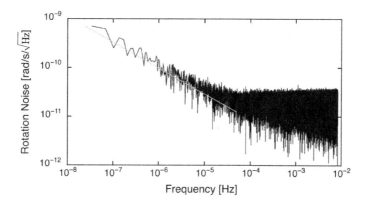

Figure 3.89 Evidence of a $1/f$ noise process at low frequencies in the spectrum of the observed rotation rates of the G ring laser.

3.10.4 Crystalline Coated Mirrors

Thermal noise associated with the mechanical loss of the mirror substrates and coatings reduces the resolution of the extracted beat frequency of a Sagnac interferometer. In a strict sense, this should not be the case, because this effect should be entirely reciprocal and cancel out as a common mode effect. Dispersion, despite a very small frequency difference, and birefringence due to mechanical stress of the coating on the substrate or other small and less obvious processes may be responsible for this observation. Moving from Zerodur to fused silica mirror substrates made a considerable difference. So it can be expected that moving from dense amorphous IBS coatings to crystalline mirror coatings will provide another reduction in the linewidth of the observed beat note. Embedding atoms in a constraining lattice has the potential to reduce Brownian motion and hence to increase the resolution of highly resolving optical interferometers [87]. This is of great importance in applications like gravitational wave detection [46], optical clocks [32] and other ultra-high precision spectroscopic applications. Furthermore, the thermal conductivity of IBS coatings is comparatively poor and typically in the range of 1 $Wm^{-1}K^{-1}$ [45]. A promising way to improve on these relevant properties is the development of a technique that uses single crystal semiconductor hetero-structures, which take the role of a Bragg stack of alternating layers of material with high and low refractive indices. These crystalline mirror coatings are grown by embedding dopants to form an epitaxial layer, which is the desired hetero-structure. These layers are subsequently transferred and optically contacted to a super-polished mirror substrate [44]. They exhibit a drastically reduced mechanical dissipation. The single crystal multilayer is grown using molecular beam epitaxy (MBE) onto a GaAs wafer and contains alternating layers of quarter-wave thickness of GaAs for the high refractive index material and $Al_{0.92}Ga_{0.08}As$ for the low refractive index layers. A typical size for the stack is 38.5–41.5 periods, resulting in a thickness of the high reflection layer of 9.5–10.5 μm and yields a transmission loss in the range 10–5 ppm. The final coating structure is then treated with a chemo-mechanical process of lapping and etching to extract the crystalline optical grating from the wafer. After cleaning, the Bragg stack is bonded onto the mirror substrate, and an annealing process, to remove friction, is the final fabrication step. The demonstrated result confirmed a very low mechanical loss angle of $\phi_{coat} < 4 \times 10^{-5}$, which is an order of magnitude smaller than that of IBS coatings (see Section 3.10.3). Absorption values in the range of 1 ppm have been found for some layers made for the near infrared regime. It is important to note that compared to their amorphous counterparts, the crystalline coatings exhibit a relatively strong birefringence. The obtained values for the linear cavity birefringence are $\theta \approx 1 - 5 \times 10^{-3}$, which compares to $10^{-6} < \theta < 0^{-4}$ in high finesse cavities employing IBS coatings. In applications where this matters, the polarization of the laser light can be matched to the grating, since the orientations of fast and slow axes on the mirror substrate are known and marked. Crystalline coatings are available for the near infrared and further out into the infrared domain. Although in principle feasible, this technology has not yet been demonstrated for the longer wavelength range of the visible spectrum. This means that crystalline coatings for the 632.8 nm transition are not yet available. We have investigated the potential of crystalline mirrors for ring laser gyro applications [196] in the near infrared at $\lambda = 1.152$ μm. The results are discussed in Section 3.11.3.

3.11 Laser Transition on Different Wavelengths

Although the $3s_2 \longrightarrow 2p_4$ at $\lambda = 632.8$ nm is very suitable for operating a ring laser gyroscope, there is a motivation to operate the laser at a different wavelength. A shorter wavelength has the benefit of an increased scale factor, which promises higher sensitivity. The $3s_2 \longrightarrow 2p_{10}$ transition at $\lambda = 543.4$ nm, for example, promises a 16% increase in sensitivity of the gyro for the same physical dimensions. Moving to a longer wavelength reduces the sensitivity of the gyro to scatter sources in the form of defects or coating roughness, which in turn promises less vulnerability for back scatter coupling. Here we have explored the $2s_2 \longrightarrow 2p_4$ transition at $\lambda = 1152.3$ nm, which also allowed the application of crystalline mirrors [9, 45]. In this section we report on the experiences with the alternative transitions.

3.11.1 Ring Laser Operations at 543.3 nm

Going to a shorter wavelength for the gyro operations means that both scatter and absorption losses in IBS coated mirrors will go up. This will cause a significantly degraded gyro performance for a wavelength shorter than 450 nm, where the losses will progressively go from 25 ppm to over 500 ppm[5] at 322 nm. The design specifications for these mirrors aimed for 0.3 ppm transmission loss and less than 12 ppm combined absorption and scatter loss for each mirror. The center of the stop band of the coating was set to 520 nm instead of the operational wavelength of 543.3 nm. The reflectivity curve is reasonably flat over a fairly wide range of $\approx$50 nm, so that the next available HeNe laser transitions at 594, 604 and 611 nm would still find sufficiently low losses to be excited instead of the desired transition at $3s_2 \longrightarrow 2p_{10}$. With the shift of the bandstop center by about 20 nm, the losses for the next available transition increased to 0.1%, so that lasing at 543.3 nm has been readily obtained. We used a stainless steel cavity in the form of the GEOsensor design [8]. The cavity was mounted vertically to a wall. It was 6.4 m long and enclosed an area of 2.56 m^2. The isotope shift for ^{20}Ne – ^{22}Ne for the 543.3 nm transition is 1 GHz, which is close to the value of 875 MHz for gyroscope operation at 632.8 nm. A real disadvantage of this system is the low laser gain of this transition, which is about 30 times weaker than the gain of the red HeNe line. Therefore, it was necessary to replace a section of the stainless steel cavity enclosure with a second gain tube, 400 mm long and 3 mm in diameter, in order to provide more laser gain. An additional 50 watt RF-transmitter, impedance matched to a set of 7 pairs of electrodes, provided the necessary gain over the entire length of the gain tube. This assembly was operated together with the pre-existing gain section, 100 mm long and 4 mm in diameter, fed by a 20 watt transmitter, matched to a pair of electrodes.

Figure 3.90 gives an impression of the setup. Maintaining continuous operation on $\lambda = 543.3$ nm for a long time proved to be difficult, since the laser extinguished quickly when the losses in the cavity went up. While the smaller diameter of the long gain tube increases the laser gain by reducing the time of the de-excitation of the neon atoms through wall collisions, it also presents a significant aperture when the borehole and the optical axis of the cavity are not perfectly aligned with respect to each other. Due to the considerable length of the gain tube, proper alignment is a delicate task. The ringdown time of the cavity was measured to $\tau = 112$ µs, which indicates a total cavity loss of $L = (\tau \times FSR)^{-1} = 190$ ppm, or 48 ppm per mirror on average for the FSR of the cavity of 46.875 MHz. The measured losses are larger than the 25 ppm specified, which has been attributed to the insertion loss of the long and narrow gain tube.

[5] R. Lalezari (Five Nine Optics; private communications).

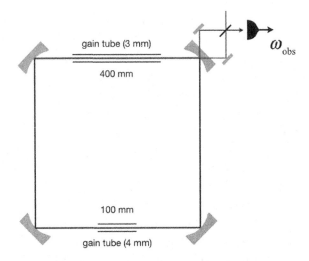

Figure 3.90 The gain of the HeNe laser transition at 543.3 nm is about a factor of 30 smaller than that for 632.8 nm. Therefore it is necessary to operate the gyroscope with a second excitation rf-transmitter unit and an additional 400 mm-long gain tube in support of the existing one of length 100 mm. The gain tubes had a diameter of 4 mm for the shorter one and 3 mm for the longer one.

Figure 3.91 Excitation of the low gain 543.3 nm HeNe laser transition.

During the initial operation of the gyro on the 543.3 nm transition, we observed pulsing of the laser radiation in the cavity at repetition rates of about 2 s. Since the transition is <2 GHz wide, the pulsing cannot be caused by mode-locking. The effect was finally traced back to the stability of the plasma excitation in the gain tube. With seven alternating pairs of RF-electrodes, generating a plasma over the full length of the 400 mm long gain tube, the plasma quickly becomes segmented and flickers noticeably when the intensity stabilization via a feedback circuit is operated, causing the observed pulsing effect. A chain of five magnets placed in the vicinity of the gain tube stabilized the plasma and made it homogeneous over the entire length.

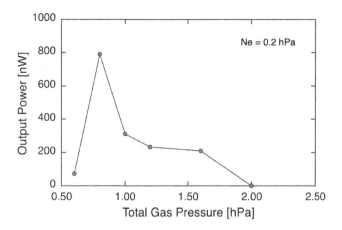

Figure 3.92 Operational parameters of the operations at 543.3 nm wavelength. Stable mono-mode lasing is only achieved in a fairly small regime, when compared to the 632.8 nm transition.

Figure 3.91 shows the arrangement of the electrodes along the plasma tube. The magnets were taped to the other side of the plastic support rail.

In terms of the rotation sensing capability, the system operated as expected. However, it did not run over a long period of time. A small increase in loss, for whatever reason, usually caused lasing to stop after a few hours. The observed Sagnac beat note was 132.8 Hz, which is 16% higher than the 114.7 Hz for the same system operated at 632.8 nm. Within the very limited mechanical stability of our test gyro and apart from the significant increase in the scale factor, we could not find any other important advantage for the operation of a laser gyro at this transition. In particular the operational difficulty of a much larger gain section is a significant disadvantage. Figure 3.92 shows the mapped out available regime for the operation of the $3s_2 \longrightarrow 2p_{10}$ transition. With 0.2 mB of neon and around 1 mB of helium, this range is a lot smaller than the operational regime of the standard red transition. This indicates that the suppression of neighboring longitudinal modes from over-pressuring in very large structures is not available. Both transitions share the same upper state; the difference in behavior is caused by a pressure dependent population increase in the $2P_{10}$ lower level and is physically attributable to temporary negative ion attachment [81].

3.11.2 Ring Laser Operations at 594–611 nm

There are three exploitable laser transitions [238] spaced closely together in the predominantly orange regime, namely the $3s_2 \longrightarrow 2p_6$ (611.8 nm), $3s_2 \longrightarrow 2p_7$ (604.6 nm) and $3s_2 \longrightarrow 2p_8$ (593.9 nm) in a helium–neon gain medium. The

Table 3.4. *Transition probability A_{ik}; measured ringdown times; and the corresponding loss per mirror L, Quality factor Q and Finesse F for number of transitions explored for rotation sensing on the same ring laser.*

λ [nm]	1152.27	632.8	611.8	604.6	593.9	543.3
ν_0	260	474	490	496	505	552
A_{ik} [s^{-1}]	1.07×10^7	3.39×10^6	6.09×10^5	2.26×10^5	2.00×10^5	2.83×10^5
τ [μs]	21	97	38	51	64	112
L [ppm]	254	55	140	105	83	48
Q	3.4×10^{10}	2.9×10^{11}	1.2×10^{11}	1.6×10^{11}	2.0×10^{11}	3.9×10^{11}
F	1.3×10^3	2.8×10^4	1.1×10^4	1.5×10^4	1.9×10^4	3.3×10^4

operation of a gyro simultaneously on two or more wavelengths is generally desirable, since it provides the opportunity to self-reference the gyro scale factor in an effort to move forward to *absolute* rotation sensing, provided that mode competition can be avoided. However, in the case here, all three transitions occupy the same upper state, thus potentially giving rise to deleterious mode competition. We used IBS coated mirrors, designed for a center wavelength of 543.2 nm. The system came up with the orange transitions, due to the 50 nm wide stop-band of the coating and the higher gain of the orange transitions.

This reflects in the measured ringdown times, where the losses increase as the optical frequency gets to the edge of the passband. Table 3.4 summarizes the properties on all available transitions that we have explored over the years. While the exact gain of each transition depends on the exact dimensions of the gain tube, the gas composition, the isotope ratio and the excitation density, the observations given here were all taken on the same instrument. The transition at 604.6 nm has about half the gain of the 611.8 nm transition, and the 593.9 nm has roughly one-third. This is in good agreement with the transition probability A_{ik} of these three transitions (given in Table 3.4), provided by the National Institute of Standards and Technology (NIST, USA). We also note that the wavelength of 543.3 nm is more than 10 times harder to excite than the transition at 632.8 nm, which acts as a reference for HeNe laser gyros. This also reflects in the fact that the pressure range over which the red 632.8 nm transition can be operated is much wider than for all other transitions [8, 9, 196, 238]. Figure 3.93 shows the observed relative gain of the HeNe laser lines in the range 594–611 nm. For each of the lines, we obtain a different scale factor. The value of the Sagnac frequency $\delta f_{611.8\ nm}$ is 117.0 Hz. As the optical frequency goes up, we obtain the values $\delta f_{604.6\ nm} = 118.4$ Hz and $\delta f_{593.9\ nm} = 120.5$ Hz.

Unlocked rotation sensing on the rate bias from the Earth itself was possible in a sense of a proof of concept on all three lines, however not simultaneously. The evidence from frequency pulling due to backscatter coupling was strong and varied

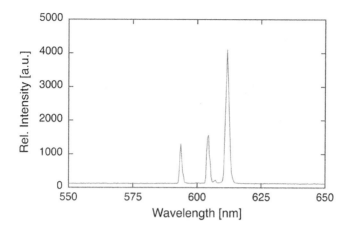

Figure 3.93 The three simultaneously excited "orange" laser lines of the HeNe gain medium, observed in a scanning Fabry–Perot analyzer. The gain of each transition decreases as the wavelength reduces. (Adapted with permission from [238], ©2022, The Optical Society.)

a lot over time [238]. The significant difference in gain for each transition prevented a simultaneous single mode operation of at least two of the three transitions. A closer evaluation of the possibility for the estimation of an absolute scale factor obtained from simultaneous unperturbed operation on two different transitions further apart than the $3s_2 \longrightarrow 2p_4$ ($\lambda = 632.8$ nm) and the $3s_2 \longrightarrow 2p_3$ ($\lambda = 635.2$ nm) could not be obtained. This is not unexpected, since the difference in gain is considerable. Both wavelength groups have in common that they share the same upper state. So mode competition will dominate the operation, even when each line runs in single mode, as outlined in Section 3.10.2. However, for the strongest line alone, a proper single mode operation is feasible, albeit at a much more limited gas pressure regime.

3.11.3 Ring Laser Operations on Crystalline Mirrors at 1.152 μm

Crystalline coatings, as introduced in Section 3.10.4 have been investigated in our GEOsensor type ring laser structure. The square cavity is 6.4 m long and encloses an area of 2.56 m^2. Since crystalline mirrors are not available in the visible domain, we have used the $2s_2 \longrightarrow 2p_4$ transition at a wavelength of 1.152276 μm. While this near infrared regime is very suitable for the new type of mirror coating, we have the disadvantage that this neon transition is a doublet with the competing $2s_4 \longrightarrow 2p_7$ transition at 1.152502 μm, only 51 GHz away. Operating on the NIR wavelength causes a reduction of scale factor of almost 45%. However, due to the longer wavelength, the scatter loss reduces as do the losses from thermal noise [196]. The mirrors consisted of a super-polished 25 mm wide fused silica substrate and an

Figure 3.94 A worst case Nomarski map of a crystalline mirror in the near infrared regime (left), and a sample mirror (right).

8 mm diameter crystalline coating wafer bonded by van der Waals forces to the substrate. The motivation to explore these coatings originated in their property of minimized thermally induced mechanical fluctuations, which makes them distinctly different from the industry standard IBS coatings. The mirrors were optimized for the operation in our gyro cavity with an angle of incidence of $45°$, a transmission of 1.9 ppm for s-polarization and losses of 15.1 ppm for p-polarization in order to ensure stable operations.

The high-reflectivity Bragg stack consisted of 38.5 periods of alternating GaAs (high index, 3.45 at 1.152 μm and 85.5 nm ideal quarter-wave thickness) and $Al_{0.92}Ga_{0.08}As$ (low index, 2.96 at 1.152 μm and 99.5 nm thickness). The absorption was established to be below 5 ppm as an upper limit. At the time of this experiment in 2014, the available technology of crystalline mirror manufacturing still had problems in keeping the scatter losses low. The micro-roughness of the multilayer wafer amounted to 1.7 Å. In addition to this, there were a significant number of growth related defects, which dominated the scatter loss. These defects are a common effect in molecular beam epitaxy (MBE), and our mirror samples showed 10 μm scatter centers with a density of the order of $1000/cm^2$. This compares to the diameter of our beam of 3.62 mm. Figure 3.94 gives an impression of the state of our mirrors, by showing an example of approximately similar quality as a color-inverted Nomarski map with the defects represented in black (left panel). The size of the image corresponds very approximately to the size of the laser beam, and the bonded coating wafer is shown attached to a fused silica substrate on the right side.

Later generations of crystalline mirrors showed a considerably reduced scatter from coating defects. The effect of their presence can be significantly reduced by flipping the coating crystals onto their reverse side in the bonding process. The number of coating defects builds up slowly during the crystal growth process. Therefore, the number of defects is very low at the bottom of the multilayer stack. However, once a defect has manifested itself, it prints through all subsequent layers. As a

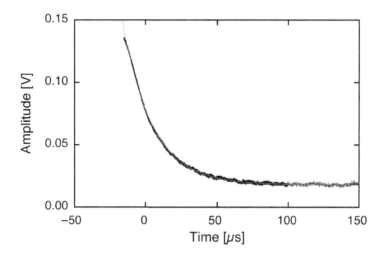

Figure 3.95 A cavity ringdown from a 0.2 mB gas mix of ^{20}Ne and ^{22}Ne at equal ratio. A value of 21 μs was typically obtained on the neon doublet at 1.152276 μm and at 1.152502 μm.

result, the bottom of the stack has considerably fewer defects than the top. In this way a reduction to about 10 defects per cm^2 has been achieved. Unlike dielectric mirror coatings, these crystalline coatings are not transparent in the visible spectrum. Therefore we pre-aligned the cavity with a set of IBS mirrors on the standard wavelength of 632.8 nm. Once lasing was obtained, we brought the laser cavity up to atmospheric air pressure, by gently adding helium from our gas supply. The IBS mirrors were swapped for a set of crystalline mirrors, taking care that the pre-alignment was not lost. The ring was then filled with a 50:50 mixture of ^{20}Ne and ^{22}Ne. Monitoring one of the output ports of the cavity with the help of a Si-based CCD camera, lasing was eventually achieved by toggling the cavity around the pre-aligned position at one of the corners. Once the alignment had been optimized, ringdown measurements were taken, as shown in Figure 3.95. Values around 21 μs, corresponding to a quality factor $Q = 2\pi v\tau = 3.2 \times 10^{10}$, were obtained, corresponding to a total loss (L) of the cavity with an FSR of 46.875 MHz of $L = 1/(\tau \times FSR) = 0.0011$. The loss was much higher than would be expected from the quality of the mirror coatings, due to the fact that some of the IR light was lost from the cavity by leaking around the edges of the 8 mm diameter coatings, as the alignment camera image clearly revealed. For successful rotation sensing, the system requires at least 1 mB of helium in the cavity in order to quench the competing second transition. When operated as a gyroscope, the IR regime shows comparable properties to the standard red transition. Figure 3.96 presents one example of a rather typical time series (left diagram) and the corresponding TDEV on the

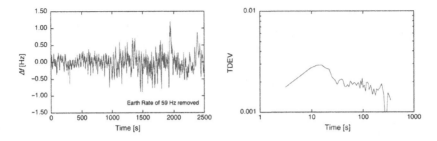

Figure 3.96 Rotation sensing at a wavelength of 1.152 μm is less tolerant than the operations at the 632.8 μm transition, due to a much smaller range of working gas pressures. A considerable number of the excursions in the Earth rotation rate are due to the mechanically unstable gyro structure bolted to a wall.

right side, normalized to the Earth rotation rate. The scale factor is almost at half the value of the standard red transition. Since the stability of the ring laser structure, mounted vertically to a wall in the old Rutherford building at the University of Canterbury in Christchurch (NZ), is low, we cannot draw meaningful conclusions from this measurement series in terms of gyro resolution. As a result of the light leakage from the cavity due to the small wafer size and the relatively large number of defects on the coating, the important properties of the crystalline coatings, namely low thermal noise and high heat conductance, were outside the regime where they can be meaningful tested. However, we note that this gyro unlocked on Earth rotation alone. The first ever laser gyro by Macek and Davies [142] required a turntable to produce rotation sensing.

3.12 Groups of Sensors; Networks

Networks of large scale rotation rate sensors can consist of several widely spaced single component sensors, which are tied horizontally onto the Earth's crust as rigidly as possible. Combining the continuous observations of all of these sensors would then allow one to establish the true rotation rate of the Earth as well as the two components of polar motion. A concept like this still has to deal with local or regional movements, independently of the global rotational signal, which can be a tedious exercise. Another approach would be the construction of a whole large scale sensor block, similar to the strap-down inertial measurement unit in aircrafts. Such a local sensor group is a demanding construction, when a sensor resolution of less than 1 prad/s is required. Such a sensor design could be integrated into an abandoned mineshaft or as part of a purpose-built structure. There is currently no precedent for a large scale sensor group in a mineshaft, although proposals have been put forward [53].

Figure 3.97 Aerial view of the construction site of ROMY. The excavation is shown on the left. The actual installation of the ring laser hardware took place when the construction was almost finished, as shown on the right. (Left side: adapted with permission from [74] ©2022 American Physical Society; right side: photo courtesy of H. Igel)

However, there is one example of a specially designed structure, which is the ROMY ring laser at the Geophysical Observatory in Fürstenfeldbruck, a small town west of Munich in Bavaria (Germany). ROMY stands for **RO**tational **M**otion in seismolog**Y**. Since the ROMY structure was particularly designed as a multi-component ring laser cluster, each sensor with twice the scale factor of the G ring laser, it could be optimized for the purpose. Arranged in the shape of a tetrahedron, each of the four independent ring laser cavities is triangular, with a length of 12 m on each side, and one of the gyros acts as a redundant sensor, to allow independent verification of consistency. In order to minimize the excavation effort at the site, the sensor block is designed as a tetrahedron standing on its tip, with the apex 14 m deep in the ground. Every corner of the tetrahedron has a massive concrete support structure, each of which rigidly ties the corner-boxes of three of the four rings together. This provides a stable geometric reference between the individual sensor components.

ROMY is a fully underground structure, so that strain from local winds has minimal loading or friction effects, despite the near top soil location. The excavation of the terrain was done in a minimally invasive way, by digging a cone shaped hole, only slightly larger than the ROMY structure. A shotcrete layer on the walls of the hole during the excavation process provided additional stability. In this way, the soil structure of the terrain around the platform of the sensor block could be maintained. Figure 3.97 gives an impression of the minimal invasive construction of the tetrahedral ROMY platform. On the left side it shows an aerial view of the excavated site with the apex point of the construction already concreted out. The beginning of the central access shaft is already visible. The three concrete arms in a radial direction will each carry one of the corners of the top of the structure. This is shown on the right side of the figure, at a time when the vacuum pipes of the ring

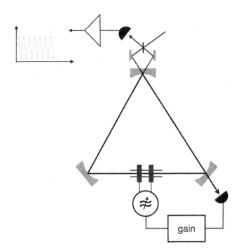

Figure 3.98 Illustration of the design of each of the ROMY ring laser interferometers.

laser structure were mounted. A couple of weeks later, the construction was closed and the top soil filled back in.

3.12.1 The ROMY Sensor Design

ROMY has one horizontal ring, which allows comparison with the G ring in order to check for regional rotational motion at about 200 km distance. All the other three rings are tilted by about 57° from the horizontal, and the apex at the bottom of the monument joins three corners of the tilted rings together. Angled granite support structures carry the mirror holder boxes, while the corner boxes for the horizontal ring are directly bolted to the concrete platform. The locations of the corner boxes define the physical size of the structure, and the corners are joined by stainless steel pipes to form the beam enclosure. Short bellows near the corner boxes reduce deformations from strain and make sure that the mechanical rigidity of the corner construction is not compromised. In the middle of one of the three sides of each ring, there is a 5 mm wide and 20 cm long capillary for laser excitation. The width of the capillary acts as a spatial mode filter and has been designed to minimize the losses on the desired $TEM_{0,0}$ laser mode. Higher order transverse modes with a larger mode volume, however, are discouraged since the gain tube acts as a soft aperture. As with all of our ring lasers, there are no Brewster windows or other loss increasing components around the entire cavity. In fact, there are only the three curved supermirrors (ROC = 12 m) as interacting intra-cavity components with a specified total loss of approximately 12 ppm per mirror (scatter, transmission and absorption) in

Table 3.5. *The projection of Earth rotation on each of the ROMY components*

Ring	Eff. Colatitude	Azimuth	Δf (calc.) [Hz]	Δf (obs.) [Hz]
Horizontal	41.8	0	545.3	554.0
North	112.3	0	302.9	305.3
West	48.2	−30	460.8	440.4
East	48.2	30	460.8	439.9

the system. Since the entire beam path is enclosed by a UHV (ultra high vacuum) compatible enclosure (pipe and mirror box housing), the resonator can be first evacuated and then filled with a mixture of (0.2 hPa) neon and (6.3 hPa) helium. Again, as usual for our gyros, lasing is achieved by RF excitation. Figure 3.98 depicts the basic sensor concept. Due to the open gain section, the laser gas is everywhere inside the cavity. Overpressuring the laser cavity increases the homogeneous linewidth and thereby avoids mode competition in a regime of ±90 MHz around the optical frequency. Therefore it is possible to operate the 36 m long cavity of the gyro on a single TEM_{00} laser mode per sense of propagation, despite a longitudinal mode spacing of only 8.6 MHz. Mode jumps are nevertheless not infrequent for any of the four lasers. Due to the substantial length of the ring cavity of 36 m, the steel structure easily contracts or expands by more than the 3 μm tolerance provided by the increased homogeneous line broadening. Another complication is the large internal stainless steel surface area of the vacuum recipient. Although each of the pipes was baked over several weeks to reduce the outgassing of hydrogen from the pipes, the residual diffusion into the cavity is still considerable. Contamination of the laser gas with hydrogen diminishes the achievable gain from the exploited $3s_2 \longrightarrow 2p_4$ transition at 632.8 nm (red). The application of a CapaciTorr D 200 getter pump in each ring reduces this effect significantly [182], thus increasing the time of continuous operation. To date it has been possible to operate each of the four cavities for several months on a single gas fill.

In order to obtain a stable beat note, the beam power in the cavity has to be stabilized. A small portion of the light leaking through one of the mirrors is detected and amplified by a photomultiplier. The resulting voltage is then fed back to drive the power of the RF transmitter, such that the laser radiation in the cavity remains constant. At another corner of the interferometer, both beams are taken out and interfered on a beam splitter. The resultant beat note corresponds to the Sagnac frequency and carries the information of the instantaneous rate of rotation that the interferometer cavity is experiencing. Each of the gyros has a different projection on the Earth rotation axis, which we can obtain by extending Eq. 3.1 by

$$\delta f = \frac{4A}{\lambda P}\Omega\cos\alpha\cos\beta,\qquad(3.118)$$

where α and β describe the north–south tilt (colatitude) and the azimuth rotation. For enhanced mechanical stability, the horizontally orientated ring on top of the ROMY sensor block is slightly smaller than the three angled rings. By making the sides 11 m long, the cavity could be placed inside the perimeter formed by the slanted rings. If all the rings were orientated with their area vectors collinear to the Earth rotation vector, we would obtain a beat note of 798.329 Hz for the three larger rings and 731.846 for the smaller one. In their true locations we obtain the beat notes as listed in Table 3.5. The observed differences from the design values are caused by some small relative misalignment in the orientation of the rings with respect to each other. These offsets were unintentional and happened during installation. They also include small offsets caused by the effective beam position inside the stainless steel enclosure after the cavity alignment. The obtained ringdown time of all four cavities ranged from 300 μs–1.2 ms, with the majority of the cavities usually operating around values of 700–800 μs. We observed a considerable variability, although none of the resonators were opened up to the atmosphere over the first three years of operation. This is apparently caused by small variations of the beam position inside the cavities, as the entire ROMY foundations exhibit small setting effects. These large cavities are far more sensitive to misalignments than their smaller counterparts. Figure 3.99, taken from the horizontal ring, shows an example of a typical beat note from the ROMY ring cluster. There is no change of laser mode visible in this eight hour time span, which means that the optical frequency of ring cavity has not varied by more than ±100 MHz (left plot) over this eight hour time window. Backscatter coupling causes a slow variation of about 3 mHz. When this effect is removed, we see no evidence from mechanical cavity instabilities. Due to the larger scale factor, we do however see a larger signal level from microseismic activity. The interference fringes have a good contrast ratio of K = 0.8. This is compatible with the astigmatism from the curved mirrors in the cavity, and the signal level is as high as 1.7 V without any apparent noise.

In the very short term we observe the expected high sensor stability, as the Sagnac beat note, integrated over 1 minute, is of the order of ±50 μHz and can be as low as ±20 μHz. The structure in the noise signal is caused by microseismic activity and actually presents a measurement signal, which is evident when the corresponding power spectrum is evaluated. This is shown in Figure 3.100. Although the power spectrum has been integrated over 10 minutes, the line indicating the beat note is very sharp, emphasizing the inherent stability of the installation. There are two fairly distinct sidebands around the Earth rate carrier at approximately

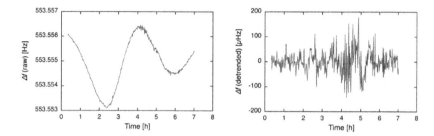

Figure 3.99 A short time series of the Sagnac beat note taken by the horizontal component of the ROMY ring laser cluster. Backscatter coupling causes the variation of the beat note (left panel). When backscatter is removed, we find no evidence of any cavity dither in the very short term (right panel). The larger noise level in the regime between 4 and 5 hours comes from microseismic activity.

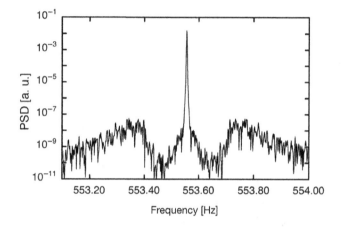

Figure 3.100 Power spectrum over 10 minutes of ring laser observations from the ROMY horizontal component. The main signal peak corresponds to Earth rotation. The secondary microseismic band at approximately 0.2 Hz around the carrier is clearly visible.

0.2 Hz distance. These are caused by Love wave-type rotational signals from the secondary microseismic band. Depending predominantly on ocean wave activity, these sidebands can become either smaller or significantly more pronounced. With all other parameters being constant, backscatter coupling reduces as $L^{-2.5}$ when the length L of the cavity goes up [97]. That means that the backscatter coupling of the ROMY rings is expected to be about a factor of eight smaller than for G, which is not quite the case. We observe a factor of two, and that could be due to more scatter losses by a much larger laser patch on the mirror. The peak to peak amplitude of one beam injected into the other beam by the backscatter process is obtained from the beat note at the Sagnac frequency from the intensity of each laser

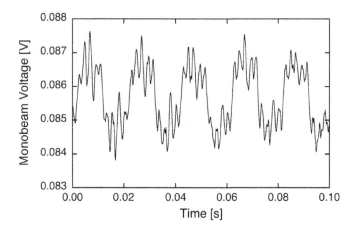

Figure 3.101 A sample of the intensity variation of one of the laser beams in the western slanted cavity of the ROMY ring laser cluster. On a constant intensity level of about 1.7 V, a small variable signal can be found. The lower frequency corresponds to a residual hum from the 50 Hz power line, while the higher frequency signal is caused by the backscatter process.

beam in the cavity. Figure 3.101 shows an example of the intensity recording of the cw beam, obtained from the slanted ring at the western side of the ROMY tetrahedron. The foremost evidence is a trace of 50 Hz hum from the power line with a peak to peak value of just under 2 mV sitting on top of the 1.7 V DC intensity voltage. On top of the hum we find the backscatter signal at a frequency of 449.6 Hz with amplitudes of 0.5–1 mV. In order to check the consistency of the measurements of the ROMY structure we can compare the signature of the extracted microseismic noise of corresponding rings, such as for example the horizontal and the slanted northern ring. They are orientated in the same way with respect to the east–west direction but have a different projection angle relative to the north–south direction. Therefore, we would expect to see the same envelope of the rotational signal, albeit at different amplitudes. Figure 3.102 depicts the results. The diagram shows a superposition of the two time series taken at the same time, each of them converted to the actually observed rotation rate, according to the respective scale factor of each ring. The thin signal record is from the horizontal component; the broader record belongs to the tilted ring. The two gyros did not quite work at the same level of resolution. This was caused by higher losses (significantly lower ring down time) in the northern slanted ring at the time of the measurement, which therefore introduced a higher noise level. For this graph we have subtracted the DC component of the Earth rotation rate from the observation. The remaining signal is typical for a windy day, where gusts of wind cause strain effects, due to friction from the wind passing over the local terrain. Most of the time the two rings show excellent

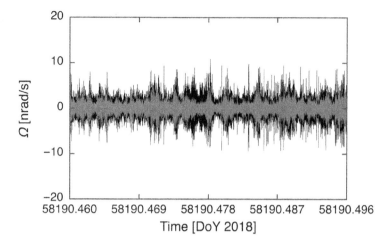

Figure 3.102 Comparison of the rotation rates of two ring laser gyros from the ROMY cluster with corresponding orientations. The signatures from the microseismic excitation, experienced by both instruments, exhibit a very similar structure.

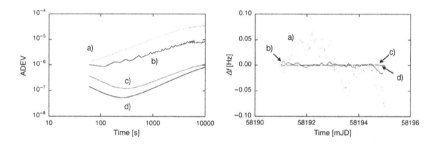

Figure 3.103 A comparison of the performance of the four identical ROMY gyros over several days reveals a large variety in quality. The Allan deviation (left panel) from the west ring exhibits a resolution almost at the level of the G ring laser, albeit not at the same long term stability. The east ring, in contrast, is nearly two orders of magnitude worse. The time series of the observations (right) indicate the same behavior.

agreement, as expected. However, there is a small section near the middle of the plot where there is a signal in the slanted ring, but only a very weak correspondence in the horizontal ring. This might indicate a rotation around the north–south horizontal axis, to which the horizontal ring is only indirectly sensitive by tilt.

The reported here, initial ROMY ring laser performance is not a final given for each sensor. As Figure 3.103 shows, the identical construction of the four different interferometers provides a wide span of performance. While the west ring and the horizontal component operate almost at the resolution level of the G ring laser, but naturally not at the same level of stability, the north ring experiences a somewhat larger drift of the beat note and exhibits a lower signal to noise ratio. Retrospectively this was found to be caused by a lower Q-factor, due to an issue in alignment where the beam was hitting the mirrors slightly off-center. In this measurement the east ring suffered from very low laser gain, as the cavity experienced a tiny vacuum leak caused by a leak in the vacuum seal in one of the cavity window flanges. This comparison illustrates the practical difficulty of keeping a set of large interferometers operating at top performance over long periods of time.

On top of the issues of maintaining the alignment, in particular that of the slanted rings, there are further deleterious effects to deal with, coming from ongoing gradual building settling, as well as a drift from laser gain variations. the latter is caused by residual outgassing from the vacuum tubes. Although each ring has a small getter pump installed, this is not sufficient for all cavities even in the absence of a leak. The exchange of laser gas has been necessary, approximately every three months in the first two years of operation. All cavities are identical in their design, so if all the small imperfections in each component have been fixed, there is no reason why ROMY could not operate at least at the same level of sensor resolution as the G ring laser. Furthermore, since ROMY has twice the scale factor of G, it is certainly possible to eventually obtain a better sensor resolution. At this point in time it is still open how well the stability can be ultimately controlled. G is intrinsically very stable, due to the monolithic Zerodur structure; therefore passive stabilization is entirely sufficient. ROMY requires active control of the cavity lengths to gain the required stability. Taking the already very low noise level on the very short term, as shown in Figure 3.99, as a reference, it is very likely that active perimeter control does not require a bandwidth higher than 1 Hz, and apparently also no control of the beam steering is required.

3.12.2 *The Orientation of a Cluster of Ring Laser Gyros*

To a first approximation, the four components of ROMY form a rigid sensor block, which is tightly coupled to the local ground at the Geophysical Observatory Fürstenfeldbruck. It is certainly possible that the structure as a whole can still deform slightly under external loading or building settling effects. Therefore, it cannot entirely be excluded that some beam-steering effects from structural relaxation processes of the ring laser laboratory can take place. The ROMY structure as a whole sensor block sets up a three-dimensional coordinate system. Due to the importance of the topic, we reproduce our work from [74] in this section. Any combination of three of the four components can be used to reconstruct the instantaneous Earth rotation vector, relative to the local inertial frame of reference, from the interferograms. The fourth component is redundant and offers

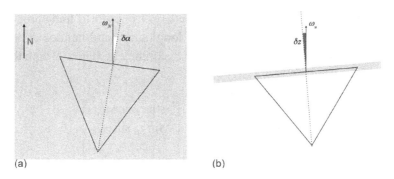

Figure 3.104 Graphical illustration of the ROMY sensor block orientation when viewed in a two-dimensional representation from the top (a) and along the longitude line from a western direction (b). The angles $\delta\alpha$ and δz indicate the misalignment errors relative to north and the local vertical.

the possibility of ensuring a consistent measurement. Figure 3.104 illustrates the assembly schematically in a simplified two-dimensional way. The entire tetrahedron is sketched as a triangle. The complete sensor block can be misaligned with respect to the north direction, and it can be tilted against the local vertical. While the length of each side is tightly constrained by the vacuum tubes, any misalignment of the corners $\delta\alpha$ with respect to the local meridian and the tilt angle δz of the entire structure is less well defined. With the horizontal ring given, the orientation of all other rings is constrained. We have used the observations from all four rings to establish the misalignment $\delta\alpha$ and the tilt angle δz of the entire ROMY structure with respect to the local horizontal plane. From several independent measurements, a misalignment angle of $\delta\alpha = 0.19205° \pm 0.00042°$ has been obtained. This indicates a very minor rotation of the entire sensor structure toward the east. The corresponding tilt angle of the sensor block is $\delta z = -0.07754° \pm 0.00009°$, with the minus sign indicating a tilt toward the north. ROMY observes a combination of global Earth rotation and local ground motion. For larger gyroscope installations like ROMY, biases from backscatter coupling are small, and we have neglected them here. In order to reconstruct the full Earth rotation vector, we have transformed the measured beat note δf_n from each of the four ring laser components of the tetrahedral structure into a global Earth-centered Earth-fixed $\omega_{X,Y,Z}$ frame (Figure 3.105) and solved the observation Eq. (3.119) for the vector $\mathbf{\Omega}_E$:

$$\delta f_n = \kappa_n \cdot \mathbf{\Omega}_E \cdot \mathbf{n}_n, \quad \kappa_n = (4 \cdot A_n)/(\lambda_n \cdot P_n), \tag{3.119}$$

with κ_n the respective scale factor of each ring laser component n, $\mathbf{\Omega}_E$ the Earth rotation vector and A, P and λ defined as before. The orientation vector $\mathbf{n}_n = D \cdot R \cdot \mathbf{n}_r$ is composed from the orientation $\mathbf{n}_r$ in the local North, East, Up (NEU, $\omega_{N,E,U}$) frame (Figure 3.105), a left-handed system, defined by the azimuth angle α_n and the zenith angle z_n (Figure 3.104). The matrix D is used to transform the local observation into the global frame with the longitude θ and latitude ϕ of the ring laser site. To obtain the true orientation, the misalignment matrix R with the offsets δa and δz is applied:

$$n_n = D \cdot R \cdot n_r, \quad n_r = \begin{bmatrix} \sin z_n \cos \alpha_n \\ \sin z_n \sin \alpha_n \\ \cos z_n \end{bmatrix} \tag{3.120}$$

with

$$D = \begin{bmatrix} -\sin \varphi \cos \theta & -\sin \theta & \cos \varphi \cos \theta \\ -\sin \varphi \sin \theta & \cos \theta & \cos \varphi \sin \theta \\ \cos \varphi & 0 & \sin \varphi \end{bmatrix}, \quad R = \begin{bmatrix} 1 & -d\alpha & -dz \\ d\alpha & 1 & 0 \\ dz & 0 & 1 \end{bmatrix}.$$

The ω_Z axis is aligned with the Earth rotation axis, while the ω_X axis passes through the Earth's surface, where the Greenwich meridian intersects with the equator. The ω_Y axis is orientated $90°$ toward the east, in order to form an orthogonal right-handed system. Figure 3.105 illustrates the orientation of the various frames of reference with respect to each other. In the global frame of reference, Earth rotation is almost entirely represented in the ω_Z component of the full Earth rotation vector,

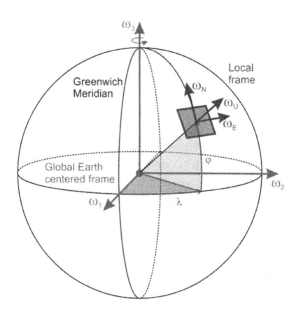

Figure 3.105 The angular rotation is measured in the local N,E,U system, which is subsequently transformed into a global reference frame. (Adapted with permission from [74] ©(2022) by the American Physical Society.)

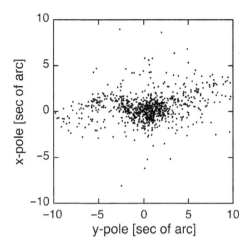

Figure 3.106 The position of the reconstructed xy-component of the Earth rotation vector from ROMY observations, based on a dataset of 47 days in length. (Reprinted with permission from [74] ©(2022) by the American Physical Society.)

while ω_X and ω_Y contain the polar motion, a multi-frequency signal that is generally small, with periods extending from one day to well beyond one year.

We used 47 days of continuous measurements on at least three sensor components and averaged them to hourly values of the instantaneous Earth rotation vector – 1125 values in total. We then took the average over all measurements as an estimate of the biased mean pole position, which comes out as (0.35 ± 0.022) arc sec in the ω_X-direction and (0.47 ± 0.07) arc sec in the ω_Y-direction. The observed small offset from zero is caused by remaining scale factor and bias errors in each ring. Figure 3.106 shows the result. The time series of the Earth rotation rate $\Delta\Omega$ is shown in Figure 3.107 with a constant value (Earth rate) of $\Omega_0 = 72.921$ μrad/s subtracted. The reconstructed vector Ω from the tilted rings is shown together with the rotation rate residuals obtained from the horizontal ring alone, with the projection angle adjusted for a minimum offset. The data gaps occur when one or more of the slanted rings lose the interferogram due to sudden changes of the operating laser mode. In terms of measurement resolution, this time series compares well with the other two components of the Earth rotation vector, indicating that this is the current limitation from the unconstrained cavity. Over the length of our measurement series, a long-term sensor stability of $\Delta\Omega/\Omega_0$ of 5×10^{-5} and less than 0.5 arc sec for the pole position has been achieved for the combined sensor design of the four large laser interferometers. After transforming the measurements into the Earth-centered Earth-fixed frame of reference, a consistent observation of the Earth rotation vector with a repeatability of about 0.1 arc sec was obtained.

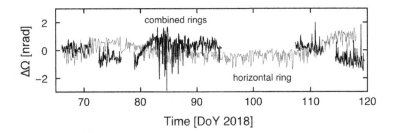

Figure 3.107 The reconstructed instantaneous Earth rotation rate from ROMY observations, based on a dataset of 47 days in length. (Reprinted with permission from [74] ©(2022) by the American Physical Society.)

3.12.3 Details on the Transformation into the Earth-Centered Earth-Fixed Frame

Each component of ROMY has a well defined orientation relative to the local horizon system (NEU frame). The orientation of each ring is given by the respective azimuth (α_r) and zenith angle (z_r) (see Figure 3.104). Their normal vectors are given by $n_r = [\sin z_r \cos \alpha_r, \sin z_r \sin \alpha_r, \cos z_r]^\mathrm{T}$, referenced to each laser beam plane. To determine the misalignment ($d\alpha$) and the tilt (dz) of the entire structure with respect to the NEU frame, we apply the transformation

$$\boldsymbol{n}_n^{NEU} = R(dz, d\alpha) \cdot \boldsymbol{n}_r, \tag{3.121}$$

with

$$R(dz, d\alpha) = \begin{bmatrix} 1 & -d\alpha & -dz \\ d\alpha & 1 & 0 \\ dz & 0 & 1 \end{bmatrix},$$

as both the misalignment and the tilt are small. Consequently, the normal vectors in the NEU frame are established by the relation

$$\boldsymbol{n}_n^{NEU} = \begin{bmatrix} \sin z_n \cos \alpha_n \\ \sin z_n \sin \alpha_n \\ \cos z_n \end{bmatrix} =$$

$$= \begin{bmatrix} 1 & -d\alpha & -dz \\ d\alpha & 1 & 0 \\ dz & 0 & 1 \end{bmatrix} \begin{bmatrix} \sin z_r \cos \alpha_r \\ \sin z_r \sin \alpha_r \\ \cos z_r \end{bmatrix} \tag{3.122}$$

and then transformed to the global XYZ frame (see Figure 3.105) with the help the actual latitude (φ) and longitude (θ) of ROMY:

$$n_n^{XYZ} = D \cdot n_n^{NEU} = \tag{3.123}$$

$$= \begin{bmatrix} -\sin\varphi\cos\theta & -\sin\theta & \cos\varphi\cos\theta \\ -\sin\varphi\sin\theta & \cos\theta & \cos\varphi\sin\theta \\ \cos\varphi & 0 & \sin\varphi \end{bmatrix} n_n^{NEU},$$

where D represents the respective transformation matrix. The observed Sagnac frequency for all rings is given by

$$f_n = \kappa_n \cdot \Omega \cdot n_n^{XYZ}. \tag{3.124}$$

κ_n denotes the scale factor of the n-th ring, and Ω is the Earth rotation vector; at this level Earth rotation variations are assumed to be too small to matter, so that the Earth rotation vector can be set to be equal to $[0, 0, \Omega_0]$. This leads to the form

$$f_n = \kappa_n \cdot \Omega_0 \cdot n_n^z, \tag{3.125}$$

with

$$\begin{aligned} n_n^z = {} & \cos\varphi\sin z_r \cos\alpha_r + \sin\varphi\cos z_r \\ & - d\alpha\cos\varphi\sin z_r \sin\alpha_r \\ & - dz\cos\varphi\cos z_r + dz\sin\varphi\sin z_r \cos\alpha_r. \end{aligned} \tag{3.126}$$

This can be rewritten in matrix notation as

$$\begin{aligned} f_n = {} & \kappa_n \cdot \Omega_0 [\cos\varphi\sin z_r \sin\alpha_r, \; -\cos\varphi\cos z_r \\ & + \sin\varphi\sin z_r \cos\alpha_r] \cdot \begin{bmatrix} d\alpha \\ dz \end{bmatrix} + f^{theo}, \end{aligned}$$

with

$$f^{theo} = \sin\varphi\cos z_r + \cos\varphi\sin z_r \cos\alpha_r, \tag{3.127}$$

which allows for the estimation of the misalignment angles angles dz and $d\alpha$. The obtained values are then substituted into Eq. 3.122 to provide the true orientation of each ring laser gyro component in the local NEU frame of reference. In order to estimate the full Earth rotation vector based on ROMY's observations, we use Eq. 3.124. At this point we have established the correct orientation vector n_n^{XYZ} for each ring. In order to obtain the components of the Ω vector, we build the system of equations

$$f_\kappa = H \cdot \Omega_{\text{local}}, \tag{3.128}$$

where f_κ refers to the observation vector divided by the respective scale factors, the matrix $H = \begin{bmatrix} n_H^{NEU}, n_E^{NEU}, n_N^{NEU}, n_W^{NEU} \end{bmatrix}^{\mathsf{T}}$ contains the normal vectors of each gyro (H - horizontal, E - east, N - north and W - west) in the NEU frame and the locally experienced rotations are $\mathbf{\Omega}_{\text{local}} = \mathbf{D}^{-1}\mathbf{\Omega} = [\omega_n, \omega_e, \omega_u]^{\text{T}}$. Next, we apply the least squares method to establish the values of the $\mathbf{\Omega}_{\text{local}}$ vector:

$$\mathbf{\Omega}_{\text{local}} = (\mathbf{H}^{\text{T}}\mathbf{H})^{-1} \cdot (\mathbf{H}^{\text{T}}\mathbf{f}_\kappa). \tag{3.129}$$

Finally, we establish the global rotation vector $\mathbf{\Omega}$ vector by transforming $\mathbf{\Omega}_{\text{local}}$ into the global XYZ frame, with the application of the transformation matrix $\mathbf{D}$. These first results from the initial ROMY operation have been published in [74].

4

Data Acquisition and Analysis

Ring lasers measure pure rotation and are entirely insensitive to translational motions. The gyroscope relates angular velocity to a beat note from two mono-chromatic optical frequencies. This beat note can reach several hundreds of kHz for an avionic application, where the platform turns rapidly. For the application in the geosciences, the system is rate biased away from zero by the angular velocity of the Earth, which is $\Omega_E = 2\pi/86164 \sin(\phi)$ rad/s for a gyroscope horizontally placed at latitude ϕ. For the G ring laser at latitude 49.14°, the experienced rate bias is $\Omega_E = 55.15$ μrad/s, corresponding to a beat note of $f_s = 348.515$ Hz. Geophysical signals modulate this bias value in several ways. Polar motion and local tilts both change the projection of the rotation vector onto the normal vector of the laser plane, as they are effectively changing the angle ϕ. Variations of the rotation rate of the Earth, due to the interaction of the geophysical fluids with the solid Earth, change Ω_E directly.

4.1 The Effect of Detector Noise in the Ring Laser

Periodic low frequency geophysical signals of interest have periods of $\frac{1}{2}$ day, 1 day, 14 days, 28 days and finally 1 year for the Annual signal and up to 435 days for the Chandler wobble. Further signals at the decadal or secular scale are currently out of scope because continuous meaningful observations over such a long time span do not yet exist. Compared to the rate bias, introduced by the Earth rotation, the diurnal polar motion signal (f_{pm}) contributes by as much as 5 μHz $\leq f_{pm} \leq$ 50 μHz, depending on the position of the moon in its orbit. At most this corresponds to 1 part in 10^7 relative to the rate bias of the Earth. Less prominent geophysical signals, like the variation in Length of Day (LoD), are a factor of about 10 below this signal level, corresponding to $\Delta\Omega/\Omega_E \approx 10^{-8}$. These stringent requirements set the ultimate goal for the desired sensor resolution to values around 1 part in 10^9. We note, however, that tests of fundamental physics would require yet another factor of 10.

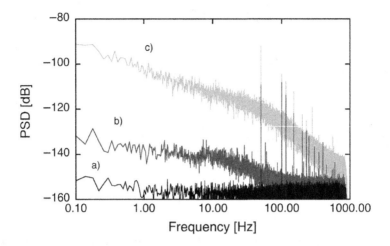

Figure 4.1 Comparison of the noise level of two detectors of different intrinsic noise levels. Both detectors are illuminated by 42 nW of cw beam power. The C10439-01 detector in low gain mode is shown with the laser switched off (a) and in operation (b). It has a noise level of up to −40 dB below the high gain HUV2000B detector from EG&G (c).

For this reason it is important to minimize detector and amplifier noise on the digitizer readout voltage. Apart from a photomultiplier, two different types of photo-diodes were used in our application over the years, namely a EG&G HUV2000B and a Hamamatsu C10439-01. Both diodes have a built-in trans-impedance amplifier. While the EG&G HUV2000B had an additional Analog Devices OP 07 operational amplifier attached, in order to achieve a signal level of 3–5 V for the 16 bit digitizer (NI PCI-MIO-16XE-50), the Hamamatsu C10439-01 was operated in the low gain mode in order to gain sufficient bandwidth, and the output signal of about 45 mV was directly fed to a Kinemetrics 3 channel 24 bit Basalt data logger at short distance for very low detector and amplifier noise. Figure 4.1 shows a comparison of different operation conditions for these detectors, exposed to the radiation of the counter-clockwise propagating laser beam. There is a significantly lower noise level in the recorded data of the C10439-01 amounting to up to −40 dB over the frequency band of interest of $0 \text{ Hz} \leq f \leq 1 \text{ kHz}$. The spectra also reveal that there are a lot of very sharp frequency components on the recorded data. With a few exceptions, these are related to the mains frequency of 50 Hz and its harmonics. At the time of writing, it was not clear whether these frequencies are injected by hum loops of the data logger system or if they are already modulated onto the optical beams via the RF-excitation of the gain medium. We find these signals everywhere with a variable power spectral density (PSD), even in the spectrum of the Hamamatsu C10439-01 diode when the laser plasma is switched off. Apart from the mains hum,

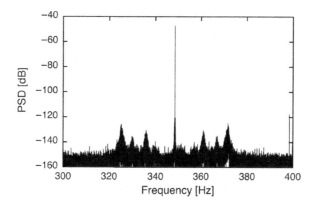

Figure 4.2 Spectral beam power of the combined beam taken with the C10439-01 detector. The spectral power density of the Sagnac beat note exceeds 90 dB.

there are no other features on this flat spectrum down to a signal floor at a level of −130 dBV. The interferogram from the two counter-propagating laser beams typically has a signal power exceeding 90 dB. Figure 4.2 shows an example. In this plot there are also a few spurious intermodulation signals visible. Due to their low spectral power densities and their wide frequency separation, they do not interfere with the rotation-induced beat note. The origin of this set of low level spurious frequency signals is not known. There are no corresponding signals anywhere else in the measurement system. Candidates are structural resonances of the gyro body and platform, causing torsional motion and fluctuations in the ionization of the plasma of some description. The former is unlikely because the seismometer on the ring laser structure did not show any indications of such structural resonances. If the suspected latter cause applies, we have not yet found the applicable mechanism. The observed spurious frequencies are not dependent on the plasma intensity. Furthermore, they are independent of the placement of the electrodes, as discussed further in Section 3.9.6.

4.2 Time and Frequency

The main output of a ring laser gyroscope is a beat note between the two counter-propagating optical frequencies in the traveling wave resonator. Depending on the size of the instrument, this beat note is in the range of $60 \text{ Hz} \leq \Delta f \leq 5 \text{ kHz}$. In order to extract rotations from the instrument, this beat note has to be detected by either a photomultiplier or a photodiode, amplified to a suitable voltage level of, say, ± 5 V, digitized and and run through a frequency estimation routine to obtain the exact frequency, corresponding to the actual rate of rotation, experienced by the interferometer. This data processing method requires a stable frequency source for the digitization process and a sufficiently accurate timing process, to establish the epoch of transient events correctly. Although it is important to capture the phase of geophysical signals in the diurnal and semidiurnal band correctly, the correct reference to the observation epoch has particular importance for higher frequency signals, such as seismic events. While an offset in time of about one minute already generates a perceptible phase shift for the extraction of solid Earth tides, the timing of seismological signals needs to be accurate to 1 ms with respect to UTC in order to properly relate the observations of different stations to each other. While a correct timescale is required to obtain the phase of the geophysical signals of interest correctly, there is also the need for high stability of the sampling frequency, to avoid the reconstructed frequencies from the periodic geophysical signals drifting. A typical quartz crystal of a free running data logger provides a stability of $\Delta f/f \approx 10^{-6}$. This is good enough for short data runs. However, applied to the long-term monitoring of polar motion, such a system would loose about 0.1 s per day. Therefore, it is desirable, if not mandatory, to use a GNSS- disciplined time and frequency source for the data logging process. The G ring laser at the Geodetic Observatory Wettzell is embedded in the local infrastructure for time and frequency distribution. The sampling frequency for the data acquisition system is derived straight from a hydrogen maser, and the time-stamping is synchronized to the local cesium clock of the observatory, which contributes to UTC by providing UTC(ifag) to the Bureau International des Poids et Mesures (BIPM).

4.3 Time Series Analysis

Diurnal and semidiurnal periodic geophysical signals have frequency bands around 11.6 µHz and 22.5 µHz, respectively. In order to distinguish the individual contributions from each other without error, the timing drift of the frequency of the digitizer clock must be about one order of magnitude smaller than the desired frequency resolution in the analysis process. In order to resolve a frequency of $f_{\text{sig}} = 100$ nHz in a straight Fourier transform, a dataset of nearly $T = 1/f_{\text{sig}} \approx 116$ days is required. This illustrates that there are two requirements for a data logging system.

- There are geophysically induced variations in the observed rotation rate of the Earth, which slowly change over a wide dynamic range with durations from several hours up to hundreds of days. These frequencies have to be captured accurately with high resolution. The coherence between the source of the signals and the timescale must not be lost over long observation times of hundreds of days.
- The unperturbed linewidth of the interferometer beat note is of the order of several µHz. In order to obtain the full measurement resolution of the sensor signal,

a frequency extraction technique is required that reaches the sensor resolution quickly and therefore operates well below the Nyquist limit.

In order to satisfy these conditions, it is important to employ data acquisition techniques that reliably reach a resolution of about 1 µHz for a digitized sine wave of 30–60 s duration. Several methods satisfy these requirements and have been widely used on our large ring laser observations. The interferogram of a Sagnac interferometer, strapped down to the Earth, contains a single spectral line with a high fidelity of the order of 90 dB or more. Due to some experimental imperfections, the spectrogram of a ring laser record may contain spurious frequency components, such as hum from the AC power line and a significant number of harmonics. We have observed small peaks of 5–10 dB at a frequency of 50 Hz and its harmonics, such as 100 Hz, 200 Hz and many more. Other deleterious effects, such as backscatter coupling, give rise to harmonics of the Sagnac frequency, most dominantly at the second and third harmonics of the interferometer beat note. The difference in signal levels between the beat note and the first harmonic is at least 40 dB for the G ring laser. With the application of a second order Butterworth filter of 20 Hz bandwidth centered around the primary peak, a very sharp single frequency component with a Lorentzian line shape remains, which must be detected with the best available frequency estimation method. Over the years, we have used several different methods, as outlined below.

4.3.1 Autoregression: AR(2)

For the estimation of a single narrow frequency component, a second order autoregression model is required. The basic concept of this frequency estimator requires that the output parameter depends linearly on its own previous values plus a stochastic term [113, 115]. It takes the form

$$x_t + a_1 x_{t-1} + a_2 x_{t-2} = \epsilon_t, \tag{4.1}$$

where the x_{t_i} are the recorded data points and the a_i are the adjustment parameters. ϵ_t represents a random process, which usually has a Gaussian distribution with a zero mean and unit variance. The line center itself is then estimated by

$$f_0 = \frac{1}{2\pi} \cos^{-1}\left[\frac{-a_1(a_2 + 1)}{4a_2}\right]. \tag{4.2}$$

The full width at half-maximum (FWHM) is then computed by

$$\delta = \frac{1}{2\pi}\left\{\cos^{-1}\left[\cos(2\pi f_0) - \zeta\right] - \cos^{-1}\left[\cos(2\pi f_0) + \zeta\right]\right\}, \tag{4.3}$$

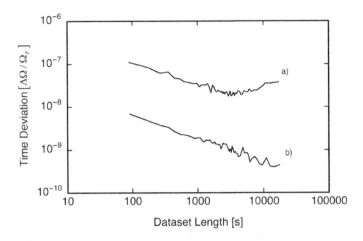

Figure 4.3 A typical time deviation plot of the G ring laser from the performance level of the year 2003 is shown (a). The lower TDEV curve (b) shows the resolution limit of the data logging system with respect to an ideal synthetic signal.

where

$$\zeta = \frac{\sqrt{(4a_2 - a_1^2)(1 - a_2)}}{4a_2}. \tag{4.4}$$

Figure 4.3 shows a comparison of two identically AR(2) processed TDEV time series. One sampled data stream contains the digitized output frequency of the G ring laser (a), and the other data stream takes the pure output signal of a signal generator, tuned to the same frequency and set to the same signal amplitude (b). One can see that the AR(2) frequency estimation of the measured rotation rates is not limited by the frequency estimation process. In the absence of laser noise, cavity perturbations and induced microseismic dither, the AR(2) frequency estimator resolves the synthesiser frequency about one order of magnitude more sharply than the interferometer beat note. This means that the frequency estimator is not the limiting component in the data acquisition chain.

4.3.2 The Buneman Frequency Estimator

The Labview program development environment, which is mostly used for the data processing of our ring laser data, contains a numerical frequency estimator, denoted as the *Buneman Frequency Estimator*. This software module is based on a Fourier transform and can be applied to the digitized data stream. It provides an exact frequency estimate for the case of a single pure sine wave [106]. The frequency of interest is obtained in some relative units from

$$\beta = b + \frac{n}{\pi} \tan^{-1} \left[\frac{\sin\left(\frac{\pi}{n}\right)}{\cos\left(\frac{\pi}{n}\right) + \frac{|F_b(x)|}{|F_{b+1}(x)|}} \right], \qquad (4.5)$$

where n is the number of data points in the array $\mathbf{X}$ of the sampled data, and F_b corresponds to the value of the Fourier transform of the input array $\mathbf{X}$ at the frequency b. The value of b is determined by using the greatest value of $|F_b(\mathbf{X})|$. β converts to the frequency of interest f_0 by

$$f_0 = \frac{\beta f_s}{n}. \qquad (4.6)$$

We have compared the performance of the AR(2) and the Buneman frequency estimator by subjecting them to a synthetic sine wave masked with normal distributed pseudo-random noise. The amplitude of the sine wave was set to 0.3 V and the noise level was set to $\sigma = 0.002$ V. This corresponds to a situation, much worse than the typical ring laser operation scenario, where the noise level remains well below 100 µV. The sampling rate had the same values as used in the data logger, namely 6.7 kHz. Figure 4.4 presents the results obtained. The x-axis shows the respective length of the analyzed dataset, where our typical integration times of 30 and 60 seconds are indicated by vertical lines. The vertical axis shows the deviation from the true frequency value in units of µHz. For the parameters used in this test, we find that the AR(2) frequency estimates drop below an error level of 12 µHz after

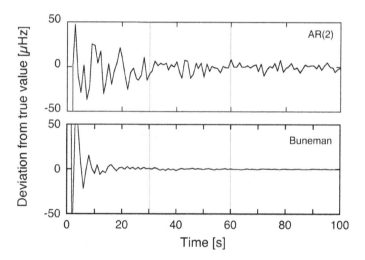

Figure 4.4 Illustration of the performance of the AR(2) and Buneman frequency estimator. The vertical lines at 30 and 60 s indicate a length of the dataset that is routinely used in our ring laser operations. After 30 s of integration, AR(2) deviates by less than 12 µHz and the Buneman estimator by less than 3 µHz from the true frequency.

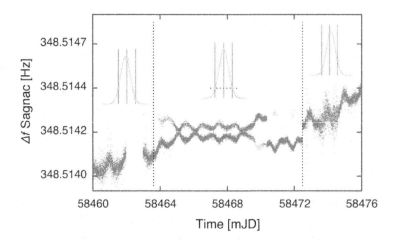

Figure 4.5 Illustration of a situation where the Buneman frequency estimator delivers ambiguous results. At the beginning and the end of this dataset, the frequency estimates are correcty derived. In the time interval between the two dotted lines, we find a double track with erroneous frequency estimates.

30 seconds of data, while the Buneman frequency estimator already stabilizes around 3 μHz after about 20 s. Our measurement resolution improved in small steps over the years, so that it became advisable to switch to a frequency estimator with higher precision.

The Buneman frequency estimator works well under normal conditions; however, one should also be aware that the method can fail, and invalid results are then obtained. We recall that the frequency is estimated by a Fourier transform, below the Nyquist limit. So the frequency of interest is derived from an interpolation process between neighboring spectral components of the Fourier transform, where the contributing spectral lines are those with the highest spectral powers. Under rare circumstances this requirement can be ambiguous, in particular when the neighbors on each side of the tallest peak have the same amplitude. In such a case it depends randomly on the noise of the measurement whether the left or the right spectral line from the tallest component is selected for the analysis. This, however, leads to a situation where the frequency estimate flips between two distinctly different values with a gap between them. Figure 4.5 illustrates this scenario with an example from our observations.

4.3.3 The Single Tone Estimator

A fast and generally more reliable module for the extraction of the Sagnac frequency is the *Extract Single Tone Information* function [106] module from the Labview development package. This subroutine evaluates the function

$$x(n) = A \cos \left(\frac{2\pi f n}{f_s} + \phi \right).$$ (4.7)

In order to achieve this, the function performs an FFT analysis over the windowed (Hanning window) time series and computes the envelope function e of the type $e = \sin x / x$, where x contains the bins with the largest amplitudes in the spectrum. For our application, where the signal to noise ratio exceeds 80 dB, the frequency estimates quickly drop below the error level of 1 ppm when the number of data points used in the calculation is larger than 1000. At the sampling rate of the G ring laser (2 kHz), this is achieved at 0.5 s of sampling. Apart from the processing of multi-frequency seismic data, all Earth rotation monitoring systems today use the single tone extractor. We usually choose a block size of 60 s, corresponding to 120,000 data points.

Apart from a verified performance of the respective frequency estimators, one should also look at the fundamental limit of the frequency estimation process. Therefore, it is important to compute the *Cramer–Rao lower bound* [112], which defines the theoretical lower limit for the variance of the estimated frequency given by

$$\sigma_f^2 \geq \frac{12}{(2\pi)^2 \eta N (N^2 - 1)}.$$ (4.8)

In this expression the number of sampled data points is represented by N, and $\eta = A^2 / 2\sigma^2$ with A the amplitude of the sinusoidal signal of interest and σ the variance of the noise on the signal. For 30 seconds of sampling at 7.6 kHz and η taken from the power spectrum of the actual measurements as 1.2×10^9, we obtain for the theoretical lower bound $\sigma_f \approx 5 \times 10^{-14}$. At the time of writing the achieved sensor resolution approached 10^{-9}, and it can be concluded that the data acquisition process is not currently limiting the achievable sensor performance.

4.3.4 Instantaneous Frequency Estimation by Hilbert Transform

The single tone estimation discussed in the previous section is practically carried out in the spectral domain by the application of a fast Fourier transform. In contrast to this approach, it is also viable to exploit the time domain. Frequency counters are an obvious choice, but they require comparatively long integration times in order to achieve a resolution below 1 μHz, which is required for high resolution Sagnac interferometry. The application of the Hilbert transform on the digitized measurement signal provides access to the envelope and the instantaneous phase of the recorded Sagnac signal. It has the shape

$$x(t) = A(t) \cos(\omega t + \varphi(t)).$$ (4.9)

An important property of the Hilbert transform is the rotation of the phase by 90°, and it is written as [41]

$$h(t) = A(t)\sin(\omega t + \varphi(t)). \tag{4.10}$$

The expression of the analytic signal is then defined as a complex quantity

$$f(t) = x(t) + ih(t) = A(t)e^{i(\omega\delta + \varphi(t))}, \tag{4.11}$$

where the real part contains information on the envelope of the amplitude $A(t)$, and the imaginary part contains the instantaneous phase of $x(t)$. After unwrapping the phase for discontinuities and differencing the phase with respect to time, the instantaneous frequency can be obtained. In order to investigate the applicability of both concepts for the exact recovery of the beat note of a ring laser at high resolution, we have compared their performance on a synthetically generated dataset under different time intervals of integration, as well as on real measurement data. The result is summarized in Figure 4.6. The first test, shown in the top panel, examined how well a predefined frequency can be recovered as a function of integration time. The frequency estimation by employing a Hilbert transform (a) showed a larger offset of up to 0.5 mHz for short integration times between 1 and 10 s. The single tone frequency estimator (b) demonstrated a much more robust performance and deviated by less than 1 mHz. For integration times of about 30 s upwards, both estimation procedures agreed well. The middle chart depicts the remaining standard deviation of the frequency estimation taken over 500 frequency estimates in the presence of a low level of uniform noise as a function of integration time. The single

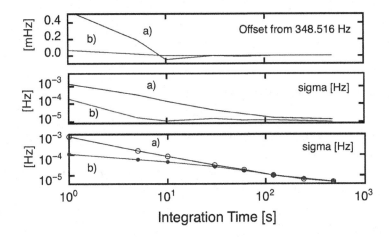

Figure 4.6 Comparison of the convergence properties of the single tone frequency estimator and the Hilbert transform approach applied to the frequency estimation of a Sagnac Interferometer measuring the Earth rotation rate.

tone frequency estimator (b) shows a significantly lower scatter than the Hilbert transform signal estimator (a). Again we find that for an integration time longer than 30 s, both concepts produce comparable results. For the given problem of extracting the G ring laser beat note from the digitized voltages, we show the results in the bottom chart. Both concepts agree reasonably well for integration times longer than 10 s. For shorter integration times, a clear advantage is found for the single tone extractor (a). This may be related to the boundary artifacts that are typical for the Hilbert transform. It was necessary to delete 3% of the available data points on either end of the extracted phase record. In order to obtain a continuous dataset for a long observation time series, an overlapping data evaluation strategy must be chosen. The application of the frequency estimation technique based on the Hilbert transform, however, has the advantage of producing a point by point phase estimate, which is useful for the study of high frequency effects in the gyro cavity.

4.3.5 The Effect of Subtle Crosstalk in the Digitizer

We discussed laser-related frequency pulling and pushing effects in Section 3.9. Here we add another very subtle potential error source to this list, namely the effect of signal crosstalk in the digitizer stage of the data logger. Many multi-channel digitizers are set up to multiplex the otherwise independent input channels to a single digitizer unit. The digitizer itself is usually designed such, that it does not present a load for the input signal. All this results in a small but non-negligible sensitivity of the digitizer to electrical crosstalk effects, albeit at a very low level. Crosstalk levels for data sample voltages in the range of 60 mV have been observed to be around 45 dB below the actual signal level, which is not sufficient. To avoid any issues from crosstalk interference, a signal crosstalk level well below −65 dB relative to the signal voltage is required. This is a concern in particular for low signal level ring laser recordings. This mode of operation can be desirable in order to avoid noise from additional amplifiers in the signal chain. Since there are situations where this matters a great deal, the effect needs to be detailed here. One prominent area where this vulnerability concerns the operation is the backscatter correction process (for details on backscatter, see Section 3.9.4).

The essence of this correction process is the clean separation of a small AC signal component ($\approx$3 mV) at the Sagnac frequency on top of a much larger DC voltage ($\approx$1.7 V). The ratio of these two signal components for each beam and their phase difference is a sensitive input quantity to the correction process. The AC voltages are within a factor of 2–4 of the undesired AC crosstalk signal. The two AC signals of the ring laser beams, as well as the beat signal of the combined beam, have the same (Sagnac) frequency but a different relative phase with respect to each other. Since the correction process is highly sensitive to this phase relationship between the two

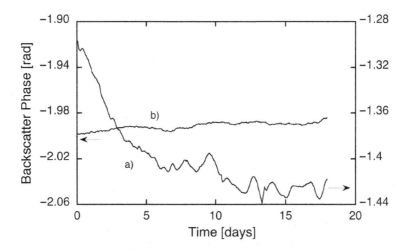

Figure 4.7 Comparison of the phase difference between the two laser beams in the G ring laser. In one case (a) the phase estimate is polluted by digitizer crosstalk. In the other dataset (b) the effect of this crosstalk problem was suppressed. The ordinate scale has the same range for both datasets.

AC signal components of the individual beams, signal crosstalk leakage from the combined beam will significantly perturb the correction process. We illustrate this with an example by comparing the established phase difference between the Sagnac modulation of the two individual beams with and without crosstalk, as shown in Figure 4.7. The G ring laser was operated pressure stabilized under two distinct conditions. In the first one, there was some small crosstalk contamination of -45 dB below the input signal level present in the system (a). In the other dataset (b) the crosstalk effect has been suppressed as much as possible (≈ -65 dB). For a gyro made of Zerodur and operated under pressure stabilized conditions, one would only expect a very small variation of the phase over longer periods. The larger excursions in graph (a) result from additional phase shift contributions, introduced by the crosstalk in the digitizer, causing perturbations of the order of ±50 μHz. This is more than an order of magnitude above the desired stability limit. This problem not only occurs with various high level data acquisition cards from the vendor National Instrument but also with high end seismic signal digitizers, such as the Kinemetrics Basalt4X/8X or the Obisdian4X/8X.

4.4 Spectral Analysis

The prime observable from a ring laser gyroscope is the waveform of the beat note between the two counter-propagating laser beams. By placing a low noise photodetector on each arm of the beam combiner and subtracting both signals from

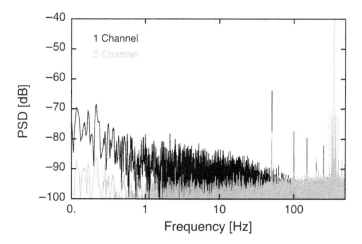

Figure 4.8 Comparison of the spectrum of the Sagnac waveform between a single detector on one port of the beam combiner and a combined detector on each of the two available output ports. Common mode noise is rejected, while the amplitude of the desired measurement signal is doubled.

each other, we recover all the available signal power. At the same time, spurious common mode signals and noise are rejected from the Sagnac waveform. Figure 4.8 illustrates the effect. As expected, there is a 6 dB gain in the spectral power for the combined detector signal, and low frequency noise is significantly reduced. Since the 50 Hz hum crosstalk from the electrical power line is coupling into our measurement system, it is much reduced due to the fact that it is almost a common mode signal contribution. While the spurious frequencies and system noise on the electrical subsystems are suppressed, noise and signals on the optical path of the Sagnac interferometer are also amplified, since these signals have opposite phase, and they add up in the evaluation process. In Figure 4.9 the effect of this extra gain is illustrated. It is important to note that it is not only the Sagnac frequency itself that improves by 6 dB in signal power; there are also spurious, ill-defined frequencies with an offset of a $\delta f \approx 12.4$ Hz, $\delta f \approx 18.3$ Hz and $\delta f \approx 23.7$ Hz. They show up like modulation sidebands around the Earth rotation signal as a carrier. What the spectrum cannot show is that the amplitudes of these apparent modulation signals are highly variable.

So far we have looked at the digitizer input signal in this section. The frequency estimator processes chunks of one minute duration, logged at a rate of 2 kHz, and determines the instantaneous Sagnac frequency as one value per minute. These values are accumulated continuously. Now we are looking at an interval of more than one year's worth of data. From this measurement series we take the power spectrum, which is shown in Figure 4.10. The corresponding spectral resolution is 34 nHz on

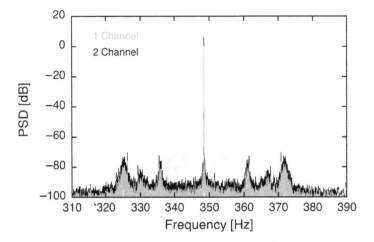

Figure 4.9 Spectrum of a small region around the Sagnac frequency. The comparison illustrates the power gain between a single detector on one port of the beam combiner and a combined detector on each of the two available output ports.

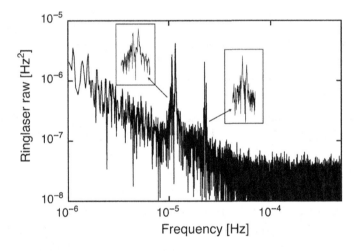

Figure 4.10 Spectrum of one year of nearly continuous observation of Earth rotation. Polar motion and solid Earth tides are resolved with a resolution of 34 nHz.

the abscissa. By taking the square root of the power spectral densities and converting the spectral components to radians per second, we get a spectrum of the observed variations of the rotation rate of the Earth as shown in Figure 4.10. The semi-diurnal tidal band is sharply defined, while the diurnal polar motion band shows up a lot broader. For periods shorter than eight hours, we observe a sensor noise floor of 5×10^{-8} Hz^2/Hz, corresponding to 65 prad/Hz. The regime of longer periods is dominated by a $1/f$ frequency dependence. The insets present the geophysical signals at higher resolution. The gravitational attraction from sun and moon leads to two distinct double peaks. The $1/f$ trend in the noise floor at long periods is thought to be caused by thermal noise from the mirrors in the cavity.

4.5 Filtering

4.5.1 Filtering the Sagnac Interferogram

The Sagnac interferogram has a fidelity in excess of 90 dB above the noise floor, and under ideal circumstances, the beat note is the only major signal in the range of the detector. However, in terms of the operating hardware, a ring laser structure, like G, is a large construction occupying two laboratory rooms, thus causing signal leads to be long and the ground potential to float between different parts of the installation. This gives rise to hum signals, with the most prominent candidates at 50 and 100 Hz and a large number of harmonics, with a spectral power of around 20 dB each.

Backscatter coupling between the two laser beams in the cavity, although small, causes some frequency pulling and pushing, which in turn leads to some distortions to the measured sine wave. As a result, we can see evidence of the second harmonic frequency of the detected Sagnac beat note, with the second harmonic at 697 Hz reaching a signal strength of about 60 dB (see Figure 4.11). This is still well over 40 dB below the spectral power of the fundamental frequency, but nevertheless is far from being ideal. Over the years, we have installed various types of detectors for the detection of the Sagnac signal and other signals in our large gyroscopes. The use of a Hamamatsu H3164-10 photomultiplier – a detector with very high gain – also provides access to the FSR frequency of 18.75 MHz. The silicon PIN diode

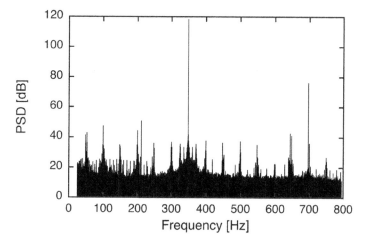

Figure 4.11 The spectrum of 800 Hz bandwidth shows evidence of many unintended signals.

module HUV-2000 B (EG & G) was used in the beginning of the project because it was convenient to use. Both detectors were chosen for their large sensitive area of more than 5 mm in diameter. While the photomultiplier offers a bandwidth of several hundred MHz, the PIN diode module has a cut-off frequency of 100 kHz. For the G ring laser and the other smaller gyroscopes, the beat note is sampled at a rate of $S = 2$ kHz, which is well above the Nyquist limit of $S = 2f_{max}$. For the larger rings UG-1 and UG-2, sampling rates of up to 10 kHz were used as the beat notes were 1512 Hz and 2177.6 Hz, respectively.

For many years, we have used analog second order Butterworth filters with a passband of 2–10 Hz around the main Sagnac beat note and a frequency counter to measure Earth rotation. These filters were substituted by digital versions of the Butterworth filter when the numerical AR(2) frequency estimator was in operation. When the logging system started to use Fourier transform-based frequency estimators, the application of spectral filters did not present any further advantage. In order to reduce potential noise issues from discrete electronic circuitry, we took all the filters out of the detector chain. The application of the digital filters was also discontinued, since they did not provide any advantage. The application of a bandpass filter can quickly introduce a phase shift to the filtered signal. For a frequency estimator this usually does not present a problem. For the backscatter correction process, however, a phase shift introduced by a filter is not acceptable. The phase sensitive quadrature demodulation technique, depicted in Figure 4.12, avoids this problem and provides access to the instantaneous frequency and the rate of change:

$$\varphi(t) = \arctan\left(\frac{\sin\varphi(t)}{\cos\varphi(t)}\right), \tag{4.12}$$

with the instantaneous frequency written as $f(t) = \Delta\varphi/\Delta t$ and $\Delta f(t) = \Delta^2\varphi/\Delta t^2$. It is a very practical approach to generate the reference frequency numerically on the computer, provided that the sampled observations are derived from a high quality clock, such as a GPS- disciplined oscillator or, as in the case of the G ring laser, from an H-maser.

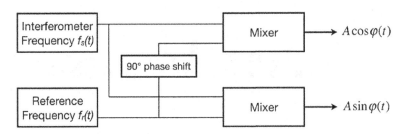

Figure 4.12 Schematic depiction of the quadrature detection process.

4.5.2 *Realtime Phase Shift Calibration*

For accurate rotation sensing, it is not enough to operate an unconstrained Sagnac laser interferometer alone. The stabilization of the scale factor of the gyroscope and the monitoring of the tilt motion are important factors, but not sufficient. Backscatter coupling pushes or pulls the optical frequencies by several parts in 10^6, as detailed in Section 3.9.4. Correcting for that corresponds to a stability improvement of several orders of magnitude; hence we have to establish the correction values carefully. We recall that the correction term Δf_s can be written as

$$\Delta f_s = \frac{1}{2} f_s m_1 m_2 \cos \varphi, \tag{4.13}$$

where f_s is the Sagnac frequency, m_1 and m_2 the fraction of the backscattered light on each of the laser beams and φ the phase angle between them. While m_1 and m_2 can be obtained from the ratio of the modulation amplitude and the intensity of each laser beam with little, if any, sensitivity to gain variations of the photo detector and the subsequently arranged amplifier, the correct estimation of the phase angle, on the other hand, is critically dependent on the bandwidth of the photo detector and the layout of the electronic subsystem, which can shift the relative phase of the optical signal substantially. It turns out that the estimated phase angle is strongly dependent on the laser intensity and also sensitive to variations of the temperature of the electronic circuitry. In order to capture the phase shifts introduced by the detection circuitry, an independent optical sine wave signal on a distinctly different modulation frequency of $f = 280$ Hz is injected into the signal path, at each of the two intensity detectors. The frequency source is a signal generator, which modulates the brightness of an LED in front of each photo detector. Figure 4.13 illustrates this process in a block diagram. Since the phase of the injected sine wave at each detector is available from the frequency source, the additional phase shifts,

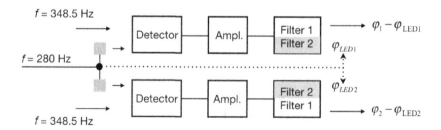

Figure 4.13 Realtime phase monitoring circuitry for the intensity detector chain. An optical sine wave signal from an LED is injected into the detector chain in parallel to the measurement signal. Recovered from the detector and compared to the phase of the optical input signal, this concept allows the phase shift induced by the detector chain to be recovered.

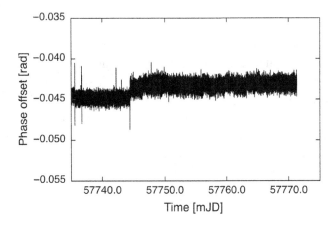

Figure 4.14 The in-situ monitoring of the phase offset between the two intensity detectors. Due to the stable environmental conditions, we usually obtain a constant value.

φ_{LED1} and φ_{LED2}, through the entire circuitry can be established for each channel from comparisons between the phase angle of the input signal and the respective amplifier output. The term $\cos\varphi$ of the backscatter correction is then obtained as $\varphi = (\varphi_1 - \varphi_{\text{LED1}}) - (\varphi_2 - \varphi_{\text{LED2}})$. In order to separate the backscattered modulation at 348.516 Hz from the electronic circuit phase shift at 280 Hz, two quadrature detection filters are employed, which are close enough in frequency to assume an identical effect on the phase shift of the amplifier and yet wide enough apart that the isolation of the signal contributions is good enough to avoid any interference. An example from a measurement series is presented in Figure 4.14. A typical value for the additional phase offset between the two intensity detector chains is −0.043 rad. Although the phase offset is very stable, small changes have occurred on occasion.

4.5.3 *The Application of Filters in Rotational Seismology*

One of the requirements for a sensor in a permanent realtime observation network is a quicklook preprocessing diagnostic feature, so that the performance of the data recording is available at a glance. At the timescales of interest for the observation of rotations in seismology, we can consider the rate bias of Earth rotation to have a constant offset. Seismic signals from earthquakes generate oscillations around this bias value. So we can consider the Earth rate effectively as the zero line of this particular application. Figure 4.15 shows such a quicklook display, covering six hours in a stacked representation. The horizontal axis covers 15 min, and the color code identifies each of the 4 quarter hour segments. The vertical axis stacks all these data segments chronologically. Excursions from the zero line, caused by induced rotations from surface waves, show up proportional to their signal

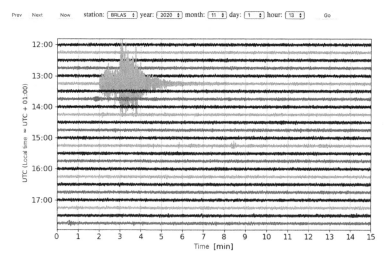

Figure 4.15 A sample quicklook image covering six hours.

strength. The example shown in Figure 4.15 was recorded on November 1, 2020 at 13:17 UTC. Microseismic noise, if strong enough, produces some structure of the stacked baseline, and anthropogenic perturbations are frequently visible as well (usually caused by a passing car), like the signal located right after 15:23 UTC in the example plot.

The preferred method for this type of data visualization is frequency demodulation, which was already introduced in Section 3.6.7. With the phase locked loop tracking the carrier "Earth rotation," the error signal is zero in the absence of any additional ground motion. Microseismic ground motion and seismic surface waves, however, are tracked by the locking mechanism, and the time dependence of the excursions shows up directly in the error signal of the feedback loop. The drawback of this approach is the low dynamic range of any available technical realization in the signal band of interest. If we exclude the domain of strong motion signals (10–200 Hz), we require high resolution in the range of $0.001\ \text{Hz} \leq 10\ \text{Hz}$. Commercial demodulation circuits usually cover the typical audio band up into the regime of several MHz. Therefore they cannot not provide the frequency resolution that we desire for this purpose. For accurate processing we therefore use a numerical method to obtain the necessary high frequency discrimination for the recovery of the earthquake signal [103]. The following steps are involved:

- Beat note digitization (24-bit, $f_s = 5$ kHz)
- Bandpass filtering to reduce aliasing and up-sampling to 10 kHz
- Convolution with a truncated time domain Hilbert filter
- Estimation of the instantaneous frequency

- Offset correction and scaling
- Compression and down-sampling
- Seismic signal estimation

The spectral resolution scales as $f_{min} = 1/T$ on the length T in seconds of the processed dataset. The ROMY ring laser, for example, forms a continuous batch of data points for each time interval of 1600 s in length. Each dataset is filtered by a zero-phase Butterworth bandpass filter to reduce potential side band noise effects, before it is upsampled to 10 kHz. For best performance of the filter, the impulse response at 20 times the time–frequency product of the bandpass filter is removed from the start and the end of the batch of data. In the next step we convolve the remaining dataset with a truncated time-domain Hilbert filter (Dave Hale, Colorado School of Mines, 06/02/89). The instantaneous frequency is then obtained by

$$ f = \frac{x(t)dH[x(t)]/dt - dx(t)/dt\,H[x(t)]}{2\pi(x(t)^2 + H[x(t)]^2)}, \tag{4.14} $$

where f is the instantaneous frequency, $x(t)$ the Sagnac waveform and $H[x(t)]$ the Hilbert transform. In order to keep the resolution as high as possible, but still within reasonable data storage requirements, the constant Earth rotation value is subtracted. The remaining numbers are scaled to form integers in multiples of 1 μHz. These integer numbers are further processed by a seedlink plugin, which applies a *Steim2* compression as well as the subsequent down-sampling to seismologically relevant sampling rates (i.e., 100, 20, 2 Hz). For the case of very low and very high frequencies, however, the raw data has to be treated in different ways, such as those introduced in Sections 4.3.1–4.3.3, and not within this automated processing procedure. Seedlink represents a robust transfer protocol for real-time seismological data streams. Details can be found at the *Incorporated Research Institute for Seismology* IRIS.

4.6 Ancillary Hardware

4.6.1 Electronic Feedback Loops

Active Sagnac interferometers are operated near the laser threshold, in order to excite only one single laser mode. Since the gain of the neon isotopes in this regime is very large, it requires an adequately tight feedback circuit that stabilizes the intensity of one of the two laser beams. When one beam is held constant in brightness, the other beam stabilizes at the same level. Typical solutions are comparators that compare the voltage of an intensity monitoring photodiode against a fixed preset reference voltage. A negative feedback with appropriate gain then adjusts the RF-power from the transmitter, so that the detected laser radiation becomes stationary at

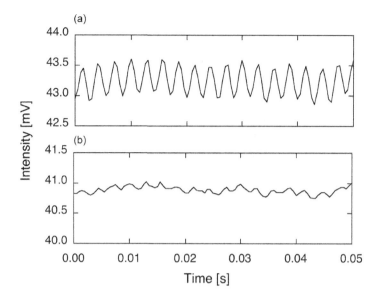

Figure 4.16 Example of a typical recording of the intensity of each beam in the G ring laser cavity. The intensity controlled beam (a) shows a 1.2% beat note amplitude from backscatter coupling on top of the beam intensity of 43.2 mV. The modulation on the unconstrained beam (b) is much smaller in this example.

the given value of the preset reference voltage. Due to the presence of beam power-dependent non-reciprocal effects on the laser beam, as introduced in Section 3.9.5, the requirements for the stability of the feedback loop are high. Furthermore, there is the problem that due to the unavoidable backscatter coupling, there is the additional requirement that the beam power stabilization must not compromise the backscatter correction process.

Backscatter coupling adds a small proportion of the respective counter-propagating optical beam power to each beam. The difference in the optical frequency creates a small modulation signal at the frequency of the Sagnac signal on top of the DC voltage on the detector output. Figure 4.16 shows a typical example. On the intensity controlled beam (a) is a signal of 0.5 mV peak to peak, while the brightness of the laser beam in the cavity has a signal voltage of 34.25 mV. The unconstrained beam (b) carries a much smaller backscatter amplitude, and the DC signal contribution is also a little less. This does not necessarily suggest that the beam power imbalance in the cavity is ≈5%. Small differences in the amplifier gain of the photodetectors or deficiencies in the respective detector alignment are prominent causes for apparently different signal levels. In fact a careful measurement with a power meter indicated an effective beam power imbalance of 1–2% in this case.

In order to keep the backscatter amplitude unperturbed, we operate a digital feedback loop, which also includes a low-pass filter with a cutoff frequency of

10 Hz. This setting ensures that the DC brightness is properly detected and appropriately stabilized. The backscatter amplitude, however, is not altered by the regulator. A high bandwidth of the feedback loop would modulate the backscatter signal inversely on to the laser gain medium. The consequence, then, is that one beam exhibits a nearly flat intensity distribution over time, while the uncontrolled beam shows a compromised signal, which is a mixture of the backscatter signal amplitude and a periodic intensity modulation from the other beam. In this situation the ring laser loses the ability to provide a valid correction for the backscatter coupling, as discussed in Section 3.9.4. Finally, the quality of the reference voltage is also very important. The voltage setting must remain very stable over time, to avoid the development of an intensity-related drift. A drift in the laser intensity may in turn cause a beam power-related bias shift.

4.6.2 Monitoring the Discharge

A tight relationship between the brightness of the discharge and the actual laser gain has already been discussed in Section 3.7.4 and shown in Figure 3.49. The approach taken in [21] uses a photodiode with a spectral filter for 633 nm in front of it, to monitor the discharge at the laser frequency (spontaneous emission) and to use the time evolution of the brightness for post-processing corrections.

A drift upwards in laser discharge brightness indicates a degradation of the laser gain as a consequence of laser gas contamination by outgassing from hydrogen. A progressive cavity misalignment caused by insufficient mechanical stability on a heterolithic sensor structure has the same effect. A third mechanism that can cause a variation of the discharge intensity comes from unwanted light from outside of the ring cavity. Coherent extra-cavity reflections from the uncontrolled beam, which feed back onto the photodetector of the intensity stabilized beam, can change the brightness level falling onto the photodetector of the controlled beam. In order to investigate this effect, we have implemented a continuous discharge monitor on the G ring laser. Furthermore, we have also recorded the RF-power going to the discharge section in order to look for any correlations.

Figure 4.17 shows some examples. Unfortunately there are no simultaneous observations available, since the diagnostic tools were not in place at the same time. The monitored variation of the RF-power going to the plasma discharge is shown in Figure 4.17a. The full power required for single mode operation is of the order of 2 W. There is some variability evident, which however remains within ± 10 mW and can be considered as sufficiently stable. The brightness monitor shown for a measurement series with strong back-reflections from a spectral filter (b) proves the high sensitivity to interference from extra-cavity signal reflections, as illustrated in Figure 3.79 (see also Section 3.9.8). The data for the diagram was taken with

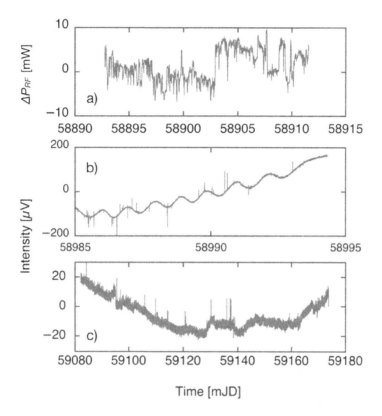

Figure 4.17 Measurements of the required RF-power (a) confirm good stability. The variation of the discharge (b) brightness is a good indicator of extra cavity interference as presented in Figure 3.79. Once consolidated (c), the level of variation reduces considerably.

the pressure vessel open and no constraint on the optical frequency. Once these unwanted reflections were contained, the variation of the brightness (c) reduced considerably and stayed within ± 20 µV, corresponding to 0.04% of the discharge monitor voltage. There was no longer a distinct upward trend detectable, which indicates stable operating conditions. We consider the application of independent out-of-loop monitoring signals as an essential augmentation for a Sagnac interferometer, but we do not recommend utilizing them for error correction purposes.

4.6.3 Tiltmeter Applications

A single component ring laser, even when the scale factor is perfectly known, cannot reconstruct the instantaneous rate of rotation of the Earth without significant uncertainty because it is not possible to obtain the orientation of the normal vector on the area circumscribed by the laser beam with sufficient resolution. Variable groundwater levels and seasonal changes cause slow tilt variations by deformations underground. Over a whole year, a relative change in tilt of 2 μrad has been observed on the platform of G. For a horizontally placed ring laser gyro with some additional north–south and east–west tilt, we can modify the scale factor to

$$\delta f = \frac{4A}{\lambda P} \Omega \sin(\varphi + \delta_N) \cos \delta_E, \qquad (4.15)$$

where φ is the latitude, δ_N a small additional misalignment of the sensor with respect to the local horizontal plane and δ_E the corresponding small east–west misalignment [118]. At mid-latitudes this means that the orientation sensitivity for a north–south tilt is high, while small east–west tilts will barely become visible, as the result of the cosine function dependence.

The G ring laser at its location on a geodetic observatory is well surveyed within the local surveying grid of the observatory. The position of the laser beams has been determined with errors smaller than 1 mm. However, even if this were one order of magnitude better and given that the effective reference plane of the gyro could be established from the nearly 3 mm wide laser beam at the same level of accuracy, this would still leave an uncertainty of 25 μrad for the orientation of the laser plane. This uncertainty translates into an orientation error of up to 7.331 mHz on the observed 348.5167 Hz beat note. In contrast, the currently achieved resolution for the instantaneous Earth rate is actually below 10 μHz. These numbers illustrate the sensitivity of the gyro technique to the orientation and the importance of tilt quite impressively. For absolute measurements of Earth rotation, the orientation has to be determined exactly to within 1 nrad.

However, there is a way out of this issue through finding a way to track variations in orientation angles and finally calibrate the initial sensor orientation to VLBI measurements of the global network. A similar approach is required for a multi-component ring laser system to solve for the initial misalignment error of the individual components with respect to each other. A gyroscope, however, with an orientation of the normal vector collinear to the Earth rotation axis, could measure the rotation rate with high accuracy, even in the presence of small orientation angle uncertainties, but it would loose the ability to establish polar motion. Nevertheless, there are applications in fundamental physics, like the estimation of the Lense–Thirring frame dragging effect, that would significantly benefit from a sensor orientation that provides the maximum of the Sagnac frequency of a given sensor.

A solution to the orientation problem for geodetic applications can be found by adding high resolution tiltmeters to the instrumentation. Apart from an initial offset error, these instruments deliver the variation of tilts with respect to the local g-vector. At a depth of 6 m, where the platform of G is attached to bedrock, we may expect tilts in the range of 0.2 μrad in the short term, induced by thermo-elastic and hydrologic deformations. Seasonal changes may cause peak to peak variations, which are an order of magnitude larger. Orientation variations, caused by solid Earth tides, are periodic signals, which are readily observed with a large ring laser. Signals like this are exclusively geometrical effects on the orientation of a ring laser. A tiltmeter is essentially a pendulum whose instantaneous distance from a fixed support is recorded. In contrast to the laser beams of a laser gyro, a plumb bob is also sensitive to mass attraction, which causes an apparent shift of the direction of the local plumb line. In order to remove this effect from the tiltmeter measurements, we have to consider the attraction potential of the bodies that cause the tides [119]. These values experience a modification due to the fact that the Earth deforms in the gravitational interaction process. The Love number k accounts for this, so that the ratio of the deformational part V_d of the gravitational attraction and the tidal component V_t is $V_d = kV_t$. For the full attraction potential V_{at}, we have $V_{at} = V_t + V_d = V_t + kV_t = (1 + k)V_t$, and with $k = 0.3$ we can compute the plumb line correction for the attraction [119]. For the geometrical part of the tidally induced tilts V_g, we make use of the Love number h, which is obtained as $V_g = -hV_t$ with $h = 2.143$. It is interesting to note that the attraction component and the deformation component act in opposite directions. Figure 4.18 shows the observed tilt from a *Lippmann* tiltmeter, which can resolve tilts of less than 0.5 nrad. The top panel shows a time series, approximately 100 days long with the periodic

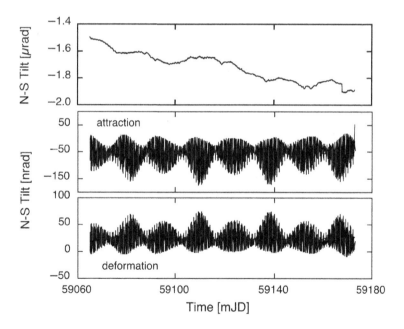

Figure 4.18 Example of a longer tilt record, decomposed into a long-term trend (top panel), the sensor response due to mass attraction (middle panel) and the true tilt from the local deformation of the subterrain (bottom panel).

tidal signals removed. The other two panels illustrate the respective contributions due to attraction and deformation separately.

Tiltmeters are basically realized in the form of a capacitor, where tilts change the separation of the two electrodes. This has the consequence that tiltmeters are sensitive to ambient temperature changes as well, which can lead to an ambiguity in observed tilt angles. The application of the measured tilts to the reduction of orientation variations of a laser gyro is essential, but this can be a very tedious process because there may be more than one temperature effect involved, with different time constants. These temperature effects may also show hysteresis, non-linearities and drift effects.

Another, however much smaller, effect on the tiltmeter is caused by mass attraction from the atmosphere. Due to the high dynamics of the weather patterns, these mass attraction variations increase the noise level on the tilt recordings. By taking a three dimensional weather model from the German Weather Service (DWD), supplemented by global data from the European Centre of Medium Weather Forecast (ECMWF), the Newtonian attraction is computed from the spatial density distribution of the atmosphere as a function of time [118]. Figure 4.19 shows one example computed over a period of 12 days. These attraction forces from the mass distribution of the atmosphere cause an additional apparent tilt on the tiltmeter. We have converted the results of this error signal (Figure 4.19) into the units of the Sagnac frequency using the scale factor of G. One can see that this effect causes perturbations of the order of 1 part in 10^8 and therefore is of the order of the effects that we wish to extract. That means that this error source cannot be neglected. In practice we see some correspondence between the residuals of the ring laser measurements and this error in tilt correction. This small variable bias is only one

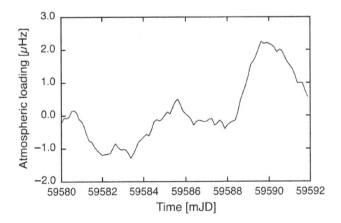

Figure 4.19 Display of 10 days of atmospheric mass attraction, acting on the tiltmeters, converted to a potential tilt-related bias for the ring laser.

contributor in the presence of several further subtle error sources at this current level of sensor resolution. Therefore we may not expect to see significant improvements from this correction alone.

4.6.4 Seismometer Applications

Tiltmeters are important for the very low quasi-stationary frequency range. Seismometers become important when the frequency band of interest is between 1 mHz and 20 Hz. The combination of ring lasers and seismometers augment each other very well. Seismometers measure translations along three linear independent directions. When they experience tilt during their measurements, for example as the result of strong motion close to an epicenter, the motion of the instrument through the event cannot be correctly reconstructed because some parts of the experienced accelerations are mapped to the wrong axis. Inertial rotation sensors can resolve this. Furthermore, we recall that ring laser gyros are entirely insensitive to translations. A combination of three single axis gyros plus one seismometer will then produce the combined measurement of six degrees of freedom of motions: three for all spatial directions in translation and three for rotations in pitch, yaw and roll.

For the conversion from the observed beat note of two optical beams to the experienced rotations, we only have to apply the scale factor as the proportionality factor for this transformation. The relationship is entirely linear. There are no zeros or poles involved. The instantaneous beat frequency of the two optical beams contain the information about the experienced rotation rate. Integrating these values once provides the angle. In contrast, seismometer accelerations have to be integrated twice to obtain the displacement. A small bias therefore introduces a significant drift. The ring laser body has to be rigid and must not deform under the experienced excitation. When it deforms it looses the alignment anyway, and lasing would stop altogether. From this point of view it is obvious that a large ring laser is not suitable for the detection of strong motion. However, for distant events, large ring lasers today are the only instruments sensitive enough, to record them. Figure 4.20 shows an example. It represents the m7.7 earthquake near Levuka (Eastern Fiji) from August 19, 2002, more than 2000 km away from Christchurch, where the C-II and UG1 ring lasers were operating simultaneously. Both rings recorded the earthquake, and since C-II was located inside the contour of UG1 with the same orientation, the measurements of the ground rotations were identical, once the measured Sagnac frequency was converted to rotation rates, using Eq. 4.16:

$$\omega = \frac{\lambda P}{4A} \Delta f - \frac{2\pi}{86164} \sin \varphi. \qquad (4.16)$$

The rate bias from Earth rotation is constant for all practical purposes in seismology, so it has to be subtracted. Both ring lasers agree perfectly, even in the fine details

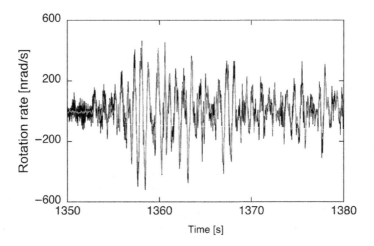

Figure 4.20 Comparison of a joint seismic observation of C-II and UG1, sited in the same place. By taking the scale factor of each instrument, we obtain the rotation rate, which agrees very well between both sensors. The only perceptible difference is the higher noise level of the smaller sensor.

of the earthquake signature. The only distinction is in the amount of noise in the observations. UG1 is far more sensitive, owing to the larger scale factor, so the noise level is considerably lower. While fiber optic gyros are ideal for rotation rates in the range of about 100 nrad/s–1 mrad/s, when deployed for example in civil engineering structures, such as bridges, large buildings, windmills, dams and so forth, large ring lasers have their place for teleseismic observations with rotation rates in the range of 0.1 prad/s–1 μrad/s.

4.6.5 Observation Files

There are several ways of recording the Sagnac frequency together with the necessary ancillary data into observation files. Apart from the strictly necessary quantities, there are also a few helpful parameters, which provide useful diagnostic information. Some of the signal sources provide observations at high sampling rates, while other signal sources, such as environmental sensors, have low sampling rates. At this point in time, only the ring laser rotation rate and the seismometer accelerations require a high data sampling rate. The processing and archiving of earthquake data is done independently and follows the procedures developed by the seismology community, using the seedlink infrastructure. All other quantities have much lower data rates.

Table 4.1 gives an overview of the multi-sensor parallel observations that are combined into a time series, which is based on the UTC timescale, and the epoch of each entry is typically represented in the modified Julian Date (mJD) so that the

Table 4.1. *Illustration of the ring laser observation data files.*

Observable	Acq. Rate	Result
Δf_{Sag}	2 – 5 kHz	Δf $\sigma(\Delta f)$
Δf_{Sag} cw-beam	2 – 5 kHz	Δf_{cw}
Δf_{Sag} ccw-beam	2 – 5 kHz	Δf_{ccw}
No extra cavity reflection issues if: $\Delta f = \Delta f_{cw} = \Delta f_{ccw}$		
Backscatter correction	1 min	systematic error reduction
Optical frequency	1 – 60 s	exact scale factor
Plasma brightness	1 min	no bias, gas integrity
Polar motion model	1 min	expectation value
Tilts from Earth tides	1 min	expectation value
Observed tilt	1 min	tilt correction
Vessel pressure	1 min	diagnostics
Ring body temperature	1 min	correction parameter
tiltmeter temperature	1 min	correction parameter
Laboratory temperature	1 min	diagnostics
Barometric pressure (Lab)	1 min	diagnostics

data stream can be related to the measurement series of other Earth observation techniques in geodesy and geophysics. With the high frequency data already extracted, from a data volume and resolution point of view it is useful to bin the observations into intervals of 1 minute and 1 hour. The structure of the data files also provides the possibility to perform cross comparisons in order to guard against systematic errors as well as possible. This increases the data integrity and is indispensable. However, it also needs to be said that this table has grown out of the C-II and G ring laser project heritage, and both of these sensors were designed with long-term operation in mind. A passive Sagnac interferometer, for example, or a system with interferometric control of the diagonals [176] will have different and additional parameters on their data acquisition unit.

The primary observation quantity is naturally the Sagnac beat note. This interferogram is sampled with a repetition rate of 2–5 kHz. From the waveform the frequency is established, as detailed in Sections 4.3 and 4.5.3, using datasets one minute long. The reference epoch for each 1 min data point is the middle of the interval. From the backscatter signals on the cw and the ccw beams, we do the same processing at the same data rates. This provides three values for the beat note related to the rotation rate. Two of them are intra-cavity signals, while the combined beam with the highest fidelity passes through the beam combiner. If all three signal sources agree well with each other, we can be sure that there is no systematic error generated by the beam combiner unit, as detailed in Section 3.9.8. Averaging the 1 min estimates of the Sagnac beat note over 1 hour provides a value for the standard deviation, which is a good indicator of the overall sensor performance as well as the respective level of the microseismic activity of the Earth. The reduction of the pushing or pulling from backscatter coupling is done during the observations by the model as introduced in Section 3.9.4. The input quantities m_1, m_2 and ϕ for the backscatter correction are not discarded after the correction process. Their time evolution is of great interest for diagnostic purposes; hence these values are added to the data file.

The evolution of the recorded plasma brightness must correspond to the measured intensity of the adjusted beam of the ring cavity. If they do not agree, there is an issue from extra-cavity light, either from some light source around the ring laser electronics, the plasma itself or from back reflections of the other beam, biasing the photo-detector by coherent interference. A continuously rising plasma brightness may also indicate laser gas degradation (see Section 4.6.2). An accurate estimate of the optical frequency in the cavity is valuable for the stabilization of the gyro cavity, as well as for the estimation of the true scale factor of the entire gyroscope. Model expression for the diurnal polar motion signal and solid Earth tides are very important for the interpretation of the gyro observations. Adding these model computations straight to the ensemble of observation data simplifies the data analysis significantly. In order to capture the long-term tilt effects, caused by the seasons and the changing groundwater table, the tiltmeter readings also provide important information. Since the tiltmeter itself is subject to temperature drift effects, the tiltmeter temperature is also recorded and available for corrections from the observation file. The tilt recordings always require extra scrutiny, since tiltmeters are temperature sensitive, and a slow drift effect may degrade their quality. By utilizing more than one tiltmeter, drifts reveal themselves more clearly. However, a proper correction of the experienced tilt will always remain a challenge.

Environmental parameters, such as the ambient pressure inside the stabilizing vessel, the barometric pressure in the laboratory, the ring laser body temperature and the temperature in the laboratory provide a valuable diagnostic tool for monitoring the consistency of the ring laser observations. Systematic errors, that relate to either pressure or temperature can be identified when the time series of the diagnostic observations correlate with the respective observations.

4.7 Sagnac Interferometry for the Geosciences

A single ring laser gyro delivers the experienced rotation rate about the sensitive axis as a function of time. In order to relate this to the geophysical processes of interest, a number of corrections have to be applied, some of which relate to the particular orientation of each sensor component, while others are related to the respective sensor biases. All of these corrections are specific to each individual gyro component in the case of a sensor array. A variety of geophysical signals with a large range of amplitudes and frequencies make up the observed signal signature. For practical purposes, we have grouped all these contributions into three different categories, to which we refer as follows:

- **Sensor Model**: All internal instrumental effects that modify and therefore bias the observation signal away from the true rotation rate are grouped in this category. Most of Chapter 3 is dealing with these effects.
- **Orientation Model**: In this category we group everything that has an effect on the relationship between the instantaneous normal vector of the respective sensor component and the rotational axis of the Earth. The orientation model transforms the **local** ring laser measurement setup to a **global** signal of interest.
- **Rotation Model**: There is a large number of small geophysical effects that cause very small variations to the measured rotation rate, starting at about a threshold of 1 part in 10^7 of the Earth rate of $15°$/h. These signals cover a range of several decades in the regime of $10 \, \text{nHz} \leq f \leq 50 \, \mu\text{Hz}$. The result is a fairly complex signal structure.

Before we discuss the data evaluation procedures in detail, we summarize the procedure here in a brief overview. In order to evaluate the observed time series, we usually compute the value of the expected Sagnac frequency from the actual scale factor of the gyroscope. Then we add the diurnal polar motion from model computations. In the next step we apply the measured tilt, which in turn is corrected for the gravitational attraction part of the tilt signal. After that we apply the correction for the Earth orientation parameters (EOP), where the Chandler wobble causes the largest contribution. This data is taken from the *finals.daily* (IAU2000) archive of the IERS website.[1] With all these corrections applied, we obtain a reference dataset, which indicates what the gyroscope should deliver.

Now we take the ring laser observations, where the backscatter correction has already been applied and also perform the null-shift correction to account for the plasma dynamics. This provides us with the measured Earth rotation time series. When we subtract the corrected measurements from the reference time series, we obtain a direct comparison between the local estimates of a single ring laser gyroscope and the global combined network of the GNSS and VLBI technique.

4.7.1 The Orientation Model

In Chapter 3 we looked at a variety of effects that bias the measurement. Starting out from the simple Sagnac equation

---

[1] www.iers.org/IERS/EN/DataProducts/EarthOrientationData/eop.html

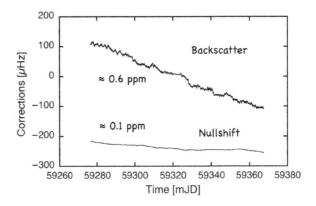

Figure 4.21 Contributions of the laser error budget of the main bias processes to the overall measurement quantity. Backscatter is the main contributor with about 1 ppm, followed by the null-shift from a non-reciprocal cavity bias, which is already half an order of magnitude smaller.

$$\Delta f = \frac{4A}{\lambda P} n \cdot \Omega_E, \qquad (4.17)$$

we have added several corrections to account for scale factor errors, a null-shift offset, backscatter coupling and many more subtle perturbation effects. Including these corrections reduced the systematic measurement errors of the instrument itself. Taking all the different effects into account provided us with a sensor model that extracts the experienced rate of rotation from the gyroscope. In order to illustrate this for comparison with the other two categories of effects, Figure 4.21 shows an example.

When we wish to relate the locally detected rotation signal to a global quantity, it is important to carefully consider how the entire sensor structure is attached to the Earth and how stable the platform itself is over time. Most of our gyroscopes are single component instruments, which means they are only sensitive about one of the three spatial axes. For practical reasons, most of these sensors are placed on the ground horizontally, with the normal vector pointing into the zenith direction. The recovery of the variation in the Length of Day requires a sensor resolution of less than 1 part in 10^8 of the rotational velocity of the Earth. Variations of the sensor orientation appear on the gyro readout in the same way as a scale factor error. As Figure 4.22 indicates, this corresponds to a requirement for knowledge of the sensor orientation (north–south tilt error of the normal vector on the plane of the laser beams) for a gyroscope at mid-latitudes of about 10 nrad.

The Geodetic Observatory Wettzell performs a full survey approximately every two years, and the marker positions are stable to better than 1 mm for all of the essential survey markers. Even the ring laser site, and most importantly the ring

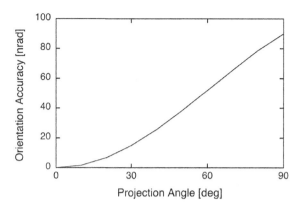

Figure 4.22 The allowable uncertainty of the orientation angle of a horizontally placed ring laser gyroscope as a function of latitude.

laser beam plane, is part of this survey network. Given the laser beam diameter of 2.5 mm and a leveling accuracy for the center of the beam of, say, 0.5 mm, which is a best case scenario, this amounts to an orientation angle uncertainty of 125 μrad for the G ring laser. G is equipped with several tiltmeters, each of which can resolve variations in orientation at the level of 1 nrad relative to local g. While we can resolve the changes in the tilt motion, the absolute orientation of the ring laser normal vector, relative to the Earth rotation vector, is still an unresolved issue. We must also note that scale factor errors and orientation errors show up in the same way. It is therefore important to realize that these orientation problems become increasingly larger as the sensor size is reduced. For a 1 m size ring laser cavity, the orientation angle uncertainty would already be as large as 0.5 mrad. From that point of view, it is an advantage to make a gyro cavity as large as possible. A single component gyroscope, placed at an angle, such that the normal vector on the ring laser plane is collinear with respect to the Earth rotation vector, does not have this problem. At 0 degrees inclination the measurement signal goes to a maximum, while the orientation error becomes negligible. At the same time, however, the gyro becomes insensitive to the polar motion signal. Apart from the sensor orientation, it is necessary to guarantee geological stability of the ring laser platform, so that the sensor orientation is rigid with respect to the Earth crust and does not change with time. This requirement is particularly critical for sensor locations near the top soil of the Earth's crust or underground platforms in caves or tunnels. Variations of the groundwater table may cause slowly varying tilts to the gyro platform (see Figure 4.23), and tunnels and caves are subject to considerable deformation as the atmospheric pressure changes. Comparing three independent tiltmeters in the same location suggests that the tilt is captured with a few percent of consistency. According to Figure 4.22, a 5% accuracy would then still be just outside the desired level of

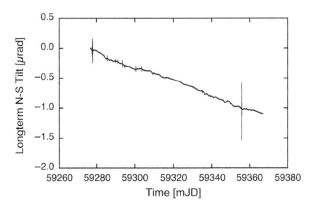

Figure 4.23 Variation of north–south orientation of the G ring laser over a period of almost 100 days. The gyro exhibits a slope in tilt in excess of 1 μrad. The measured tilts are temperature corrected, so that the overall drift of about 1 μrad is real, showing a seasonal hydrological effect.

uncertainty. Atmospheric loading and the cavity deformation effect would come on top of all and cause additional errors in the gyro orientation. For very large rings the induced strain effects may even affect the scale factor and cause beam steering (see Section 3.7.3). From these considerations we may safely conclude that it requires more than one gyro in different locations to obtain a reliable orientation of the measurement devices. By operating a network of such gyroscopes and correlating the obtained rotation rates among them, it is possible to extract the local or regional orientation changes from the global rotation signal. A sensor array like the ROMY structure can reconstruct any tilt motion without the need of tiltmeters, provided it remains rigid. A single ring laser array, however, cannot distinguish between local and global rotation unambiguously.

4.7.2 The Rotation Model

Here we distinguish between two different classes of signal, namely the orientation-related signals and variations in the rotation rate of the Earth. The former are either caused by a change of the position of the rotational axis of the Earth with respect to the body of the Earth or by some local tilt effect. The gravitational attraction of Sun and Moon gives rise to the diurnal polar motion signal, which causes a tilt in the rotation axis. A second effect from the attraction of celestial bodies is the deformation of the Earth's crust, and that induces the solid Earth tides. These signals do not change the rotational velocity of the Earth. Instead they change the projection of the instantaneous Earth rotation axis onto the ring laser normal vector. This means that we can measure the tilt component from the tidal signal but not the corresponding uplift, although that is a much larger effect. Variations of the Earth rotation rate itself are caused by a momentum exchange between the solid body and the fluids of the Earth. These effects are related to variations in the integrated mass transport, most prominently in the ocean currents, and this is known as the variation in Length of Day (LoD). Although this effect is only of the order of a millisecond in a day, it is not predictable and it adds up with each revolution. In 2003, some 40 years on from the historical experiment of Macek and Davis [142], ring laser measurements of geophysical signals having a period of the sidereal day were clearly discriminated against temperature and pressure effects acting on timescales of the solar day [184]. This demonstrates that the large laser gyros today routinely deliver on the optimistic expectations of the early days. They map out the orbital motion of the Earth in great detail.

Solid Earth Tides

Gravitational attraction from the Moon and the Sun cause large scale deformation of the Earth's crust [111]. A ring laser tied to the crust will not be sensitive to the periodic uplift at semi-diurnal frequencies; however, it will experience the corresponding tilts as the dominant deformation pattern passes the sensor location twice per day. The signal pattern depends on the position of the Moon along the lunar orbit and on the relative angle between the Moon and the Sun with respect to the Earth. While the semidiurnal component is the strongest, there are also significant diurnal, fortnightly and semiannual signal components present. Figure 4.24 shows the contribution of one month of tidal signal on the rotation rate of the Earth. This tilt signal is converted to the corresponding modification of the Sagnac beat note of the G ring laser at the Geodetic Observatory Wettzell at 49.1444° North. The date is

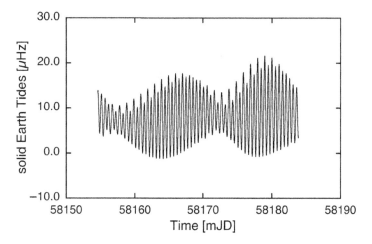

Figure 4.24 Display of the modeled contribution of the Solid Earth Tides to the observed Earth rotation velocity of the G ring laser at a latitude of 49.1444° North.

given in the modified Julian date, and day number 58150 corresponds to February 1, 2018. The tilts of the solid Earth tides change the projection of the ring laser normal vector in Eq. 4.17 on the rotational axis of the Earth. Relative to the rate bias from the Earth itself, the body tides of the Earth become visible at the level of 5 parts in 10^8.

Polar Motion

Apart from the deformation of the Earth's crust, the gravitational attraction from Sun and Moon also exert a torque on the Earth body due to the inhomogeneous mass distribution on an oblate Earth. As a result, one can observe a forced oscillation of the rotational axis of the Earth. In a celestial frame of reference, this motion corresponds to the precession and the nutation of the Earth's axis. It varies slowly as the relative positions of the Sun, the Moon and the Earth change with respect to each other. Due to the Earth rotation itself, the otherwise slow forced motion of the Earth's rotation axis causes an additional diurnal oscillation signal in the Earth fixed frame of reference. These Oppolzer terms change in amplitude from one day to another with a dominant period between nodes of 13.66 days [119]. Figure 4.25 shows the magnitude of this tilt related signal on the Sagnac beat note of the G ring laser at the Geodetic Observatory Wettzell at 49.1444° North and covers one month in length. The date is given in the modified Julian date, and day number 58150 corresponds to February 1, 2018. The maximum excursion is of the order of 40 μHz. From all geophysical signals this component has the largest amplitude variation over one day. Therefore it is the first geophysical signal that becomes observable at around 1 part in 10^7 as the ring laser development progressively gains

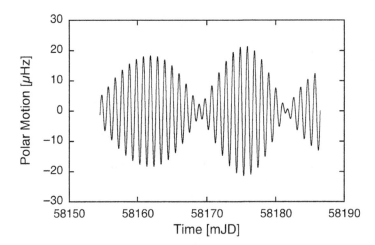

Figure 4.25 Display of the contribution of the polar motion (Oppolzer terms) to the observed Earth rotation velocity of the G ring laser at a latitude of 49.1444° North.

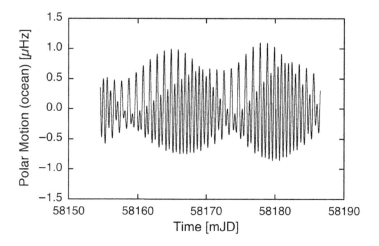

Figure 4.26 Display of the contribution of the oceans to the polar motion of the G ring laser at a latitude of 49.1444° North.

in resolution and stability. A ring laser with its sensitive axis aligned to the rotation pole cannot observe this signal. The other techniques of space geodesy, namely VLBI and SLR, cannot resolve this quantity, since their measurements obtain the full rotation matrix. It is therefore not possible to make a distinction between the nutation offsets $\delta\Delta\epsilon$ and $\delta\Delta\psi$ and the polar motion terms δx and δy. Another contribution to the polar motion originates from the oceans, responding to the external gravitational forces. Figure 4.26 illustrates this effect, which is here derived from a model. The contribution of the oceans to the inhomogeneous mass distribution of the Earth is small; hence the modeled effect is about an order of magnitude below the signal of the Oppolzer terms, and it also has a strong semi-diurnal component to it. The variability of the amplitude of the signal is again a consequence of the relative positions of the Sun and the Moon with respect to the Earth.

Loading Effects, Local Tilts and Variation of Latitude

Atmospheric loading and ocean loading cause the crust of the Earth to deform. Models suggest that loading effects from high or low atmospheric pressure patterns may cause an angular tilt of up to 10 nrad. This effect has not yet unambiguously been observed in any of our large ring laser structures. Ocean loading effects on the coast are both more predictable and larger in angle. The distance between the North Sea coast and the site of the G ring laser exceeds 600 km. The corresponding north–south tilt from ocean loading converted to a modification of the Sagnac frequency for one month is represented in Figure 4.27. The date is given in the modified Julian date, and day number 58150 corresponds to February 1, 2018. The loading signal is obtained from a loading model, and the corresponding tilt angle is converted to

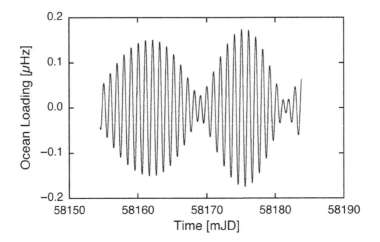

Figure 4.27 Display of the north–south tilt induced by ocean loading as it contributes to the rotation signal obtained by the G ring laser at a latitude of 49.1444° North. In 2018 this signal was still too small to be derived from the measurements.

a variation of the observed ring laser beat note. In 2022 this signal is still too small to be extracted from the ring laser measurements. Measurements with the C-II ring laser located in the Cashmere Cavern at 43.575° South on the Banks Peninsula in New Zealand naturally show a much larger tidal signal contribution in the range of 0.1–0.2 μrad [184]. The model used for the computation of the ocean loading is contained in the SPOTL-package [4]. The tilts from the ocean loading and the solid Earth tides were separated by evaluation of the phase-lag between the two signals in the time series.

Unlike a tiltmeter, which senses both the displacement due to the solid Earth tides and the corresponding change of the plumb line, a ring laser gyroscope is only sensitive to the variation of the projection of the normal vector on the beam plane and the axis of rotation. However, tilt signals cause a small local displacement, which causes a variation in latitude [222], and this is given by

$$\delta\phi_{\text{tilt}} = -\frac{(1+k-h)}{gR}\frac{\partial v}{\partial \varphi} \tag{4.18}$$

and

$$\delta\phi_{\text{RLG}} = \frac{(h-l)}{gR}\frac{\partial v}{\partial \varphi}, \tag{4.19}$$

where h, l, k are the respective Love numbers, g the mean equatorial gravity and R the Earth's equatorial radius. Here v is the tidal potential and φ the latitude. The differences between the ring laser-derived orientation changes and the tiltmeter measurements are small and of the order of 0.5 μHz on the measured Sagnac frequency for the most dominant tidal terms but cannot be neglected for a ring laser strapped down to the Earth and required to resolve frequency variations of the order of 1 part in 10^9 or better.

Local tilts of the entire ring laser platform, for example as a consequence of changing groundwater table after a strong rainfall, can reach levels of several hundred nrad in each direction within a few days. This has already been observed at the site of the G ring laser. However, such strong events are rather rare. More typically we observe seasonal variations of the orientation of the ring laser platform with rates of the order of 10 nrad per day and sometimes less. In order to compensate for this effect, we constantly operate several tiltmeters on the G ring laser structure and correct these local tilts in the post processing. These measured tilts have to be corrected for gravitational attraction of the atmosphere and the Moon. Since the tiltmeters are also susceptible to temperature variations, they have to be corrected as well.

The Chandler and the Annual Wobble

The combined contribution of the Chandler and the Annual wobble corresponds to a roughly circular motion of the rotational axis of the Earth's with a radius of up to 6 m [82, 200] at the poles. The Annual wobble has a period of 1 year and is caused by the ellipticity of the Earth's orbit, while the Chandler wobble is a free oscillation of the Earth, excited by wind and waves on the ocean. It has a period in the vicinity of 435 days, so the actual polar motion, caused by the combined signal contributions, exhibits a beat note for the amplitude of the actually observed polar motion. In Figure 4.28 we present the combined contribution of the Annual and the Chandler wobble, as measured by GNSS and VLBI, converted to a correction of the measured Sagnac frequency for the same time interval of all the other signals presented in this chapter. The ring laser performance in this period was unfortunately not stable enough to recover this signal. In Figure 4.29 we present another dataset, where the stringent requirements for stability were met.

The time series started on November 4, 2021 and lasted for 72 days. The high frequency signal signatures from the solid Earth tides ($f \approx 23$ μHz) and the diurnal polar motion ($f \approx 12$ μHz) have been subtracted from the time series, so that only the combined low frequency components from the Chandler ($f \approx 26$ nHz) and the Annual wobble ($f \approx 32$ nHz) remain. In the same way as for the diurnal polar motion signal, a single component inertial gyroscope, placed on the ground in the local horizontal plane at the mid-latitudes, is only sensitive to the north–south component of the Chandler motion. Since the G ring laser is located in the Bavarian Forest at a longitude of 12.878° E, the sensor location is almost aligned with the x-pole direction, and the necessary reductions of the full x-pole excursion for the

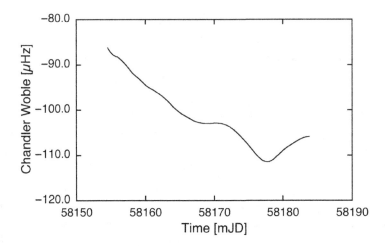

Figure 4.28 Display of the effect that the Chandler and Annual wobble has on the ring laser measurement time series. The signal was taken from the c04 time series of the IERS and then converted to a correction signal for the ring laser measurements for the same time interval as the other geophysical signals in this chapter.

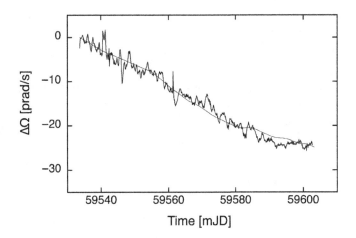

Figure 4.29 Display of the measured Chandler wobble over a period of 70 days relative to the first value. The solid line corresponds to the IERS observations, extracted from the c04 data series.

actually observed tilts are small. With this measurement example we concentrate in particular on sensor stability, which is the important property in Earth rotation sensing. Depending on the phase of the combined Chandler and Annual wobble, it takes about 30 days before the signal sticks out from the noise clearly. At the time of the measurement, the sensor stability of the single ring laser gyroscope component was less than a factor of five worse than the combined results from the GNSS and VLBI network.

Variations in Length of Day

All the observations of geophysical signals discussed so far have been created by either a change of orientation of the ring laser normal vector or a variation in the attitude of the Earth rotation vector. They become visible because of a variation of the projection angle of the inner product of Eq. 4.17. Variations in the Length of Day (LoD) are most prominently caused by momentum exchange between the solid Earth and the motion of the Earth fluids, mainly the atmosphere and the ocean currents. This signal directly corresponds to a variation of the measured rotation rate Ω_E. Since the fluids of the Earth make up less than 1 ppm of the mass of the Earth, these perturbations are only very small. They barely exceed a value of 1 ms over the length of a day of 86,400 s. Interactions with the gravitational forces of the moon give rise to a periodic component with a 14 day period, and this makes up the dominant feature of Figure 4.30. The signal in this frequency range has an amplitude of about 1 µHz, which when compared to the noise band of the measurements in Figure 4.29 is entirely masked. Other predominantly seasonal components act over several months and have larger amplitudes. The El Nino Southern Oscillation (ENSO) anomaly usually lasts for more than a year and is probably the most prominent example. More typical are, however, frequency components of around half a year. They are associated with oceanic angular momentum and atmospheric angular momentum. Although this seasonal component is larger in amplitude than 1 µHz, it is much harder to detect. With a corresponding frequency in the range of 65 nHz, this signal is much more demanding on sensor stability.

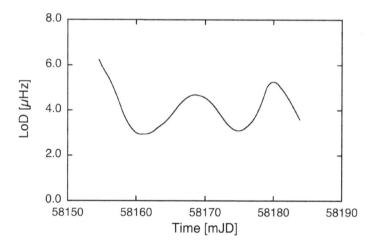

Figure 4.30 Example of the signal signature of the variation in Length of Day. The signal was taken from the c04 time series of the IERS and is converted to units of the G ring laser Sagnac frequency. So far it has not been directly observed with a laser gyroscope.

The greatest obstacle for all of these low frequency component signals is a $1/f$ increase in the sensor noise, which exceeds the resolution limit set by the shot noise for all signals with periods longer than one day. It is our current understanding that this is associated with thermally induced noise at the mirrors, as discussed in Section 3.10.3.

The Fully Corrected Observation

In the previous section we have discussed the main contributors to the geophysical signals on the large stable ring laser structure G, individually. This list of signals is not yet complete. There are more subtle effects; however their amplitudes are much smaller, barely exceeding the value of 1 µHz on a diurnal timescale or even much less. When we merge all the signals together and look at the combined effect of the known geophysical signals over the period of one month, we obtain a rather complex pattern as presented in Figure 4.31. This is what the ring laser observation time series should reproduce. In order to extract these signals of interest, we have to correct the raw measurements for sensor-related errors, namely backscatter, null-shift and tilt. Furthermore, we have to use only observation data that is not perturbed by ground motion during teleseismic activities or caused by larger storms over the Atlantic Ocean or the observatory.

In the top panel of Figure 4.32 we have presented the obtained raw measurements. It is interesting to note that this curve slopes steeply upward while we expect a gentle downward trend from Figure 4.31. This indicates that the sensor-related errors are larger than the geophysical signals of interest. Once all the known error sources and the relevant geophysical models have been removed from the observation data,

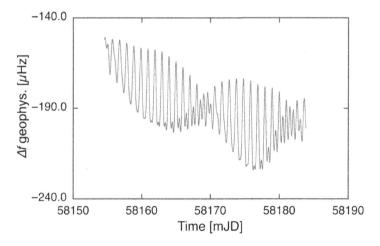

Figure 4.31 The combined contribution of the geophysical signals as shown in Figures 4.24–4.30 cause a significant change to the raw measurements.

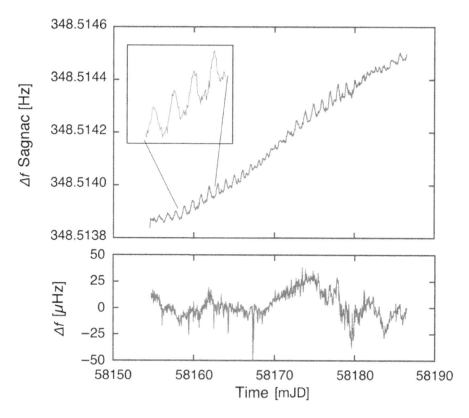

Figure 4.32 Comparison between the continuously measured raw Sagnac frequency (top panel) and the fully reduced measurement signal (bottom panel). The inset provides a more detailed representation of the polar motion signal. The remaining error is now more than one order of magnitude smaller.

we are left with a noise floor of about 10 μHz and some systematic excursions in the range of ±25 μHz. All corrections from this dataset taken together cover a range of more than 700 μHz. With these corrections applied, the resolution of G reached a regime of $\Delta\Omega_E/\Omega_E < 10^{-7}$, more than an order of magnitude short of our goals. The current noise level of the measurements and the presence of more, albeit small, systematic errors does not yet allow the direct extraction of the LoD signal from the ring laser data. In order to achieve this, the noise level and the systematic errors have to drop below a level of 1 μHz. What are the potential candidates for the remaining errors? The irregular shape of the frequency fluctuation clearly indicate that the respective error sources are purely instrumental, rather than a deficit in the application of the model corrections. For the particular set of measurements presented here, our data acquisition system did not provide a continuous and reliable record of the laser beam intensities. The signal to noise ratio was not high enough and therefore compromised the backscatter correction process. Furthermore, since the beam powers of the two laser beams in the ring cavity are not identical, which in turn causes a bias in the beat frequency (see Section 3.9.5). The third important candidate for the remaining systematic errors is the tilt correction. If the long-term stability of the tiltmeter is insufficient, it causes a characteristic drift, like the signal trend shown here. A walk off of more than 100 µHz has been observed in some bad cases.

Today, the existing large ring lasers have demonstrated that they can resolve small perturbations of the Earth's rotation with a resolution of up to 1 part in 10^8. The limiting factor for the instrument performance is not the sensitivity but the stability. We have seen that the orientation of a single component gyro causes an ambiguity in the effective scale factor. This, however, is not a fundamental limitation because the geophysical signals in a comparison with the Earth orientation parameters of the IERS can be used to disentangle this ambiguity. If the low frequency noise can be further reduced, for example by improved mirror performance, a continuous observation of the Chandler wobble and the LoD signal in realtime is within reach. On the other hand, there is still the persistent issue of controlling the attainable accuracy of a laser gyroscope. On average we find a considerable offset, of the order of ~500 µHz, between the measured and the modeled Sagnac frequency for our flagship gyro G. The estimation of the null-shift error depends critically on the correct estimation of the Lamb parameters in the rate equation, and these are not well enough established, limiting the achievable accuracy currently to 1 part in 10^6.

4.8 Ring Laser Analysis

Over the years, the ring laser analysis program has grown to quite a complex program suite. As outlined in several previous sections, there is a large set of auxiliary system monitoring parameters recorded along with the actual rotation rate. Not all of them are really relevant in the end. However, since the possible mutual interactions can be very subtle and even indirect, it is important to have the respective continuous time series of as many system and environmental parameters available as possible. At the same time, we also collect and evaluate the daily observation of polar motion, the Chandler wobble and the variation in Length of Day from the IERS. This allows us to compare the gyro measurements to the other techniques of space geodesy, namely GNSS and VLBI and enables indispensable validation tests. The parameters currently in use are listed in Table 4.2.

The observation files are generated on a daily basis. They contain the measured data points as listed in Table 4.2 at one minute intervals. These daily files are appended to the main master file. Figure 4.33 provides a block diagram of the analysis process. When the master file is read, it scans the file for gross outliers

Table 4.2. *Ring laser and environmental parameters of the G ring laser, as used in the analysis program*

Parameter	remarks
Time	in units of mJD
Δf_S	frequency estimate over 60 s (combined beam and mono beams)
$\sigma(\Delta f_S)$	standard deviation of Sagnac frequency estimates
AC, DC and ϕ	monobeam parameters for backscatter correction
ϕ bias	phase offset due to the detection electronics
Tilt in N/S and E/W	observed from several tiltmeters
f_{opt}	optical frequency derived by a optical freq. comb
I_{plasma}	plasma brightness
p_v	ambient pressure inside vessel
p_{atm}	atmospheric pressure
T_v	temperature inside vessel
T_L	temperature in underground laboratory
pm_{theo}	polar motion model from [36]
et_{theo}	theoretical Earth tides model from [4]

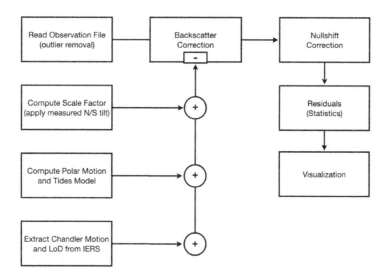

Figure 4.33 Block diagram of the analysis program structure. It combines the sensor model, the rotation model, the orientation model and auxiliary parameters in a way that the effect of each aspect can be tested separately.

and removes them. The analysis program includes a sensor model of the ring laser hardware; by default this is the G ring laser. It uses a set of a priori known system parameters along with the measured optical frequency and the empirically established sensor orientation in latitude and longitude along with the long-term trend in local tilt and generates a corresponding reference time series. This dataset also includes the scale factor corrections as derived from [98].

In the next step, the known signals of the rotation model that are not contained in the IERS data are computed. These are the diurnal polar motion model according to [36] and the solid Earth tides from [4]. The values for the observed Chandler motion and the variation in LoD are taken from the IERS archive [168]. They are converted into units of the Sagnac frequency (Hz) with respect to the orientation and the sensor model, to allow direct comparison. The result is a synthetic time series, where each individual correction can either be switched on or off. From the auxiliary measurements of the input observation file we compute and apply the necessary backscatter corrections, either according to the procedures outlined in Section 3.9.4 or alternatively based on the formalism discussed in Section 3.9.7. The synthetic dataset is then subtracted point by point from the backscatter corrected observation, such that the remaining residuals become available for further inspection and investigation. For an increased resolution it is useful to average the residuals over a longer time, which can be freely chosen in the program. The default value is two hours, which corresponds to the minimum of the Allan deviation. The final output is a display of the residuals, together with individual geophysical signals, the backscatter and null-shift correction, the tiltmeter measurements, the laser beam intensities and their phase offset, the atmospheric pressure, the temperature, the optical frequency, a power spectrum of the residuals and the Allan deviation. Apart from providing an immediate consistency check, this procedure also allows a quick identification of systematic dependencies between the individual input parameters and the obtained reduction result.

Although we consistently obtain an agreement between our measurements and the synthetic model at the level of 1–5 parts in 10^8, we can nevertheless identify some general shortcomings that still need to be addressed.

- **Auxiliary sensor drift and bias:** In this procedure we use tiltmeters to infer changes in orientation. The three tiltmeters on our system do not agree to better than ± 50 nrad in the long term. Each of them experiences a different sensitivity and trend in the presence of very small temperature variations.
- **Contamination of observables:** We monitor the intensities of the two counter-propagating laser beams with a photodetector. The backscatter correction only requires the ratio of the modulated part relative to the intensity. However, this intensity measurement is contaminated with residual plasma light for one beam more than the other, despite the application of spectral filters. This may cause a small drift in the backscatter correction.
- **Limitations in the plasma dispersion correction:** The corrections as discussed in Sections 2.6 and 3.9.7 are based on quite a large number of physical quantities,

Table 4.3. *Parameters for the HeNe transition at 632.8 nm compiled from the literature, namely [16, 56, 227]. The table is augmented by the actual experimental settings used for G.*

Parameter	G ring	[16]	[227]	[56]
γ_a	15.9 MHz	12 MHz	8.5 MHz	$(8.35 + p)$ MHz
γ_b	310 MHz	127 MHz	$59.5 p$ MHz	$(9.75 + 40p)$ MHz
γ_{ab}	163 MHz	234 MHz	$(\gamma_a + \gamma_b)/2$ MHz	$(\gamma_a + \gamma_b)/2$ MHz
p	7.52 Torr	3.95 Torr	7.52 Torr	3.5 Torr
T_p	360 K	450 K		360 K
A_{ik}	3.39 MHz			3.39 MHz
μ_{ab}	3.19×10^{-30} Cm	3.2×10^{-30} Cm		3.19×10^{-30} Cm
Losses	53.4 ppm			
Γ_D	8.442×10^8			8.442×10^8

which are not available at the necessary accuracy. Some of the values, taken from the literature, go back as far as the 1960s. Our model includes provisions to deal with some detuning of the transition by making use of the frequency comb measurements. The obtained results so far do not appear to be entirely consistent.

Table 4.3 lists the values of the respective parameters for the laser excitation used in the analysis model for the G ring laser together with values used in [16, 55, 226].

4.9 Micro-seismic Background of the Earth and Wind Shear

The Earth is not a quiet body. Due to the the highly dynamic local motion of the Earth's atmosphere and oceans, there is a constantly present vibrational noise detectable in the frequency range of 0.05–0.3 Hz. This micro-seismic activity of the Earth essentially limits the resolution of seismometers. Large ring laser structures will be eventually limited by the rotational component of the micro-seismic background, once the self-noise from the laser process has been sufficiently reduced. Until the year 2009 all of our ring lasers used super-polished ULE mirrors with high reflectivity Bragg stack coatings. As the substrates were part of the geometric length reference of the gyro cavity, a low expansion material was chosen. In 2010 we substituted these mirrors for fused silica mirror substrates. Due to the higher mechanical Q factor of the fused silica material, the vibrational noise in the interferometer reduced significantly. Figure 4.34 shows this effect for a time series of ring laser noise estimates. The data acquisition system continuously provides one frequency estimate of the Sagnac frequency every 60 s. Every hour, 60 of these estimates are averaged, and the mean and standard deviation of the

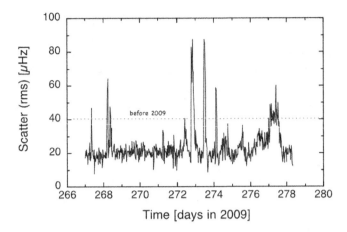

Figure 4.34 The variation of background noise as a function of time. The improved mirrors lowered the noise level from an rms scatter of 40 μHz to about 18 μHz. The short transient excursions are caused by earthquakes or local wind gusts. The broader structures are caused by storms in the Atlantic.

measurements are logged in the system database. For the current fused silica mirrors, the lower limit for the scatter of the measurements is around 18 μHz, as the flat zone between days 269 and 272 indicates. Whenever the scatter goes above this limit, it happens in response to a rotational excitation of the gyroscope. Usual causes are rotational signals from earthquakes or local wind gusts, which show up as sharp needles in the diagram. Enhanced noise levels that are persistent over more than a day are caused by storms in the Atlantic. They reflect a higher level of micro-seismic activity. The dotted line at the level of 40 μHz marks the regime where the lower noise limit of the G ring laser was prior to the fused silica mirrors. The G ring laser structure is strapped down to the Earth body. Since buildings are sources of local noise, G was placed in a dedicated underground facility and it is directly tied to the bedrock, where the laboratory is not connected to the ring laser platform. However, local sources of noise are still not decoupled entirely from the ring laser installation. Small tremors caused by passing cars, 100 m away from the ring laser platform, are the most obvious signs of anthropogenic disturbance. Luckily the signal levels are infrequent and small, the events are very short and the excursions have a mean value of zero, so that these perturbations do not (yet) matter. Much larger are the effects from gusty winds on the observatory. Potentially, winds can interact in two ways. They can cause a loading effect on the hill and the local terrain under which G rests. Wind gusts may also cause a small torsional deformation of the terrain around the ring laser due to friction effects from the resistance of obstacles, such as nearby trees. For Figure 4.35 we have set up a detailed finite element model of the local terrain, based on a digital terrain model of the laboratory environment, including the

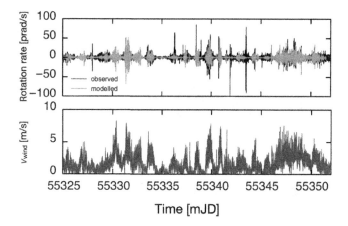

Figure 4.35 Local winds cause torsional ground motion, most prominently from friction on topographic obstacles, such as trees. The top panel displays the comparison of the modeled wind torsions (gray) and the ring laser response (black). The bottom panel represents the corresponding measured wind velocities. (Adapted with permission from [73] ©2022 Springer Nature)

local topography and land use. This model was subjected to wind patterns that we obtained from a local wind sensor, feeding wind direction and wind strength into the modeled terrain. The comparison of this simulation and the measured response from the gyro reveal that wind pressure effects are negligible, both for the ring laser location and for the nearby mountain range. However, torsional shear noise excitation from wind friction on the surrounding terrain does show up, and these effects are more than an order of magnitude larger than wind loading. While the noise level of the instrument is significantly increased during periods of gusty wind, the effect on the ring laser measurements is small but not negligible. The torsional excitation quickly becomes incoherent with increasing distance from the ring laser site, as the individual wind cells are rather compact. So far we have not observed a continuous laminar flow pattern near the gyro site, which would give rise to a larger perturbation.

4.10 Transient Rotation Signals in Ring Laser Measurements

All of the geophysical signals that we have discussed in this chapter so far are either periodic or change slowly over several days or even months. In contrast to these, earthquakes generate transient events that require at least a bandwidth of about 100 Hz in order to capture their full signature [99, 100, 145, 157, 235]. While seismometers cannot distinguish between translations and tilt, ring lasers are only sensitive to rotations. Figure 4.36 provides an example for the rotational

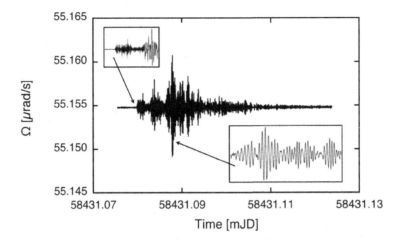

Figure 4.36 The M 6.8 Jan Mayen event is an example of a large teleseismic event recorded at the G ring laser at a distance of 2790 km in the early hours of November 4, 2018. The insets show a blow up of the onset of the earthquake and the Love wave arrival.

signal obtained at the G ring laser from the M6.8 Jan Mayen earthquake from the November 4, 2018 at a distance of 2790 km northwest of the ring laser location. The insets present a more detailed view of the onset of the earthquake (p-SH conversion) and the Love wave arrival around the maximum amplitude of this earthquake example. Superimposed on the signature of the earthquake ($\delta\Omega_s$) is the rate bias generated by Earth rotation (Ω_E), which for the G instrument has a value of 55.15 μrad/s. The diagram only contains the angular motion of the seismic field around a vertical axis. It is also important to note that a ring laser does not contain any moving mechanical parts. So the transfer function is unity and does not have any poles or zeros to consider. As a result the observed beat note can be directly converted to a rotation rate, by dividing the measured Sagnac beat note by the scale factor $K = 4A/\lambda P$. For the G ring laser at the Geodetic Observatory Wettzell at $\varphi = 49.1444°$ N, we obtain

$$\delta\Omega_s = \frac{\Delta f}{K} - \Omega_E = \frac{348.514\,\text{Hz}}{6318839.208} - \frac{2\pi}{86164\,\text{s}}\sin(49.1444°)\,. \qquad (4.20)$$

For the Jan Mayen event, shown in Figure 4.36, the measured ground rotation rates peak at 5 nrad/s.

4.10.1 Eigenmodes of the Earth

The body of the Earth has a surprisingly large Q factor; therefore, large earthquakes can excite eigenmodes of the Earth body, which usually persist for several days before they are damped away [101, 239]. Assuming for simplicity that the Earth is ideally spherical, all relevant properties only depend on the radial distance from the geocenter. From the solution of the equations of motion for an isotropic homogeneous elastic Earth for small displacements around an equilibrium position, one obtains two types of oscillations, namely toroidal and spherical modes. Toroidal modes have no component in the radial direction, and according to [239] the displacement in spherical coordinates as a function of t is given by

$$u_\theta = \frac{_n W_l(r)}{\sin\theta} \cdot \frac{\partial}{\partial\phi} \left[P_l^m(\cos\theta)e^{im\phi} \right] e^{-i\omega(_n T_l^m)t} \tag{4.21}$$

and

$$u_\phi = -_n W_l(r) \cdot \frac{\partial}{\partial\theta} \left[P_l^m(\cos\theta)e^{im\phi} \right] e^{-i\omega(_n T_l^m)t}. \tag{4.22}$$

The expression $_n T_l^m$ describes the eigen angular frequency, $_n W_l(r)$ the radial eigenfunctions and P_l^m represent the respective Legendre polynomials: where l and m are integers. In a similar fashion one can find the corresponding terms of displacement for the spheroidal modes $_n S_l^m$:

$$u_r = _n U_l(r) \cdot P_l^m(\cos\theta)e^{im\phi}e^{-i\omega(_n S_l^m)t} \tag{4.23}$$

$$u_\theta = _n V_l(r) \cdot \frac{\partial}{\partial\theta} \left[P_l^m(\cos\theta)e^{im\phi} \right] e^{-i\omega(_n S_l^m)t} \tag{4.24}$$

$$u_\phi = \frac{_n V_l(r)}{\sin\theta} \cdot \frac{\partial}{\partial\phi} \left[P_l^m(\cos\theta)e^{im\phi} \right] e^{-i\omega(_n S_l^m)t}, \tag{4.25}$$

where $_n U_l(r)$ and $_n V_l(r)$ represent the respective eigenfunctions. When these modes are excited by a strong earthquake, the displacements can be detected by a broadband seismometer, and the rotational component can be picked up by a sensitive gyroscope. To date, only the G ring laser has been able to measure the spectrum of the local rotations from the eigenmodes of the Earth, experienced at the location of the instrument. In Figure 4.37 we represent the observed spheroidal and toroidal eigenmodes obtained from the M 9.0 Tohoku earthquake in 2011. The data shown here were recorded by an STS-2 broadband seismometer and the G ring laser at the Geodetic Observatory Wettzell, where both instruments are collocated. The top panel presents the spheroidal eigenmodes, recorded on the Z (vertical) component of the seismometer, while the middle panel contains the toroidal eigenmodes obtained from the transversal horizontal direction. The lower panel provides the pure rotational component, obtained from the ring laser. Theoretically one would expect the latter two spectra to be similar, since the transversal horizontal displacement and the rotations caused by the Love surface waves correspond to each other. Comparing the two lower spectra of Figure 4.37, however, shows a certain disagreement between the obtained amplitudes from the two sensors. This is caused by the fact that the

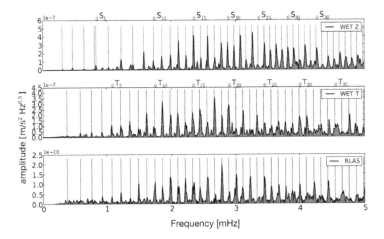

Figure 4.37 Spectrum of the eigenmodes of the Earth excited by the 2011 Tohoku earthquake, taken with the broadband STS-2 seismometer. The top plot shows the spheroidal modes visible in the Z (vertical) component, the middle plot displays the seismometer displacements caused by the toroidal modes of the horizontal transversal component and the bottom diagram represents the rotational component, from the ring laser, contained in the toroidal modes.

physical positions of the nodes and antinodes for the sinusoidal patterns of the translations and the rotations of the Earth's crust do not coincide.

4.11 The Effect of Magnetic Fields on Large Ring Lasers

Without exception, all of the large HeNe ring lasers operated by our collaboration in New Zealand, Germany and Italy exhibit a difference in brightness between the clockwise and counter-clockwise beams, despite an entirely reciprocal design of the ring laser cavity. One possible cause for this appears to be a combination of birefringence of the mirror coatings with a non-reciprocal mechanism, such as a tiny rotation of the polarization due to Faraday rotation. In order to investigate this, we have placed a strong magnet behind the mirror of one corner of the G ring laser in order to study the effects from magnetic fields on the ring laser mirrors. Figure 4.38 illustrates the design of the mirror holder and the way the magnet is attached to one of the corners of the G ring laser, far away from the gain medium. In this arrangement, the magnet, a neodymium rare earth disc with a field strength in excess of 10,000 gauss, is located 35 mm away from the actual mirror coating. Early experiments on the Canterbury ring laser have already indicated that a magnet close to the gain medium could push the observed Sagnac frequency up by several Hz. Therefore we have designed the G ring laser structure with a mu-metal cage around the excitation capillary, in order to shield the plasma from potential variations of the

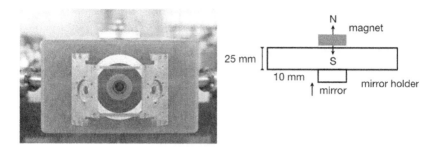

Figure 4.38 Display of the mirror holder and the location of the neodymium magnet on the outer face of the ULE plate.

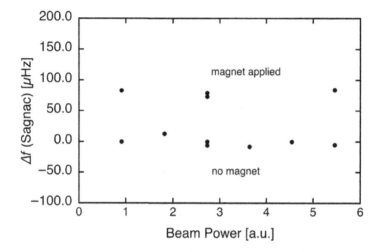

Figure 4.39 The placement of a strong rare earth magnet behind the mirror holder (see Figure 4.38) caused an offset of about 80 μHz, which was independent of the actual laser beam power.

Earth magnetic field. The first test looked at the dependence of the laser gain on the presence of the magnetic field. In the absence of a magnet, the gain of the laser was varied from a very low level right next to the laser threshold up to the regime where multiple longitudinal laser modes can readily be observed. After returning to the point of origin, the same experiment was carried out with the magnet attached. There was no obvious functional difference observed; nor does the amplitude of the Sagnac beat note depend on the presence of a magnet. However, for the measurement of the Sagnac beat note, a distinct and constant offset of +80 μHz has been readily observed between the cases with and without a magnet present (see Figure 4.39), which again was independent of the applied laser power. Magnetically induced or modified birefringence, together with a small difference or change in polarization,

Table 4.4. *Observed shift of the*
Sagnac frequency as the result of the
presence of a strong magnetic field at
one mirror of the ring laser.

Experiment	Δf_{Sagnac} [Hz]
No magnet	348.517062
Forward polarity	348.517133
Reverse polarity	348.517037

caused by Faraday rotation, are the most likely candidates for the appearance of this additional non-reciprocity. Since the laser beams are penetrating into the stack of the thin film layers, some sort of interaction is not unlikely. The generation of this offset turned out to be very reproducible. The available magnet was unfortunately unsuitable for another experiment with reversed polarity of the magnetic field, due to an obstruction on the backside. So another magnet of comparable strength had to be found, and the experiment was carried out at a later time and yielded slightly different numbers. Table 4.4 summarizes the results.

Magnetic fields have long been known to produce null-shift effects in a HeNe gas laser [37]. This is caused by Zeeman splitting of the laser transition in the presence of a magnetic field. An example, presented in [11], shows a bias of $4.25°$/h for a HeNe laser gyro at $\lambda = 1.15$ μm exposed to a magnetic field of 5 gauss applied to the plasma. This is about 100 times higher than the Earth magnetic field at mid-latitudes. Converting the external magnetic field to the range of the Earth magnetic field of, say, 49,560 nT, will then generate a possible bias of 206 nrad/s at the most. Since the short-term variation of the magnetic field rarely exceeds ± 25 nT, the variable part of the effect would then be 2000 times smaller. In comparison to the long DC excited gain section of the experiment reported in [11], the G ring laser only has an approximately 3 cm long gain section, encapsulated in mu-metal. This suggests that the effect of an external magnetic field does not play a role under normal operating conditions. However, in order to explore the sensitivity to magnetic fields in more detail, we have placed the strong neodymium magnet right over the electrodes, 2 cm away from the RF-excited gain medium, very close to the capillary and inside the shielding. The effect of the magnet near the gain medium over the entire range of the single mode laser powers produced a constant offset again, but approximately 25 times larger. The result is illustrated in Figure 4.40. This time the Sagnac beat note was shifted upward by 1.845 mHz. The effect is considerably larger than at the corners, since the magnet was placed much closer to the laser beams. Furthermore, the Zeeman splitting will dominate the observed offset. The effect of Zeeman splitting is absent at the mirrors with no

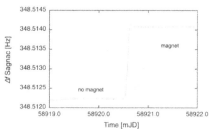

Figure 4.40 The magnet was placed with about 20 mm clearance over the plasma in the gaintube section. This caused an offset in the observed beat note of more than 1.845 mHz.

plasma within several meters of the beam path. The influence of the magnetic field and the induced non-reciprocal extra losses from Faraday rotation on the two laser beams will, however, still be there and will also contribute. When the magnetic field at the plasma rotates the polarization of the optical beams slightly away from pure s-polarization, it obviously generates an effect equivalent to increasing the losses differently to both beams and might explain the observed unequal beam powers in the laser cavity. Since the plane of polarization is rotated in opposite directions with respect to the direction of propagation of each laser beam, Faraday rotation is a likely candidate for a difference in the losses between the clockwise and the counter-clockwise cavity.

In Section 3.11.1 we introduced magnets for the stabilization of the size and shape of the plasma in a long gain tube, to facilitate a stable gain medium. In general, this appears not to be necessary for our laser gyros on the 633 nm transition with their small plasma length of 3 cm. However, at a level of resolution below 1 part in 10^8 of Earth rotation, this might no longer be the case. Currently we cannot exclude the possibility that subtle plasma fluctuations are contributing to the observed $1/f$ noise level behavior, shown in Figure 3.89. A longer test with the strong magnet placed right over to the plasma did not show any evident improvement in stability. Therefore we discontinued using a magnet. The obvious disadvantage of the pushed beat note due to the Zeeman splitting outweighs any possible subtle advantage, at least at this point in time.

4.12 Effects on Open Ring Resonators

A small variation in the length of the resonator does not matter in a completely unperturbed ring gyro because the corresponding change in the optical frequency is entirely reciprocal and disappears in the beat note, as long as the longitudinal mode index does not change. In this book we have often seen that this ideal condition is never met. Here we present an example of how this asymmetry can even be exploited for a pressure sensing application. The measurements have been carried out by the ring laser group in Conway (Arkansas) [65, 67]. It is based on several realizations of an equilateral triangular gyro in an open configuration. The sides are between 13 m and 15 m long. An alternative square configuration with each side 5.65 m long have also shown the same phenomenon.

All the cavities were set up with a free space propagation path between highly reflective (99.8%) mirrors. Closed corner boxes and aluminum tubes insulate the cavity from the ambient atmosphere. The cavity itself is slowly flushed by evaporating liquid nitrogen, to keep water vapor out of the cavity, which would otherwise quench the laser process. Laser excitation is facilitated by a 75 cm long plasma tube, sealed with Brewster windows and providing a free boresight of 7.7 mm. Unlike most of the systems covered in this book, a DC discharge is employed. The gas pressure in the plasma tube was set to 4.41 hPa in order to increase the homogeneous line width for the suppression of neighboring longitudinal laser modes.

Figure 4.41 illustrates the setup of the square system as an example. All open cavity systems already showed a high sensitivity to atmospheric pressure variations

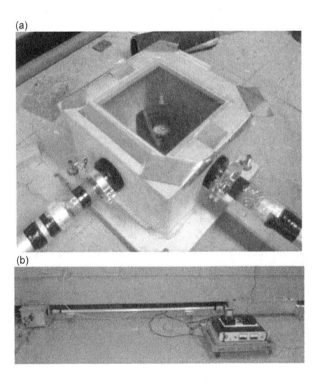

(a)

(b)

Figure 4.41 Impression of the open architecture of the Arkansas ring lasers. The corner boxes enclose the mirrors (a) without sealing them against the local atmospheric pressure. Aluminum tubes connect the beam lines (b) in order to prevent water vapor from quenching the laser beam. (Photo courtesy: R. Dunn)

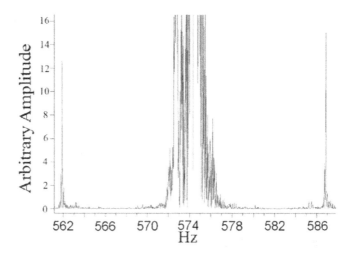

Figure 4.42 Display of a sample spectrum from the large Arkansas ring laser. The Sagnac frequency is obtained as 574 Hz. Strong backscatter coupling and a high sensitivity to ambient pressure changes cause significant instability. Please note the weak sidebands ±12.5 Hz away from the main carrier. They were obtained from the application of a portable Helmholtz acoustic generator. (Reprinted with permission from [66] ©American Institute of Physics)

early on in these experiments. Particularly noteworthy is the sensitivity to infrasound in the frequency range well below 20 Hz. Figure 4.42 shows a spectrum of the observed Sagnac beat note of 574 Hz of the large triangular structure, provided by courtesy of the PI Robert Dunn. Due to the unstable nature of the large resonator, significant atmospheric pressure changes and the high level of backscatter coupling, the Sagnac signal is not so well defined and jitters by about ±3 Hz about the true rate bias of Earth rotation. This is a common experience with this type of construction. This particular measurement demonstrates the high sensitivity of the system to low acoustic frequencies, because it was sensing the 12.5 Hz frequency radiated from a portable Helmholtz resonator placed in the vicinity of the ring laser structure. The Helmholtz resonator consists of a 55-gallon drum with a half-inch thick aluminum top connected to an aluminum duct 1 m long and 10.2 cm in diameter; the duct is open to the external environment. Air in the cavity (drum) acts as a spring, while the air in the duct acts as a mass; for a given geometry, a resonant frequency is created. An audio woofer between the volume and duct drives the combined system at resonance. The radiated infrasound from the duct opening is small but sufficient for these tests. Care was taken to isolate the Helmholtz resonator as much as possible from the floor by suspending it from the ceiling rafters so that the vibrations initially traveled through the air. The output of the resonator was calculated to produce pressure variations of the order of ≈2 mPa [65]. This well defined infrasound signal causes a weak but very distinct frequency modulation of the Sagnac beat signal, visible as the two sidebands ±12.5 Hz away from the main carrier. The system shows particular sensitivity to infrasound frequencies in the range of 0–20 mHz as are radiated by large scale geophysical events like hurricanes and erupting volcanoes, as we will discuss in Chapter 6.

Closed ring laser systems like G are not sensitive to these kind of signals, and the origin is not external ground motion but is a result of backscatter coupling inside the cavity. The losses from the mirrors and the Brewster windows give rise to significant backscatter, which perturbs the Sagnac frequency. Since the experimental details are no longer available, we use some rough but realistic calculations and apply the model described in Section 3.9.4 to the Arkansas gyro. A horizontally placed equilateral triangular ring with 15 m on each side at the latitude of Conway (AR) of 35.09° North has very crudely a Sagnac frequency of 547.6 Hz. Observed was 574 Hz, which suggests frequency pushing by 26.4 Hz. When we set the AC backscatter amplitude to about 25% of the mono-beam DC amplitude and assume an asymmetry between the two beam intensities of 7%, this roughly reproduces the required offset of the Sagnac frequency. These are very typical values for a ring laser with significant back reflections, even for systems with (bad) super-mirrors. Figure 4.43 illustrates the dependence of the pushed frequency on the effective backscatter angle. If we consider an externally induced change in cavity length by 1 nm, this would change the phase angle by 0.57°. It would also change the optical frequency by 10.5 kHz. In order to experience an effective change of length of 1 nm, the ring would only require a change in ambient pressure of 23 mPa, which

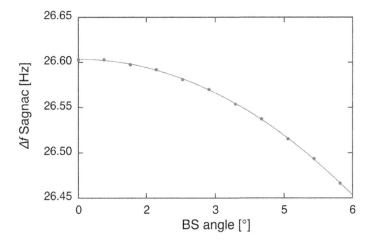

Figure 4.43 The expected variation of the pushed frequency of the Arkansas ring laser with variation of the effective backscatter angle.

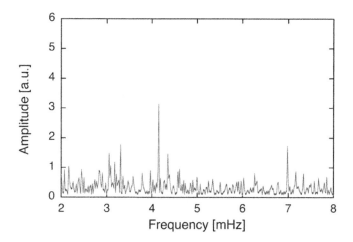

Figure 4.44 A sample spectrum of the infrasound regime, observed by the Arkansas ring laser. The peak at 7 mHz was caused by the super Typhoon Huiyan coming ashore at the Philippines on November 7, 2013. The simultaneous peak at 4.2 mHz is in response to the Volcano Shiveluch eruption in Eastern Russia. (Adapted with permission from [67] ©American Institute of Physics)

is an incredibly small value. The experiment described above with the Helmholtz resonator indicates that a periodic cavity length variation of 0.1 nm is already enough to produce a distinct frequency signature.

Infrasound with well defined frequencies from geophysical events can therefore frequency modulate the open cavity gyro, with the very long wavelength of the excitation frequency acting on the entire cavity at once. This simplified discussion neglects possible additional perturbation effects from an unequal beam path length external to the cavity, before the two exiting beam are mixed. There is no information on this available from the experiment anymore. However, it is very realistic to compare this to the effect of a free space beam combiner, treated in Section 3.6.6. Figure 4.44 shows a simultaneous observation of several events obtained from this system as an example [65, 67]. The signal at 7 mHz is detected when hurricanes or typhoons come ashore. The signal in Figure 4.44 was recorded from the super typhoon Huiyan, which came ashore in the Philippines on November 7, 2013. The eruption of the volcano Shiveluch in Eastern Russia generated the signal at 4.2 mHz at the same time. Finally, we can also see some signal in the buoyancy frequency range of 3–3.3 mHz.

5

Alternative High Resolution Rotation Sensing Concepts

Optical interferometers are not the only types of high precision rotation sensing devices with the potential of probing variations in the Earth's rotation. Atom interferometry and the exploitation of the Josephson effect with superconducting helium passing through a small orifice of about 0.1 μm diameter are also very viable candidates. Table 5.1 summarizes available results from the literature for such rotation sensing devices. All these techniques have similar challenges in common. Sensitivity is the most obvious parameter of interest, but aspects like scale-factor stability, sensor orientation and sensitivity to horizontal acceleration are serious error sources and common to all techniques. The success in the mitigation of such errors eventually defines the applicability of a monitoring device for geophysical perturbations. Finally, if it is a single component sensor, the problem of sensor orientation needs particularly careful consideration. Large area sensors are much easier to align with the instantaneous rotation axis of the Earth. Furthermore, there must be the possibility to access and precisely monitor the variation of the orientation that the rotation sensor experiences. Due to their size of several meters on a side, active laser gyroscopes can be considered advantageous, although this is still subject to a significant challenge. These considerations eventually motivated the development of large ring laser gyroscopes in our group. In this chapter we briefly review the alternative rotation sensing instrumentation and their prospects for applications in geophysics and geodesy.

5.1 Externally Excited, Passive Cavities

Early versions of passive gyros were reported as early as 1977 [71]. In this era before the availability of IBS-coated super-mirrors, they already provided very promising results. The Sagnac effect for a passive gyroscope in the sense of the classical experiments of George Sagnac and Michelson–Pearson–Gale exploits the phase

Table 5.1. *A comparison of high resolution rotation sensing technologies and their best performance parameters.*

Technology	Sensitivity [prad/s/ $\sqrt{\text{Hz}}$]	Source
Atom interferometry	600	Class. Quant. Grav. **17**, 2385–2398 (2000)
Josephson effect	2000	Rep. Prog. Phys. **75**, 016401 (2012)
FOG (strategic)	72,000	L-3 Space & Navigation Datasheet
FOG (BlueSeis3A)	20,000	SRL, **89**(2), 620–629, (2018)
FOG	≈1200	2 km Fiber around 4.2 m diameter disc RSI, **84**(4), 041101–041101–26, (2013)
Passive opt. interferometer	24,000	Optics Lett. **6**(11), 569 (1981)
Passive opt. interferometer	10,000	Class. Quant. Grav. **33**(3), 035004, (2016)
Passive opt. interferometer	1000	Sensors **20**(18), 5369, (2020)
Large ring laser gyroscope	12	Phys. Rev. Lett. **107**, 173904 (2011)

difference that the beams experience on their way around the cavity. The rate of rotation then corresponds to the phase shift

$$\delta\phi = \frac{8\pi A}{\lambda c}\mathbf{n} \cdot \omega, \tag{5.1}$$

with ω the rate of rotation acting on the gyro and all other symbols defined as in previous chapters. It is interesting to note that already the early attempts on the investigation of passive laser gyroscopes had switched from the difficult phase measurements of the pre-laser era to frequency measurements. While passive laser gyros carry two promises, namely the absence of anything but the mirrors in the cavity and the potential of going to very high laser powers in order to reduce the shot noise to insignificance, there are also disadvantages that need to be taken into consideration and which we discuss in this section.

At first we will look at the challenges that are common to both, the ring laser gyro and the passive laser gyro. The unperturbed Sagnac frequency is the measurement quantity of interest and this frequency, is entirely a property of the cavity. So the actual scale-factor, the orientation and the mechanical stability of the gyro body itself, together with the platform on which it is operated, present the same challenges in both concepts. The same applies to the Q-factor of the cavity, as well as the backscatter properties and the Brownian and substrate noise of the mirror coatings. In [121] it has been asserted that the operation on different longitudinal mode indices for the two counter-propagating laser beams avoids the backscatter coupling, but this is not the case. neither the active nor the passive gyros make a difference in this matter, as discussed in Section 3.9.4. As a word of caution, it needs to be emphasized that the absence of an entirely reciprocal setting of the interferometer will unavoidably reduce the common mode rejection of mechanical cavity noise and the scale factor instability (see Section 3.6.4 for details).

In the laser gyro, a useful (common) cavity mode for both senses of propagation is automatically selected by a cavity resonance close to the maximum gain of the laser transition. Once oscillating, the linewidth of the laser reduces to incredibly small values of the order of about 10 µHz (see Section 3.6.5). The oscillation frequency dithers around by up to 100 Hz, but this does not matter as it is in common mode for both beams. The width of the resonance of the empty cavity is of the order of 200 Hz, which requires a truly elaborate PDH locking scheme in order to be rigidly tied to an external laser, which has to be carefully mode-matched for both beams on the same cavity mode. A significant perturbation is presented by the easily introduced residual amplitude modulation (RAM), and this needs to be suppressed. A sensor resolution of $\Delta\delta\omega < 10$ nrad/s at around 1 Hz is reported in [121] for a 1 m^2 size installation. An improved value of $\Delta\delta\omega < 2$ nrad/s for the regime of 5–100 Hz has been obtained by [137, 138] on a system of comparable size. None of these structures has yet been built from zero expansion material, and neither operates on the same longitudinal mode index in order to remove non-common mode cavity effects. The latter system employs an ultra stable laser to track the variable size of the cavity. This is responsible for the enhanced regime of high resolution, despite the fact that the mechanical cavity properties and the drift of the injection laser combine in this procedure. The measurement scatter of the 1×1 m system in Wuhan is of the order of 100 mHz for Sagnac frequency estimates at 1 minute measurement intervals. In comparison, we obtain 100 µHz for G. A careful analysis of all noise contributors is given in [138]. Among the most serious noise sources are the cavity variations due to fact that the system is not working on the same mode and the discriminator noise from the PDH locking scheme. Issues due to backscatter coupling have not yet been addressed in these systems. Until now, they have been masked by the other dominant noise sources. Passive resonant cavities will also require much longer runs for theses effects to become apparent. The promise of a lower shot noise limit is there, but not yet exploitable. Ring Lasers have their disadvantages as well. They have to be operated at a much lower optical power. In contrast to passive gyros, active laser gyros require proper handling of the non-reciprocal frequency shift induced by the presence of the plasma in the cavity. At this point in time it remains to be seen which of the two concepts will eventually carry further.

5.2 Fiber Optic Gyroscopes

Passive rings and fiber optic gyroscopes (FOG) are comparatively similar. Both feature an external light source. However, the fiber optic gyroscope does not form a resonant cavity and hence does not exploit a frequency difference. From that point of view, the FOG is much closer to the original phase detection interferometer of George Sagnac than the passive resonant cavity discussed in Section 5.1. Single turn Sagnac interferometers at rest require a path length of nearly 2 km in order to measure Earth rotation to an accuracy of 2% [77]. When the HeNe gas laser finally became available, optical laser gyros, essentially representing a multi-turn rotation sensor, became practical "small size, high resolution" instruments for aviation. Guided light around a low loss fiber has very similar properties. Instead of enclosing a large area A in one turn, a circular coil made of I turns of a fiber of length L and diameter D can accumulate a large scale factor ($L = I\pi D$) in a matchbox-sized lightweight package:

$$\delta\phi = \frac{2\pi LD}{\lambda c}\mathbf{n} \cdot \omega. \tag{5.2}$$

Proposed in 1967 [160] and demonstrated in 1976 [223], the fiber optic gyro has a lot of operational advantages. It is a small, solid state device without the need for highly symmetrical plasma excitations. When we look at the performance of the various types of gyros, the ring laser gyro populates the high resolution, high stability end of the spectrum, suggesting itself not only for applications in geophysics and geodesy but also for the highly accurate navigation application with a sensor drift of the order of 0.01–0.001°/h. While FOGs can reach the high demand end of navigation, most of these devices are built for the 1–10°/h range followed by devices for the 0.1–0.01°/h range [129]. Apart from the Shupe effect, where non-reciprocal refractive index changes cause a variable non-reciprocal bias, one of the main problems with the use of FOGs is Rayleigh backscattering, which yields a readout error and, since the optical path lengths are not completely stable, it gives rise to noise. Current high quality fiber optic gyros exhibit a typical sensor resolution of $\delta\varphi = 0.1°/\sqrt{h}$ or slightly less. The low power consumption, large dynamic range and data rates of up to 4 kHz make them ideal for uses in aviation. Furthermore, they suggest themselves for civil engineering or any other application that requires a small packaged unit for rotation rate sensing, apart from navigation. The entirely integrated, hermetically sealed units have dimensions around $20 \times 50 \times 80$ mm. In the context of this book we do not explore the properties of FOGs as navigation devices. Here we look at some applications in earthquake structural engineering monitoring as well as geophysics. For an excellent treatment of the theory of fiber optic gyros, we refer the reader [129].

5.2.1 Inertial Rotation Sensing in Structural Engineering

In order to monitor changes in larger buildings, for example due to a slowly developing deformation as a result of structural weakening, we not only need sufficient sensitivity but also stability. For our tests we used a Northrop Grumman LITEF GmbH fiber optic gyro type μFORS-1 (serial no. 1376) with a specified stability of 1°/h. This approach offers several inherent advantages, which can become important in health monitoring applications, where false alarms or even worse, missed alarms, are not acceptable. When the structure suffers strong motion impacts during an earthquake, there is no longer a reliable reference, since both the building and the ground are moving at different rates and in different directions. So the sensor needs to produce absolute measurements.

Torsional motions during an earthquake is very destructive, so these should be captured well. Inter-story drift from shear forces causes significant weakening of a structure. Absolute inertial rotation sensing around a horizontal axis via FOGs can capture this. The excursion angle experienced by a structure to which the FOG is strapped is obtained by a single integration, unlike for accelerometers, which have to be integrated twice. Therefore the results usually diverge quickly. Last but not least, if a system like the μFORS-1 is strapped to some structure, it always senses the a priori known projection of the Earth rotation vector as a constant background signal. This fact can be readily exploited for permanent sensor health monitoring purposes. Figure 5.1 shows a typical static observation signal from a strapped down FOG, along with the corresponding normalized Allan deviation estimate.

The FOG was operated with a sampling rate of 1 kHz. A linear regression over the measurements shown in the left panel yields an offset from zero of $\delta y = 3.1946 \times 10^{-3}$°/s, which is the experienced Earth rotation rate of 11.5°/h for a sensor flat on the floor at the Geodetic Observatory Wettzell, where the test was conducted. The drift was small. When integrated up to obtain angles and with the slope of the Earth rotation rate removed, we can see the noise and the drift of the sensor, as shown in Figure 5.2 for a short period of time. The noise level of this particular sensor can be taken from the right panel of Figure 5.1 as ~0.08°/(h $\sqrt{\text{Hz}}$) and the

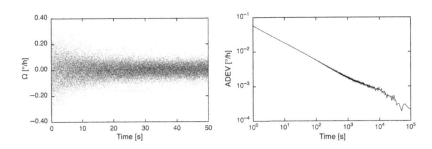

Figure 5.1 Time series of the measured rotation rate, taken with a sampling rate of 1 kHz (left). After switch-on, the sensor takes about five seconds to optimize the observation parameters. The ADEV plot shows white noise behavior over almost one day (right).

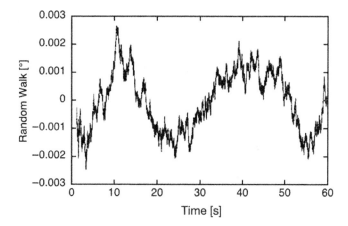

Figure 5.2 Display of the integrated angle of the μFORS-1 measurements taken over one minute with the Earth rate removed.

Figure 5.3 Display of a single story building rigidly mounted to the shake table. By shifting heavy metal sheets sideways on the top of the structure, an unsymmetrical mass distribution results, which gives rise to substantial torsional oscillations.

angular random walk as $10^{-3}/\sqrt{h}$. An FOG of the tactical grade class ($1°/h$–$10°/h$) is well suited to studying the response of civil engineering structures under medium or strong motion conditions. We have subjected our sample device to simulation tests on a shake table installation. The important results of these tests have already been reported in [72, 190]. So we limit the discussion here to the sensor qualification test.

Before any structures were tested, we operated the FOG alone, strapped down on the bare one-dimensional shake table (see lower right section of Figure 5.3). There were only small perturbations recorded, not exceeding values of $\pm0.2°/s$, which is just barely above the sensor noise level of $0.1°/s$ (see the top panel of Figure 5.4). This confirmed of that the linear motion shake table itself did not induce

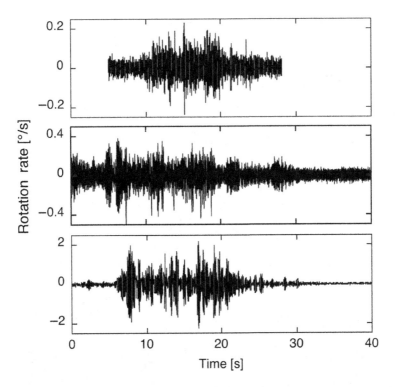

Figure 5.4 Observations of rotations around the vertical axis on a one-dimensional empty shake table (top). The same experiment with a one-story building model and symmetrical mass distribution shows slightly enhanced exursions (middle). The measurements with a moderate unsymmetrical mass distribution shows a considerable increase in torsional motion (bottom).

any significant amount of rotational motion around the vertical axis on any test structure. When a symmetrical single story model of a building structure was rigidly mounted onto the shake table, as depicted in Figure 5.3, it was subjected to the same synthetic ground motion as before (see middle panel of Figure 5.4). Heavy metal shields in the plane of the top of the building model could be moved sideways in order to make the mass distribution unsymmetrical in the direction orthogonal to the excitation.

The shake table was then set into motion, simulating the signature of a real historical earthquake (mag. 8.0 Valparaiso, 1985) scaled down to 10% in amplitude. Figure 5.4 presents a comparison of FOG recordings of the measurement of the shake table alone (top panel), the shake table with the building model mounted (middle panel) and the shake table with an asymmetrical load excitation with respect to the vertical axis. Although the applied earthquake signature was always the same, the measured angle excursions do not properly line up in this comparison, since the time tagging was not synchronized and started at random for each test. There is a factor of two between the bare shake table and the fully symmetric load around the middle of the structure for a linear excitation. However, with a moderate asymmetry in the mass distribution, the torsional excursions of the structure increased by another factor of five, as shown in the lower panel of Figure 5.4. It is important to note that apart from possible deformations in the structure of the building, the measured rate of rotation around the vertical axis would be the same, independent of the chosen sensor position; in analogy to ring lasers, the FOG is an all optical device and therefore entirely insensitive to translations. So these two components of motions do not mix in the observations.

In another experiment, a model of a four-story building on the same shake table provided excellent agreement for the measurement of inter-story drift between the FOG, mounted such that was it sensitive around the horizontal axis orthogonal to the direction of motion, and a set of transducers attached to a rigid frame, referenced to the fixed laboratory floor [72, 190]. Figure 5.5 illustrates the applied installation, and Figure 5.6 compares the measurement results from the shake table sensor tests. Applied to real world buildings, FOGs can still produce valid measurements of similar quality, while the transducer method is not applicable, due to the lack of a permanent stable ground reference. The data gaps in Figure 5.6 are the result of a buffer update problem in the data logging routine of the FOG readout and have no general significance. A closer inspection of the collocated observations show that the transducers slightly underestimate the amplitude of the displacement. This is caused by some bending in the transducer arms. Further experiments have shown that these FOGs are not only suitable for comparatively small structures, like these models on a shake table. They also suggest themselves for tall structures.

Figure 5.5 Display of a model of a four story structure, placed on a one-dimensional shake table. Transducers tied to a metal frame at the back of the laboratory are used to establish the excursions of the model during the test run. The position of the FOG is indicated by the arrow, and the installation is shown on the inset.

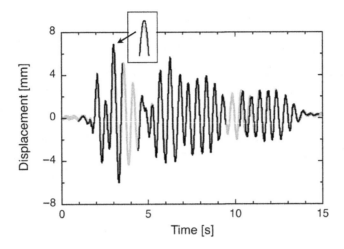

Figure 5.6 Time series of the observed displacement from inter-story drift of the building model during a simulated earthquake. The two independent measurement concepts show excellent agreement.

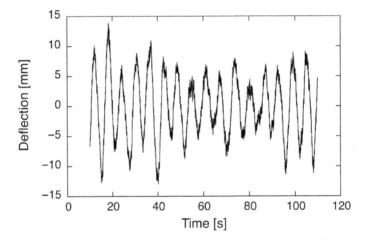

Figure 5.7 Display of the rocking motion of the Sky Tower in Auckland with a period of a little more than 6 s.

As an example, we have measured the deflection of the Sky-Tower in Auckland (New Zealand) and the torsional motion of a wind turbine in northern Germany. On a calm day we obtained the rocking motion with a deflection of 5–15 mm and a period of oscillation of slightly more than 6 s. Figure 5.7 depicts one of these measurements. For more details we refer the reader to [190]. Another example is the torsional motion around the vertical axis of an operating wind turbine, measured just below the main axle of the turbine approximately 70 m above the ground.

The experiment was carried out in the presence of moderately gusty wind and while the blades were turning. Within the short period of the measurements we obtained torsions of ± 100 seconds of arc. In general the measurements of displacements in tall static and dynamic structures yielded consistent results with good fidelity and for frequencies from as low as 0.015 Hz to well in excess of 100 Hz. Even FOGs with moderate sensitivity are useful for the analysis of civil engineering structures. The typical application approach is similar to that of navigation. The advantage is true inertial rotation sensing in an environment that lacks a reliable reference system. Measurements of inter-story drift or torsion become viable. Furthermore, we find that the excursions obtained by a FOG are widely independent of their placement. This cannot be said for rotations inferred from finite differences obtained from an array of accelerometers. However, from a commercial point of view FOGs are not ideal. The construction is rather complex, which makes them expensive, and one of our sensors failed after about 12,000 hours of continuous monitoring operations.

5.2.2 Large Fiber Optic Gyroscope

The utilizable length of the fiber ultimately limits the scale factor in an FOG, most prominently due to the unavoidable losses that are introduced. Scattering from inhomogeneities within the fiber reduces the contrast further and is another limitation on the sensor resolution. So the sensor design has to deal with two opposing goals. While the total enclosed area should be made as large as possible, the fiber ideally should be as short as possible. These and some more subtle issues limit the ultimate sensitivity of an FOG to around 1 nrad/s with a coil diameter of the order of 20 cm. For the envisaged application in space geodesy, we can, however, make the diameter as large as 4.25 m by utilizing the Zerodur disc of the G ring laser, while keeping the fiber length still around 2 km. This increases the scale factor already by a factor of 21 over the commercial 20 cm diameter device. The temperature stability in the underground laboratory is typically 2–5 mK/day and therefore suitably stable [182]. In order to explore the potential of the enlarged scale factor of the FOG concept, we have used about 2.2 km of mono-mode fiber and wound it around the G ring laser Zerodur disc structure in 162 turns. Figure 5.8 illustrates how the fiber is attached to the low expansion material. After an initial warm up phase of several minutes, the Earth rotation rate is recorded at a rate of 400 Hz, resulting in a sensor resolution of approximately $\Delta\Omega \approx 0.5 \times 10^{-6}$ rad/s. As for the smaller gyros, the signal primarily shows white noise behavior, which averages down to $\Delta\Omega/\Omega_e \approx 5 \times 10^{-5}$ over three hours. Most of the observed scatter in the measurements of this large fiber optic gyroscope has the characteristics of white noise, as one can see from the Allan deviation (a) shown in Figure 5.9. At one hour of integration, G can already resolve

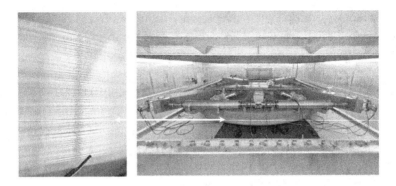

Figure 5.8 A large fiber optic gyroscope was built by winding 2.2 km of mono-mode fiber around the Zerodur disc of the G ring laser. (Photos © A. Heddergott of TU München (2012).)

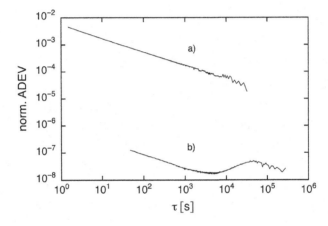

Figure 5.9 Allan deviation of the large fiber optic gyroscope (a) in comparison to the G ring laser (b). Although the scale factor is much larger than that of the G ring laser, there is a significant discrepancy in sensitivity of more than three orders of magnitude. (Adapted with permission from [193] ©2022 by IOP Science)

rotation rates in the femto-rad/s regime, which means that there is a difference in performance of more than three orders of magnitude between the active ring laser and the large FOG sensor concept. Figure 5.10 illustrates this by a short sequence from a collocation measurement. It is expected that some functional modifications, such as a brighter light source and some enhancements to the electronic circuitry, could reduce this gap. Fiber optic gyros, including the large G-FORS experiment, operate in a closed loop configuration. While this is of great advantage in aircraft navigation, because of the enlarged dynamic range of the sensor, it presents a significant disadvantage for the application of FOGs in the geosciences, where small perturbations of the nearly constant rate of rotation of the Earth are examined.

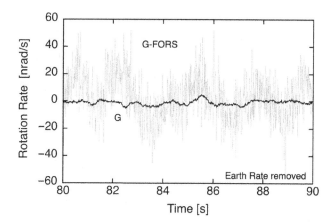

Figure 5.10 Comparison of the collocated operation of the G ring laser and the G-FORS large scale fiber optic gyroscope. The ring laser measurements are significantly more stable.

Open loop FOGs, optimized for applications in the geosciences, would improve the sensor performance somewhat, but not as nearly as much as required.

Alternatively, one can take the entire concept to an extreme by utilizing an urban fiber network and construct a loop with 20 km² area and a total length of the mono-mode fiber of 47 km, as reported in [42]. As extended optical fiber networks are progressively coming into existence, it is natural to explore even further upscaling. With a fiber loop around the city of Torino (Italy, colatitude 45°), this approach has been comparatively simple to realize [42]. While we obtain a phase shift from the rotating Earth for the G-FORS with an enclosed area of 2298 m² of 6.85 mrad from Eq. 5.1, the ring in Torino accumulates approximately 55 rad for $\lambda = 1.542$ μm. However, since there is no way to obtain the exact number of cycles, it is only possible to observe the variation of the phase shift. The linewidth of the applied fiber laser is 10 kHz, with the coexisting internet data transfer on the same lines approximately 2 THz away. The laser frequency of the two beams is shifted by 40 MHz before it is injected into the fiber loop with randomized polarization. On top of this, there is a $f_m = 990$ kHz phase modulation at one side, which modulates the cw beam before it travels around the fiber, while the ccw beam is modulated after the beam has been around the loop. The phase difference caused by the Sagnac effect is φ_{nr}. A photodiode sampling the recombined beam detects the signal [42, 51]

$$I = I_0 + I_1 J_1(x) \sin \varphi_{nr} \cos \left[2\pi f_m \left(t - \frac{\tau}{2} \right] + f_m \text{ harmonics}, \quad (5.3)$$

where I_0 and I_1 present the DC voltage and the first harmonic, J_1 is the first order Bessel function, $x = 2\phi_0 \sin (2\pi f_m \tau)$ and ϕ_0 is the phase modulation depth. From this expression φ_{nr} can be extracted. Closed loop operation is possible by compensating φ_{nr} by a frequency offset Δ_f between the two beams [42, 51]. In this way the phase difference is converted to a frequency difference, similar to that of the ring laser application:

$$\Delta_f = \frac{4vA}{nLc}\mathbf{n}\cdot\Omega = \frac{4A}{n\lambda L}\mathbf{n}\cdot\Omega. \tag{5.4}$$

In this expression L represents both the length of the perimeter of the gyroscope and the length of the fiber. The Allan deviation of this gigantic gyroscope levels out in the interval of 100–1000 s before it starts to diverge, and it reaches a minimum for the detection of a rotational signal at 2 nrad/s, taking it into the same domain as the G-FORS installation performs in phase detection mode.

Taking a rather global look at Sagnac interferometers, it is fair to say that there are in the end two most important factors that define the ultimate performance, namely the scale factor and the losses in the beam path. Applied to ring lasers this means that the linewidth, which depends on the losses, and the area enclosed by the cavity are the dominant factors with respect to the ultimate sensor resolution. Backscatter coupling and plasma dispersion are independent complications that need to be addressed separately. For large gyros the scale factor is not a problem; however, as the examples shown here suggest, the losses in fiber optic gyros grow at the same rate as the scale factor increases. It is interesting to note that the commercial fiber optic gyro *BlueSeis3a* [24] for applications in the geosciences exhibits a flat sensitivity response of $\delta\omega \approx 20$ nrad/s in the range of 1 mHz to 100 Hz, with a diameter of the coil of the order of 20–30 cm. G-FORS reaches $\delta\omega \approx 2$ nrad/s after more than an hour of integration, and the giant FOG around Torino is in the same regime of $\delta\omega \approx 2$ nrad/s after about 1000 s. Although very different in their specifications, all three systems have very similar sensitivity limits.

5.3 Helium SQUID Gyros

A gyroscope based on superfluid helium [156] consists of a thin torus, cooled down to 2.17 K in a cryostat. Under these conditions the ^{4}He inside the torus flows without friction. The quantum coherence of superfluid helium is exploited here, since very small rotation rates can significantly change the flow pattern in the toroidal measurement compartment. Superfluid systems are characterized by a macroscopic quantum-mechanical wave function $\Psi = \Psi_0 e^{i\phi}$, whose complex phase ϕ contains the kinematic description of the system and whose amplitude Ψ_0 is proportional to the superfluid density. So the gradient of the phase is related to the superfluid velocity field v_s by

$$\nabla \phi = \frac{2\pi m}{h} v_s, \tag{5.5}$$

where m is the mass of ^{4}He in this case and h is the Planck constant. Therefore, it is possible to detect very small changes of absolute rotation, experienced by the entire cryostat container in which the superfluid torus is submerged. The relationship between macroscopic quantum phase and velocity leads to a restriction on the flow states that are accessible to the quantum liquid. When the function $\nabla \phi$ is integrated around any closed path inside the fluid, the result must be an integer multiple of 2π. In other words, the circulation is quantized as

$$\oint v dl = \frac{nh}{m} = n\kappa \quad n = 0, 1, 2, 3, 4 \ldots \tag{5.6}$$

In order to exceed the critical velocity of the helium inside the torus, the cross-section is blocked with a membrane, which has a tiny hole of 0.1 μm in diameter. This acts like an amplifier, and in practical solutions a gain factor of 10^5 is readily achieved from the ratio between the torus cross-section and the diameter of the hole. When the critical velocity is exceeded, vortices are generated, and the system becomes dissipative. The onset of the vortices is very sharp. Therefore, it is possible to exploit this concept to set up a sensitive gyroscope. A membrane pump equipped with a SQUID sensor pumps the helium periodically back and forth through the torus as schematically indicated in Figure 5.11. The SQUID sensor senses the position of the membrane with a resolution below 1 nm. When the critical velocity in the orifice is exceeded and dissipative vortices are generated, they are detected by a sudden change in the membrane position. These vortex breakdowns occur at different excitation amplitudes for a system that is at rest compared to a system that is in rotational motion, and this is the essential effect for the gyroscope

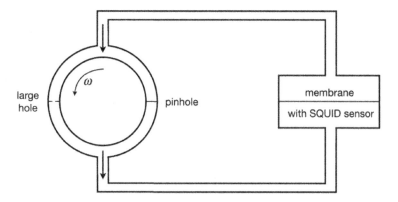

Figure 5.11 Schematic depiction of the operating principle of a superfluid helium gyroscope. The ratio of the holes in the torus define the sensitivity of the sensor.

functions [178]. When the enclosing dewar is rotating with an angular velocity $\mathbf{\Omega}$, the fluid in the connecting tubes moves with it, and this causes a phase difference $\Delta\phi_e xt = 4\pi \mathbf{A} \cdot \mathbf{\Omega}/\kappa$. The vector $\mathbf{A}$ on the area enclosed by the loop is the superfluid equivalent of the beam path in an optical Sagnac interferometer. The phase shift $\Delta\phi = 4\pi f \mathbf{A} \cdot \mathbf{\Omega}/c^2$, where the effective photon mass hf/c^2 is now replaced by the mass of the superfluid atom. Since the atomic mass of helium is about 10 orders of magnitude larger than the photon mass, this provides a strong incentive to explore this type of sensor for high resolution rotation sensing.

As a proof of principle, a rotation sensor with an area of 6 cm^2 was demonstrated. Instead of one pinhole, there were two weak links implemented, each with a matrix of 65 × 65 small holes [178]. In these experiments a sensitivity for rotations of $\delta\omega \approx 2 \times 10^{-7}$ rad/s/$\sqrt{\text{Hz}}$ was achieved. This is certainly not the lower limit of this concept, but there is still a long way to go to make this device comparable to the ring laser gyro. Apart from the technical challenge of improving the sensitivity by five orders of magnitude, there are also very practical problems to solve, such as optimizing the size of the cryostat, the proper isolation of vibrational noise and, last but not least, the issue of how to ideally orientate the sensor with 1 nrad of accuracy with respect to the Earth's surface. Finally, good long term stability and reliability has yet to be achieved. External vibrational noise, for example, from the helium compressor or the ambient microseismic activity, cause serious perturbations in this experimental setup.[1]

5.4 Atom Interferometry

Another promising technique for inertial rotation sensing is also a quantum mechanical device, based on the interference of atoms [83, 132]. Cesium atoms are laser cooled in a 2D optical molasses before they are optically pumped into a magnetic field-insensitive hyperfine state. The interferometer exploits Raman transitions between the $6S_{1/2}$, $F = 3$, $m_F = 0$ and the $6S_{1/2}$, $F = 4$, $m_F = 0$ acting like beam splitters, while the atoms move through a Mach–Zehnder interferometer setup, as sketched out in Figure 5.12. The atoms have an initial momentum $\mathbf{p}_0$ when they enter the interferometer after the preparation of the $|F = 3, m_F = 0; \mathbf{p}_0\rangle$ state. The first $\pi/2$ pulse puts the atoms in a superposition of $|F = 3, m_F = 0; \mathbf{p}_0\rangle$ and the $|F = 4, m_F = 0; \mathbf{p}_0 + 2\hbar\mathbf{k}\rangle$ states. The two-photon recoil momentum $2\hbar\mathbf{k}$ introduces a transverse velocity component on one of the wave packets, thus acting as a matter wave beam splitter. At the next stage the wave packets reach the π Raman beams, taking the equivalent role of a mirror in optics, where the ground

---

[1] R. Packard, private communication

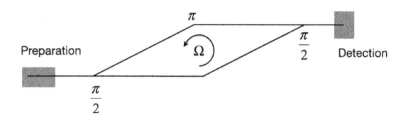

Figure 5.12 Schematic illustration of the operating principle of a thermal atom interferometer gyroscope. A Mach–Zehnder layout is used, where the interactions with Raman states are taking the roles of beam splitter and mirror.

states and momenta are exchanged. When the converging beams meet again, another $\pi/2$ pulse ends the superposition. Atoms that have made the transition from the $F = 3$ to the $F = 4$ state are detected by a resonant probe laser, tuned to the $6S_{1/2}, F = 4 \rightarrow 6P_{3/2}, F = 5$ cycling transition, where the resulting fluorescence is detected by a photomultiplier. The inertial sensitivity of an atom interferometer arises because the freely propagating atoms form fringes with respect to an inertial reference frame. Half of the wave packets are traveling on the co-rotating path, while the other half are traveling on the anti-rotating path. The inherent high sensitivity of the atom interferometer comes from the fact that the de Broglie wavelength $\lambda_{dB} = h/mv$ is depending on the mass (m) and the velocity (v) of the particles. This puts atom interferometry into the same category of sensor as superfluid helium gyroscopes, exploiting the Josephson effect. Consequently, one could expect an increased sensitivity of these sensors over optical Sagnac interferometers by

$$\eta = \frac{mc^2}{hf} \approx 10^{11}. \tag{5.7}$$

In practice this resolution has not yet been demonstrated. Optical mirrors and beam splitters are far more effective, so that large enclosed areas are simple to achieve for light beams. Thermal cesium beams as reported in [83, 132] have achieved enclosed areas of the order of $A = 22$ mm^2, while our optical Sagnac gyros have been built as large as 834 m^2.

The next generation of atom interferometers uses cold [87]Rb atoms instead of the thermal cesium beams. Since it reduces the velocity of the atoms from $v = 290$ m/s [83] to $v = 2.79$ m/s [215], the apparatus becomes substantially smaller, namely 13.4 cm in length for the interferometer zone as opposed to 2 m in the earlier design. In both cases the enclosed area remains at about the same value. However, this requires much stricter conditions for the parallel alignment of the Raman lasers, because the temperature-determined beam divergence is much larger in the slow cold atom beams as opposed to the faster thermal beams. The achieved sensitivity for rotations for the cold atom interferometer is reported to be $\delta\Omega = 6.1 \times 10^{-7}$ rad/s/ $\sqrt{\text{Hz}}$, with some room for moderate improvements.

Both types of alternative sensors, the Josephson and the atom interferometer, are substantially more complex in sensor design and operation than optical laser gyroscopes, and this concerns not only the setup and handling of the interferometers. A rotation sensor for geophysically relevant observations needs to operate unattended and permanently. Furthermore, the sensor and the measurement approach must be long-term stable, which is an important conceptional advantage of radio interferometry (VLBI). Atom interferometry and superfluid helium gyroscopes both have an enormous potential sensitivity. As a result, the corresponding gyroscope areas are small, and this has several significant consequences. Since the area scales linearly with the sensor resolution, a high sensitivity will also induce a high susceptibility to potential fluctuations in the effective area. While the geometry is well defined in a large ring laser gyro, we still have considerable difficulties in relating this large area to the body of the Earth itself in a representative way. The discussion of the orientation of the single component G ring laser in Section 4.6.3 illustrates this problem. In much smaller areas, such as in the range of 22 mm² to 6 cm² as discussed in this chapter, this becomes a real problem. In the end it is necessary to control the north–south orientation of any inertial sensor with respect to the instantaneous rotational axis of the Earth to a level of 1 nrad, which is much more than any location near the surface of the Earth can support anyway, due to the presence of local deformations from atmospheric loading and ground water variation.

While large gyros can be set up in a tetrahedral arrangement, so that changes in orientation of one component show up as rotation in the other sensor components (1 nrad = 12 nm in ROMY), this becomes even more difficult for a system with a baseline of, say, 2 cm, where a tilt of 1 nrad corresponds to a change of corner positions of 20 pm. Atoms have mass and are subject to gravity. Is it possible to stabilize the trajectories to sub-nm levels over a long time? Finally, secondary issues also need to be addressed. Cryogenic and ultra-high-vaccum systems usually require pumps, which have the side effect that they cause vibrational noise. How well can this be separated from the measurement quantity of interest? Here we need to note that none of the inertial sensors are currently in the position to fulfill all the demands at the time of writing. However, with this said, the optical ring laser gyroscope is currently the most promising candidate for an inertial rotation sensor for applications in the geosciences.

5.5 Coriolis Force Gyroscopes and Microelectromechanical Systems

In 1851 Jean Bernard Léon Foucault used his famous pendulum to demonstrate inertial Earth rotation sensing. When properly set up, the effect of gravity is compensated by the wire, and no other forces are acting on the proof mass. In this case the path taken by the pendulum is fixed in space, and the observer can monitor how the Earth is rotating underneath the trajectory of the pendulum. However, since the rotating Earth is not an inertial reference system, an observer on the body of the Earth at the latitude φ sees the effect of the Coriolis force ($\omega_c = \Omega_E \sin \varphi$), which acts perpendicular to the pendulum motion and therefore perturbs our experiment. This effect, however, can be exploited in tuning fork gyros for inertial rotation sensing. While ring lasers and FOGs cannot practically be made much smaller, microelectromechanical systems (MEMS) fill this gap. Apart from a much smaller size and low power consumption, they can be mass produced. Today, low resolution rotation MEMS sensors are contained in every cell phone, and this opens new perspectives for scientific applications. Traditionally seismic studies depended on a low number of high resolution observation points. However, with an MEMS in every cell phone connected to the internet, it is possible to obtain earthquake measurements from a very large number of observation points at the same time, albeit with comparatively low resolution. Nevertheless, for strong motion events in densely populated regions, this capability provides valuable novel inputs.

The sensing element in a gyro MEMS structure is usually a small tuning fork or similar resonating element, etched lithographically from a quartz crystal [128]. A tuning fork, where both arms are oscillating in the same plane but in opposite directions (anti-phase), is sensitive to the Coriolis force when the system is rotated around the mounting post of the tuning fork. This is illustrated in Figure 5.13.

The device is sensitive to rotations about the mounting post of the tuning fork. The Coriolis force causes an out of plane vibration, whose amplitude is proportional to the experienced rate of rotation. In order to make the system more sensitive, the mounting post has dimensions that are resonant with the Coriolis force-induced vibration. The mechanical Q-factor of the entire system defines the resolution of the device. One of the technical difficulties is the symmetry of this arrangement.

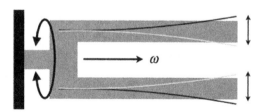

Figure 5.13 Schematic illustration of the operating principle of a tuning fork MEMS device. The two arms of the tuning fork are oscillating in the plane of the page. When the device is rotated about the mounting post, the Coriolis force exerts an out of plane deflection proportional to the rate of rotation.

It is important that the two arms of the tuning fork are balanced so that there is no force acting on the base of the fork in the absence of rotations. The output amplitude A is obtained as [128]

$$A = \frac{4v\omega}{K},\tag{5.8}$$

where $v = v_0 \sin \omega_f t$ is the velocity of the oscillating arms of the tuning fork, ω the experienced rate of rotation and K the torsional stiffness constant of the mounting post. It is important to note that the quantity of interest, namely the torque induced by the rotation of the device, is very small, which makes the system very sensitive to any imperfections in the geometrical arrangement. Designing a more complex oscillatory structure reduces the susceptibility to non-axial perturbations. The Quapason[2] is an example of such a structure and comprises a four-beam tuning fork with one oscillatory arm at each corner of a common base plate [131]. The driving design criterion, apart from the proper insulation of the sensing elements from perturbing forces and high symmetry, is low frictional loss. By and large this is the same as we have always found for ring lasers. The need for perfect symmetry is a common feature of highly sensitive rotation sensing devices. The Quapason resolves 0.1°/s over a measurement range of ±100°/s at an angular rate bandwidth of 100 Hz.

The hemispherical resonator gyro HRG [172] is a great example of the illustration of the requirement of high mechanical Q. They go back to an observation by G. H. Bryan reported to the Cambridge Philosophical Society in 1890, that a wine glass rotated around its stem and struck to give a ringing tone exhibits a beat note [35]. Hemispherical resonators operate on acoustic eigenmodes on fused silica quartz material. This provides the necessary high mechanical Q in excess of 25×10^6. Figure 5.14 illustrates the concept in a rough sketch. A bell-shape semi-sphere is excited in the lowest order bending mode with four nodes and four anti-nodes as indicated on the right side of Figure 5.14. In closed loop operation, the oscillation pattern is kept stationary by a feedback system. During open loop operation, the mode pattern moves along the perimeter with the experienced angle of rotation. In order to obtain high accuracy, it was important to balance the mass distribution around the 58 mm hemispherical resonator very carefully. In the early 1990s, the sensor design had matured to the point that a angular random walk of $ARW = 0.0006°/\sqrt{\text{Hz}}$ and a bias of $\delta\phi = 0.005°/h$ had been achieved [13], together with high reliability, albeit with a comparatively large drift of the scale factor of 48–100 ppm/day.

---

[2] trademark by Sagem

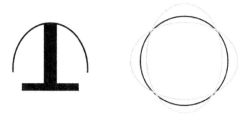

Figure 5.14 Rough schematic illustration of the hemispherical resonator gyro (left). The lowest order bending mode is excited during operation viewed from above (right). The black line indicates the neutral position, while the shaded lines indicate the positions of the nodes and antinodes of the standing wave pattern of the oscillation.

Building on the high operational reliability, a reduction in size to 30 mm and the introduction of a force-rebalance mode for closed loop operation made this type of gyro very suitable for space requirements, such as maintaining the orientation of communication satellites. A *Space Inertial Reference Unit* based on the hemispherical resonator was part of the CASSINI mission to Saturn. The system meets the required bias stability of $\delta\phi = 0.015°$/h and an $ARW = 0.001°$/h, with the angle white noise not exceeding 0.02 seconds of arc/Hz. The excellent performance characteristics of the HRG was then pushed to satisfy the requirements as a rate sensor for the Hubble space telescope. This resulted in a demonstrated performance of the bias stability of $\delta\phi = 0.00008°$/h and an $ARW = 0.00001°$/h, with a residual angle white noise not larger than 0.00015 seconds of arc/h at a maximum rate of 0.5°/s. Despite meeting the required specifications, the HRG did not end up on the Hubble space telescope, since the technological readiness level was not high enough to meet the required tight timeline.

Numerous units of the final product version, the *Scalable Space Inertial Reference Unit* (SSIRU), have been flown successfully on a large number of challenging space missions, such as the MErcury Surface, Space ENvironment, GEochemistry and Ranging (Messenger) mission and the Deep Impact mission, where a SSIRU guided impactor was set to hit comet Tempel 1. The specifications for the SSIRU are given as $ARW = 0.0006°/\sqrt{h}$, with an angle white noise of 0.001 seconds of arc/$\sqrt{Hz}$. For the bias stability, a value of 0.0005°/h at 1 s is achieved. The scale factor stability and the scale factor linearity are both estimated to be as good as 1 ppm for 1 s of integration.

5.6 Bi-directional Solid State Ring Lasers

Active HeNe laser gyros perform well, but the necessity for the confinement of the low pressure laser gas and the plasma excitation are inconvenient, in particular for space applications. Solid state active laser gyros promise a much simpler sensor design and have therefore been quite intensively studied, almost since the early days of the laser [219, 229]. The demonstration of rotation sensing with a solid state ring laser was demonstrated in 1974 by Klochan et al. [116, 117]. A comprehensive review paper covering this development up to 1992 was written by Kravtsov, et al. [122]. Most of this earlier work was carried out in the former Soviet Union. Solid state ring lasers differ from gas laser gyros by developing a gain grating, which causes the two counter-propagating laser modes to couple and eventually to degenerate. Since solid state lasers inherently have substantial gain, the gain medium can be made very small. This reduces the coupling of the counter-propagating beams, but this reduction is not enough for sensitive rotation sensing. Vibrating the gain crystal of a diode pumped Nd:YAG ring laser back and forth in the longitudinal direction smears out this grating and significantly reduces this effect [198].

This leaves gain competition as the other complication. In a gaseous HeNe ring laser, the absence of mode competition is due to inhomogeneous gain broadening, achieved by a 50:50 mixture of the two isotopes ^{20}Ne and ^{22}Ne. Such an approach is not available for solid state lasers, which require the control of the differential loss between the two counter-propagating laser modes. An electronic feedback system provides the necessary functionality [60]. A non-reciprocal Faraday rotator cell and a polarizer are employed for the purpose. On the downside, these are additional intra-cavity elements. By changing the current through a coil around the Faraday rotator, the plane of polarization is slightly rotated, which is aligned such that it introduces larger additional losses for the beam with higher intensity, while the losses for the weaker beam are smaller. The feedback system is based on the measured difference in intensity between the two counter-propagating laser beams. In order to reduce the overall losses of this concept, one of the cavity mirrors of the gyro is substituted by a polarizing mirror, while the cavity is made slightly non-planar in order to make the polarization very slightly elliptical. Finally, the length of the gain crystal can be reduced from 10 mm to 3 mm due to the more efficient laser diode pumping. The frequency of vibration needs to be around 168 kHz, to be large against the inverse of the population inversion lifetime [198]. It is important for the atoms to be sensitive to the average rather than the instantaneous intensity. For the discussion of the ultimate rotation sensing properties of a solid state ring laser cavity against any other concept of a Sagnac interferometer, we note that we have introduced a number of additional losses to the cavity, namely the crystal containing the gain medium and the differential loss mechanism for the suppression of mode competition between the counter-propagating laser modes. However, in Chapter 3 we have seen that the sensor resolution for any high end device critically depends on the minimization of the overall losses.

5.6.1 Design and Operation of a Solid State Gyro based on Nd³⁺:phosphate Glass

The incentive to look for a different gain medium comes from the desire to reduce the insertion loss of the crystal and at the same time to allow even shorter crystals for the gain medium, which will reduce the effect of spatial hole burning further, so that the additional dithering of the gain medium eventually may become obsolete. Phosphate glass can tolerate a much higher level of doping and has an insertion loss of around 0.1 m^{-1}, while a YAG crystal by comparison has a loss of more than 0.2 m^{-1}. Furthermore, the absorption band for the laser pumping is much wider and less structured, due to the amorphous structure of the glass material. Therefore, we may expect less stringent operation conditions for the pump diodes [165]. Phosphate glasses offer the best overall combination of properties, such as refractive index homogeneity, small cross-section for seeds and inclusions, good chemical durability, low attenuation for the laser wavelength and little or no optical birefringence when compared with other glasses [68]. For our experiment we used a 1 mm thick phosphate glass from Molecular Technology GmbH (Berlin, Germany), doped with a concentration of 4×10^{20} Nd³⁺ ions/cm³. In order to minimize the losses, the glass was antireflection coated on both sides. We used an AlGaAs laser diode from Intense™ (3030-HHL-TEC-FAC- 80803) to pump the glass crystal. The diode has a maximum power rating of 3.16 W at a wavelength of 807.2 nm and a FWHM of 0.2 nm. The cavity was set up as a square, 7.5 cm on each side. Figure 5.15 sketches the essential part of the ring cavity layout and shows a photo of the setup.

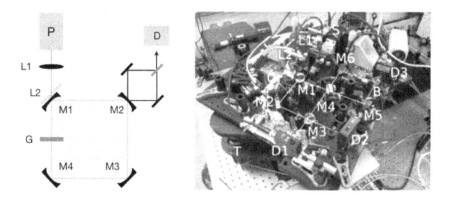

Figure 5.15 Schematic illustration of the solid state laser gyro (left). P is the pump diode, and M1–M4 are the mirrors of the ring cavity. L1 and L2 are necessary to focus the pump beam into the crystal. Interference is obtained from the standard beam combiner arrangement behind M2 and detected on the photodiode D. The photo (right) gives an impression of the full setup [165]. (Photo copyright N. Rabeendran)

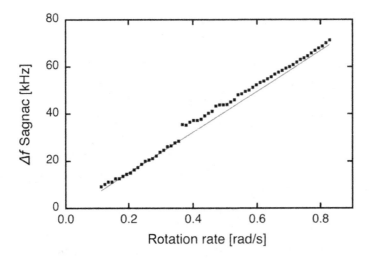

Figure 5.16 Comparison between the expected (solid line) and the experimentally obtained (squares) beat note as a function of the rate of rotation, as reported by [165].

Once set up and aligned, lasing was readily obtained. The effects of the gain grating in the solid state gain medium and the mode competition between the two counter-propagating beams were obtained in a similar fashion as reported by [198]. As a result of the additional backscatter on such a small cavity, the laser gyro was locked up most of the time when the setup was at rest. This was no surprise since the rate bias from Earth rotation alone amounted to only 3.5 Hz. The system showed an occasional beat note at frequencies between 6 and 22 kHz. These oscillations were accompanied by strong intensity variations, often in anti-phase at the same frequency on both laser beams, and were caused by strong mode competition. When the entire setup was placed on a rotary table, the ring cavity was made slightly larger, so that each side increased to 9 cm. When rotated, the two laser beams started to unlock at a rotational velocity of 0.1 rad/s. Figure 5.16 shows a comparison between the theoretically expected and the experimentally obtained relationship of the Sagnac beat note with respect to the experienced rate of rotation. The measurements were obtained without a dither applied to the gain medium, so evidence for backscatter coupling is clearly visible in the data. The measurements were carried out when mode competition was small. Mode competition effects in general reduced with higher rotation rates.

5.6.2 Design and Operation of a Solid State Gyro based on Er^{3+}-Yb^{3+}:phosphate Glass

We also investigated the properties of an Er^{3+} phosphate glass co-doped with Yb^{3+}. Erbium glass phosphates are promising because they exhibit a 24 times longer

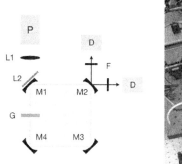 

Figure 5.17 Schematic illustration of the essential parts of a solid state laser gyro (left). P is the pump diode, and M1–M4 are the mirrors of the ring cavity. L1 and L2 are necessary to focus the pump beam into the crystal with minimum astigmatism. The beat note is taken directly from any of the two laser beams. The photo (right) gives an impression of the full setup on the turntable [165]. (Photo copyright N. Rabeendran)

lifetime of the electrons in the upper state, compared to Nd^{3+} phosphate glass. This leads to a much longer buildup time for the detrimental gain grating to form. The combination of erbium and ytterbium in a phosphate glass host material lowers the lasing threshold sufficiently to enable convenient room temperature operation [165]. Measurements were made with two sample crystals, of 1 mm and 1.5 mm length, and the concentrations of the dopants were 1.4×10^{21} ytterbium ions/cm^3 and 0.5×10^{21} erbium ions/cm^3. However, the absorption of the pump laser diode power inside the crystal proved to be difficult. Figure 5.17 shows the essential part of the resonator layout and the system itself during the test measurements on the rotary table. The dimensions were identical to the ones presented for the Nd^{3+} dopant gyro, discussed in the previous section. There was no beam combiner installed because we did not have a suitable IR camera system to overlap the output beams properly at a 1.534 μm wavelength. Due to the much lower gain of the shorter crystal of about 1.5%, compared to the neodymium doped crystal of 175%, and the considerable losses in the cavity of the order of 0.43%, it was difficult to obtain continuous operation. This also meant that rotation sensing could only be done at low SNR. The difference in operation wavelength between Figures 5.16 and 5.18 is responsible for the apparent reduced sensitivity of the erbium system. As the rotation rate induced by the turntable approaches zero, the measurements become progressively more unreliable due to lock-in. When we compare the overall performance of the Nd:YAG system [198], the Nd:phosphate and the Er-Yb:phosphate system, we find that none of these systems are serious alternatives to the HeNe gas laser gyros.

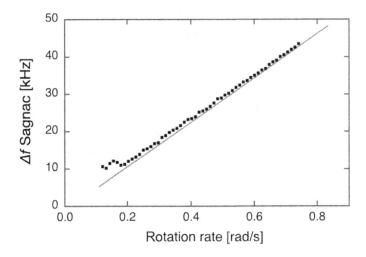

Figure 5.18 The example shows a linear dependence between the expected (solid line) and the experimentally obtained (squares) beat note. Systematic effects become dominant at rotation rates below 0.2 rad/s. Reported by [165].

Apart from the higher insertion loss of the gain crystal, there are the detrimental effects from the spatial hole burning and the mode competition between the two counter-propagating beams. The gain grating increases the backscatter coupling significantly, and the perturbations from the mode competition can be serious enough to render rotation sensing impossible. Vibrating the crystal with the gain medium at high frequencies to reduce the spatial hole burning and balancing the intensities of the two interferometer beams with an intra-cavity Faraday rotator are approximate remedies for effects that are entirely absent in the HeNe gas laser. For the two types of phosphate glass lasers introduced here, we have not applied any control of the mode competition, nor was the gain medium vibrated. As Figure 5.19 illustrates, we find that the deviation of the measured beat note from the expected value is rather large for the Nd:YAG system [198], amounting to about 50% for the case without vibration of the gain medium and a rotation rate of 0.5 rad/s. The erbium system exhibited an error of around 5%, and the neodymium phosphate glass produced an error of around 16%. With the vibration of the gain medium in operation, the bias error of the Nd:YAG system dropped to below 5%.

Is the solid state ring laser gyro a competitive alternative to a HeNe laser gyro, an FOG or an MEMS gyro? At the time of writing, it does not appear to make sense to investigate these alternatives any further. The electromechanical requirements to overcome the spatial hole burning and the mode competition are demanding and not necessary in the alternative concepts. Furthermore, both methods introduce additional losses to the gyro cavity, which reduces the ultimate sensitivity of the gyroscopes. The solid state gyros also need to be rather small in order to allow

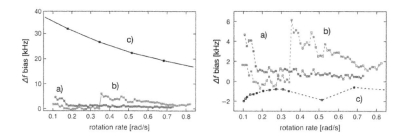

Figure 5.19 Offset between the measured and the computed Sagnac frequency of the Er^{3+}-Yb^{3+}:phosphate (a), the Nd^{3+}:phosphate (b) and the Nd^{3+}:YAG from [198] (c) for the non-dithered case on the left and the for the case when (c) was dithered on the right. The data is reproduced from [165].

efficient pumping. However, the FOG and MEMS technologies provide systems that are still significantly smaller than a ring gyro and deliver much higher sensitivity as well as far better long-term stability. Furthermore, they already represent fully developed market products.

6

Applications

Inertial sensors for acceleration and angular velocity provide the very general capability for tracking motion in real time. Navigation can be achieved by integrating the measured accelerations twice and the registered angular velocity once. In a very simple application, an accelerometer in a cell phone in the pocket of a user counts the steps in a health monitoring application. A more involved example is the motion control of robotic objects. Finally, submarine or deep space satellite navigation mark the demanding end for the application of these devices. The cost range of these devices is correspondingly large. While a cell phone accelerometer is available for a few dollars in the mass market, the cost for the high end devices may easily exceed millions of dollars per piece. This ratio of price is also reflected in the respective sensor performance, which differs by about five orders of magnitude as well, namely $0.009°/s$ for a MEMS gyro of a cell phone and $0.001°/h$ for a high performance inertial rotation sensing IMU based on a laser gyroscope.

Looking back over the last six decades, one can identify four major types of rotation sensors coming into fashion one by one. The first group comprises the mechanical gyros, where the conservation of momentum of a spinning mass was used to infer rotation rates. These were complicated mechanical devices, requiring significant maintenance. In order to be resilient against failure, structures like the dynamically tuned gyroscope (DTG) were designed [152]. The development of mechanical gyroscopes peaked in the 1960s.

The era of the mechanical gyroscope was followed by the development of ring laser gyroscopes, based on a traveling wave 'ring' cavity [142]. Most of the development for these devices happened in the 1970s. In order to reduce the cost, weight and size of IMUs, passive Sagnac gyroscopes like the fiber optic gyroscope (FOG) progressively substituted the ring lasers during the 1990s. After the turn of the millennium, MEMS structures started to take over the market for IMUs. The development was driven by the same demands as before, namely, cost, weight, size and reliability. Since the achieved specifications of all these newer devices are comparable and meet the requirements of the aviation industry, such an evolution is natural.

Inertial rotation sensors for the application in Earth observation do not follow this trend of the commercial world. They are tuned for maximum sensitivity and in particular for extremely high stability. Furthermore, there is no mass market for such devices. At the time of writing, rotation rates of $\delta\Omega_E \approx 0.6$ prad/s after 1000 s of integration have been achieved routinely, with a value of $\delta\Omega_E \approx 0.2$ prad/s after a full day of integration still marking the best achievement for a single axis 16 m^2 gyro, placed horizontally on the ground at mid-latitudes. For applications in fundamental physics, these gyros additionally require an extremely high level of accuracy of at least 1 part in 10^{10} for the measured rate of rotation, which has yet to be achieved. For Earth observation, the rotation sensors are strapped down to the body of the Earth. This means that a local sensor has to be sensitive and stable enough to observe small perturbations of a global measurement quantity, namely the variations in the rate of rotation of the Earth, while local effects acting on the scale factor or the orientation of the sensor have to be established and removed from the measurements. These perturbations include ground motion from seismic and microseismic activity, tidal deformations of the solid Earth and loading effects from ocean tides. In order to eventually detect tiny variations of the rotation rate of the Earth, the motion of the instantaneous rotation axis of the Earth as the result of the Chandler wobble and the diurnal polar motion signal have to be extracted. Some signals, like ground motion as a result of earthquakes, require update rates of up to 100 Hz. In contrast, geophysical signals, like the Chandler wobble, have periods of about 435 days ($\approx$25 nHz). For clarity, we distinguish between high frequency applications, which we discuss in Section 6.3: Rotational Seismology, and low frequency signal components, which are treated in Section 6.5. Finally, applications requiring ultra-high sensor stability are introduced in Sections 6.7 and 6.8.

6.1 General Considerations

The monolithic G ring laser on the Geodetic Observatory Wettzell is the most stable and most sensitive large ring laser structure at the time of writing in 2021. A typical Allan deviation (ADEV) sensor performance is shown in Figure 6.1. Theoretical models for the diurnal polar motion and the semi-diurnal Earth tides have been applied to reduce these effects from the observed instantaneous rate of rotation of the Earth, as have local tilt effects, obtained from the high resolution tiltmeter. One can see that the actual measurement of the gyroscope closely follows the expected

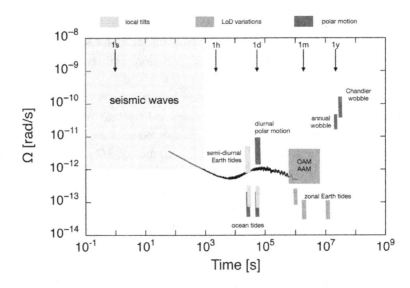

Figure 6.1 A typical Allan deviation of the G ring laser overlaid on the known important geophysical effects acting on the Earth. The diagram illustrates that most geophysical signals are already within reach of the sensor resolution, provided that the stability of the measurement series improves.

sensitivity slope for white noise but shows a distinct deviation from this trend toward periods around one day. This could have two obvious reasons. A small misalignment of the orientation due to an error of the estimated north–south tilt will cause an incomplete reduction of the diurnal and semi-diurnal geophysical signals but should generate a spectrally distinct pattern. The other potentially more diffuse and hence more probable signal contribution could come from local deformations as a result of variable atmospheric loading effects, which would then raise the question why this is not fully captured in the tiltmeter correction. Since the gyro is isolated from the ambient atmosphere by a pressure stabilizing vessel, instrumental effects with periods around one day are unlikely. In terms of stable gyro operation, this is the current bottom end sensor performance of the G ring laser over periods of 100 days. Relativistic terms, such as Lense–Thirring frame dragging and de Sitter geodetic precession [221] become detectable when the ring laser can resolve signals smaller than 10^{-14} rad/s. These relativistic effects, however, cause a tiny constant offset value, and unlike the mostly periodic geophysical signal, they require the exact scale factor of the Sagnac interferometer to be precisely known. Furthermore, the orientation of the sensor or the sensor array has to be established exactly ($\delta\varphi < 1$ nrad), and there must not be any unknown non-reciprocal optical frequency bias ($\delta f/f < 10^{-22}$) in the interferometer.

6.2 Microwave Frequencies Generated from a Large Ring Cavity

The cavity of the G ring laser has a Q factor of 3.5×10^{12}, and since it is monolithically made from Zerodur, it is interesting to explore the properties of a 16 m cavity, when it is utilized as a stable resonator for optical frequencies. In Section 3.6.5 we have compared the frequency stability of the laser cavity against the frequency comb and found a relative stability between the H maser, transferred into the optical regime by a Menlo FC1500-250 fs-frequency comb, and the gyro cavity in the range of several hundred Hz. In order to investigate the performance of a stable microwave frequency, generated from the optical frequency of the ring resonator, we have utilized the experimental setup shown in Figure 6.2. A HeNe transfer laser was locked to the ring laser cavity via a PLL circuit, utilizing a free space beat unit in the ring laser laboratory. Between the control room, where the transfer laser is located, and the ring laser laboratory, we used a compensated fiber link to ensure unperturbed signal quality. Since the available beam power from the ring laser is only in the regime of 20 nW, it was necessary to evaluate the stability of this part of the setup. The relative stability of this lock started at 10^{-17} after 1 s of integration and reached 10^{-20} with 1000 seconds of averaging. This ensured that this part of the experiment would not limit the quality of the microwave signal generation. Another portion of the transfer laser output frequency was in turn used to lock the 250 MHz repetition rate of the fs-frequency comb to the transfer laser. We then used the 40th harmonic of the repetition rate frequency at 10 GHz to compare the ring laser cavity-conditioned microwave frequency to an ULISS cryogenic sapphire oscillator [76], which has been verified to operate at a stability level $\delta f / f$ of 1 part in 10^{-15}.

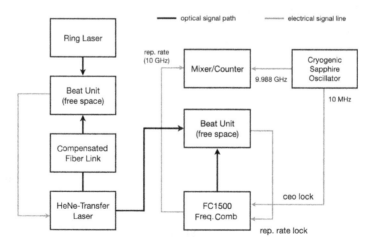

Figure 6.2 Block diagram of arrangement that utilizes a pressure stabilized ring cavity to generate a microwave frequency from an optical beam.

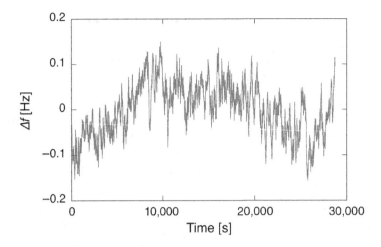

Figure 6.3 Nearly nine hours of the optical frequency of the G ring laser, down-converted to a 10 GHz microwave signal and compared to a cryogenic sapphire oscillator as a reference. A constant slope of -375 µHz/day has been removed.

The result is displayed in Figure 6.3. Over the nearly nine hours of the experiment, we observed a constant drift of the beat note of -375 µHz/day in this comparison. The ring laser cavity was slowly expanding as a result of the annual temperature cycle of about $\pm 0.4°$ in the laboratory. The measured relative frequency stability is 2×10^{-13} at 1 s with this drift removed and diverges for higher integration times in the fashion of a random process, evident by a $\sqrt{t}$ slope in the Allan deviation analysis. Optical reference cavities in high precision metrology are usually much smaller and much better isolated, both from temperature variations and mechanical vibrations. Furthermore, they are evacuated and used by an external laser coupled with the Pound Drever Hall (PDH) locking technique. In contrast we have the active laser gain medium inside the reference cavity. The obtained stability for a 16 m long traveling wave cavity is quite respectable. This includes a residual gas pressure of 10 mB around the entire cavity as well as a 3 cm long plasma segment for laser beam excitation. Although it does not seem to impact the result, there is also the second counter-propagating laser beam present in the cavity. The gain in stability from the down-conversion of the optical frequency at 473.612683465664 THz to 10 GHz gives almost a factor of 500, meaning that the optical frequency of the laser beam dithers by only 100 Hz, which corresponds to the expected width of the cavity resonance.

6.3 Rotational Seismology

The Earth has quite a considerable inertia, and mass transport phenomena inside the Earth only affect a very small proportion of the entire mass at timescales between hours and several days. Therefore, Earth rotation is perceived as a very constant motion. So it was very natural to initially use Earth rotation rate – that is, the length of the day – for the definition of the duration of the second. With the advent of highly accurate atomic frequency standards operating in a stability regime surpassing astronomical observations, the irregularities in Earth rotation could no longer be ignored. The largest of the perturbations with a period as short as one day, the diurnal polar motion, becomes visible when a rotation rate of approximately 6 prad/s can be resolved with an inertial sensing gyro. Unpredictable signals, like the variation of the length of day, which are caused by mass transport phenomena of the fluids of the Earth, are still three orders of magnitude smaller than the diurnal polar motion and present a significant challenge for a quantitative detection by inertial sensors.

Ground motion induced by seismic activities causes systematic movements to the foundation of the gyro, thus adding a local rotational component as a perturbation on top of the Earth rotation rate. The latter can be considered as constant in this time interval of interest. Since highly sensitive ring laser gyroscopes are not small, it is therefore important to attach them rigidly to the ground and to construct the platform of the gyro in such a way that it does not deform under these excitations. Large ring lasers populate the weak signal regime of seismic motion. This is illustrated in Figure 6.4. Thus, they are suitable for the detection of teleseismic signals. Since these gyroscopes have to be built as large and rigid structures on a stiff platform, they are not deformed or otherwise compromised by the source of the observation signal. The strong motion regime, on the other hand, requires an entirely different sensor type. Fiber optic gyros or MEMS structures are small, lightweight and rigid devices.

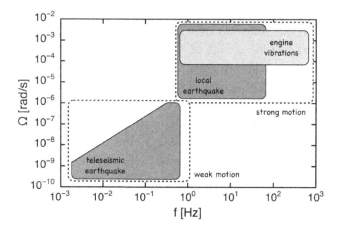

Figure 6.4 The relationship between spectral content and respective amplitudes of rotational signals is illustrated schematically over a wide dynamic range. No instrument can cover the entire dynamic range, which requires specialized sensor designs for particular purposes.

They can be strapped to the structures of interest and cope with large rotation rates. Furthermore they usually sample at much higher data rates, owing to their heritage from the field of guidance and navigation.

Rotational signals in seismology exhibit a large variety of different signatures. While earthquake signals show up like transients with a large range of amplitudes, microseismic activity is constantly observable, albeit at a very low signal level and with variable intensity. It is excited by the activities of wind and ocean waves, which are forcing these ground vibrations. Finally, effects like static rotations as a consequence of large earthquakes require superior sensor stability and resolution in order to be observed, beyond current capabilities. In the extreme case of a highly sensitive, accurate and long-term stable gyro, these inertial sensing instruments would eventually also reveal rotations from tectonic plate motion as well as relativistic effects like the Lense–Thirring frame dragging [34, 221] but are yet to be built. For the next section we limit the discussion to earthquakes and the microseismic activity.

6.3.1 Rotational Signals from Earthquakes

The rotational ground motion caused by Love waves from not too distant earthquakes are the first candidate signals that a moderate sized single component ring laser gyro strapped down to the Earth crust will detect, apart, of course, from the nearly constant rate bias created by Earth rotation itself. Our large ring lasers detected teleseismic signals from remote earthquakes from early on in their development. The first observation of Love wave signatures was already obtained with the Canterbury ring laser (C-I) in 1994 [209] from a set of regional earthquakes less than 100 km away from the gyro. The first quantitative studies on true teleseismic signals followed soon after that, with earthquake records of C-II [145, 157]. The monolithic design of C-II allowed nearly continuous operation, which was novel at that time for such a large gyro. Since C-II was located in Christchurch, it was at a convenient distance to the many earthquakes around the western side of the Ring of Fire around the Pacific ocean. The observed signals, corresponding to an oscillation around the vertical axis, were not small compared to the resolution of the gyroscope. The M7.3 Vanuatu earthquake from February 6, 1999, for example, generated Love wave excursions at rates of the order of up to 8 nrad/s, which could be detected with good fidelity on the C-II ring laser at an angular distance of 31°, corresponding to about 3450 km of separation. Since its installation in 2001, the G ring laser records large numbers of distant earthquakes on a routine basis.

Today, with the ROMY ring laser operational, rotations around all of the three spatial directions are observable. Since each component of ROMY is much larger than C-II, the ring laser is also much more sensitive, and we can now

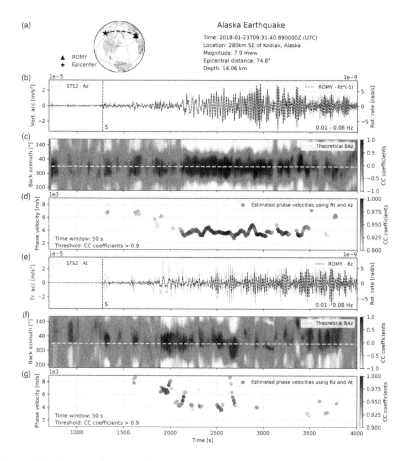

Figure 6.5 Observed translational and rotational motions for the M7.9 Alaska earthquake, January 23, 2018. (a) Earthquake details. (b) Superposition of the vertical acceleration (Az) and the transverse rotational rate (Rt). (c) Estimated back azimuth (BAz) for each 50 s sliding time window. The intensity scale indicates the cross correlation coefficient (CC). (d) Estimated phase velocities. (e)–(g) The same as (b)–(d), respectively, but for transverse acceleration (At) and vertical rotational rate (Rz), which focus on Love-type waves. (Diagram courtesy of H. Igel)

quantitatively observe the rotations around the two orthogonal horizontal axes as well, which additionally provides access to Rayleigh surface waves from a distant earthquake. As an example we present the joint analysis of an M7.9 earthquake, which took place on January 23, 2018 in a region 280 km south-east of Kodiak in Alaska. The epicentral distance was 74.8°, corresponding to approximately 8320 km along the surface. Translations were recorded by an STS-2 broadband seismometer, operated in the vicinity of the ROMY structure. In Figure 6.5b–d we compare the vertical component of the bandpass filtered (0.01–0.1 Hz) ground acceleration Az with the transverse component of the ground rotation rate Rt. The latter is multiplied by -1 in order to properly match the phase. The traces of the acceleration (gray line) and the rotation rate (dashed line) are presented on top of each other, indicating a nearly perfect match for most parts of the trace. The arrival time of the S phase is marked on the plot (Figure 6.5b). This means that the wave fronts are close to planar and that body wave $P - SV$ motions and Rayleigh wave motions are correlated. This is expected from simple plane wave theory [135]. The match of the amplitudes is achieved by scaling with an apparent horizontal phase velocity with values of ≈ 3 km/s.

In order to further examine the teleseismic ground motions, we calculate the correlation coefficient between vertical acceleration and rotation rate as a function of corresponding back azimuth in a sliding time window of 50 s duration. The maximum correlation is denoted by markers, which scatter around the theoretical straight line back azimuth, indicated by the dashed line in Figure 6.5c. Apparent phase velocities are estimated from the ratio of the rotation rate and the vertical acceleration, when the correlation coefficient exceeds a value of 0.95. The result is presented in Figure 6.5d. The high phase velocities around the SS arrival are caused by the steep incidence angles of $P - SV$ body-wave phases. The highest correlation (intensity coded) is observed in the time window containing the Rayleigh wave energy with phase velocities between 2.5 and 4.5 km/s. The lower part of Figure 6.5 in sections e–g contains the corresponding analysis of the vertical component of rotational motions Rz and the associated transverse acceleration At. This method has already been used for quite some time for earthquake observations from the G ring laser [100]. In this case, the ring laser is sensitive to SH-type motions only. Prior to the S arrival, there is substantially less energy in the vertical rotation than in the horizontal component (discussed above), as expected for a predominantly spherically symmetric Earth. The back azimuth estimation is stable almost along the entire seismogram. The indication of surface wave dispersion in the window containing the Love waves is more apparent than for the Rayleigh wave case.

The motivation to study rotational motion in seismology is driven by several deficits and open questions. First of all one has to appreciate that standard seismological observations are potentially contaminated by rotations through tilt effects. Obvious examples are strong motion events. Sensor tilts and horizontal accelerations create the same type of measurement signal, and one cannot be discriminated against the other without reliable rotation sensing [100]. Ring laser gyroscopes have a great advantage here because they are entirely insensitive to translations and only record rotations. The combined integration of displacement and rotation in all six degrees of freedom may capture the true motion of the instantaneous sensor position during an earthquake [102, 180, 205]. Rotations are considered to be drivers in co-seismic structural damage, in particular when the center of stiffness of a structure does not correspond to the center of rotation. Using rotational measurements as an independent additional data source may provide relevant additional wavefield information, such as phase velocities, propagation directions and anisotropy. Furthermore, the inclusion of rotations in the analysis may further constrain rupture processes. A specific project on rotational seismology, funded by the German Ministry of Education and Research (BMBF) within the GEO-technology program, made the construction of a large ring laser for seismological studies possible. ROMY essentially grew out of this precursor experiment. In 2006 a workshop initiated by the scientists of the United States Geological Survey (USGS) led to the formation of the International Working Group on Rotational Seismology (IWGoRS, http://www.rotational-seismology.org). A more detailed report on the application of ring lasers for rotational seismology is given in [43, 99, 187, 188, 189]. With improving sensor stability and more and larger ring lasers on better monuments, rotational seismograms of many more earthquakes have been observed to date. The construction of the ROMY ring laser finally added the observation of tilt effects associated with Rayleigh waves to the list of available measurement quantities.

6.3.2 Comparison with Array Measurements and Determination of Phase Velocity

The study of seismic rotational signals over the entire range of interest requires a sensor capable of detecting angular velocities in the range $10^{-14} \leq \Omega_s \leq 1$ rad/s, with a required frequency bandwidth for the seismic waves in the range of 1 mHz $\leq f_s \leq 50$ Hz [186]. Currently, large ring lasers are the only rotation sensors available that meet a pair amount of these demands, in particular for weak signal amplitudes. Four such devices mounted in the shape of a tetrahedron are contained in the ROMY underground ring laser structure located at the Geophysical Observatory Fürstenfeldbruck, approximately 20 km west of Munich, Germany. The observation of the individual rings must be transformed to a local North, East, Up system (see Section 3.12.1) and provide the quantitative detection of rotations from shear, Love and Rayleigh waves. Errors in the orientation of each ring laser component relative to all others have to be taken into account as misalignments during this transformation. From this point of view, inertial rotation sensors provide additional information and can also be very useful for the separation of tilt and horizontal acceleration, a persistent problem in seismology and of particular importance for seismic measurements on the ocean floor. With the availability of large ring lasers with sensitivity better than 10^{-11}rad/s/$\sqrt{\text{Hz}}$, rigidly attached to the Earth in a 'strapdown' approach, the quantitative detection of rotations from seismic signals became feasible. Since rotations can be expressed as the curl of the wavefield (Eq. 6.1),

$$\begin{pmatrix} \omega_x \\ \omega_y \\ \omega_z \end{pmatrix} = \frac{1}{2} \nabla \times \mathbf{v} = \frac{1}{2} \begin{pmatrix} \partial_y v_z - \partial_z v_y \\ \partial_z v_x - \partial_x v_z \\ \partial_x v_y - \partial_y v_x \end{pmatrix}, \tag{6.1}$$

the rotational signal component obtained from each spacial axis then corresponds to

$$\omega_x = \frac{1}{2}(\partial_y v_z - \partial_z v_y),$$

$$\omega_y = \frac{1}{2}(\partial_z v_x - \partial_x v_z), \tag{6.2}$$

$$\omega_z = \frac{1}{2}(\partial_x v_y - \partial_y v_x),$$

which can be approximated by measurements from seismometer arrays by a finite-differencing approach. One of the early verification tests for the quantitative study of ring laser-established rotational ground motion was the comparison between the array-derived and the ring laser-derived rotations from the observation of earthquake signals. In a dedicated experiment, a comparison between the G-ring and an array of nine seismometers, arranged around the ring laser site, provided collocated measurements that were remarkably consistent both in waveform and amplitude [213]. With complete 3D synthetic seismograms, calculated for the M 6.3 Al Hoceima, Morocco, earthquake of February 24, 2004, it was shown that even low levels of noise may considerably influence the accuracy of the array-derived rotations when the minimum number of required stations (three) is used. However, with all nine available stations the overall fit between direct and array-derived measurements was surprisingly good. In a similar way it is desirable to relate the measured rotations to the acceleration of a plane wave [99]. With the displacement $u_y(x,t) = f(kx - \omega t)$ and the phase velocity defined as $c = \omega/k$, the acceleration can be written as $a_y(x,t) = \ddot{u}_y(x,t) = \omega^2 f''(kx - \omega t)$. For the component of rotation around the vertical, the curl of the wavefield (Eqs. 6.1 and 6.3) reduces to

$$\dot{\Omega}(x,t) = \frac{1}{2} \nabla \times [0, \dot{u}_y, 0] = [0, 0, -\frac{1}{2} k \omega f''(kx - \omega t)]. \tag{6.3}$$

Comparing the acceleration with the curl of the wavefield around the vertical, one obtains

$$a(x,t)/\dot{\Omega}(x,t) = -2c, \tag{6.4}$$

which means that the phase velocity can be established reliably from a point measurement from a seismometer and a ring laser. The application of this relationship also produced the phase velocities already discussed in Figure 6.5d and g. From plane-wave propagation, it therefore follows that the rotation rate and the transverse acceleration are proportional and have the same phase relationship.

Although the discussion above was applied to horizontally mounted ring lasers, sensitive to rotations around their vertical axis, similar results were established for the vertically orientated G0 ring laser [157] and in more detail also for G-Pisa [19] before the ROMY structure became available and impressively confirmed the earlier results [58]. Furthermore, it was also shown that strain measurements from the PFO long baseline strainmeters produce consistent results in comparison with the array derivation method [58]. These strainmeters are realized as unequal arm Michelson HeNe interferometers, where each measurement arm is 730 m long, and the reference arm less than 0.5 m. The phase velocities can be estimated in practice through best-fitting waveform adjustments by sliding a time-window of appropriate length along the seismic signal. It was shown without exception for many earthquakes that the ring laser technology provides the required resolution for consistent broadband observations of rotational motions induced by earthquakes. Furthermore, it is important to point out that optical interferometer gyroscopes, through the absence of any movable mechanical part, have a completely linear transfer function with a value of unity. For measurements at the Geodetic Observatory Wettzell, it was possible to estimate Love wave dispersion curves with good signal-to-noise ratio, using the amplitude ratio between the transverse acceleration and the rotation rate about the vertical [123]. Since then, the ROMY array has provided more general results, which also include rotations around the horizontal axes.

6.3.3 Detection of Toroidal Free Oscillations of the Earth

Giant earthquakes radiate mechanical energy into and around the Earth, leading to worldwide oscillatory ground displacements, sometimes in the centimeter range. Due to the high Q-factor of the Earth's deep interior, they are usually observable for several days, occasionally even weeks after the event [158]. The resulting global standing wave patterns that form after the constructive interference of the seismic wave field – the Earth's free oscillations (or normal modes) – are characterized by discrete frequencies that depend primarily on the motion type (toroidal or spheroidal modes) and the structure of our planet. This implies that observations of free oscillations provide some of the most important large-scale constraints on a variety of elastic parameters, attenuation and density of the Earth's deep interior. In addition, low frequency normal modes allow us to put strong constraints on the energy, geometry and duration of large earthquake sources. Widmer-Schnidrig and Zürn [235] stated that the study of free oscillations of the Earth is one of the areas in which measurements of rotations could be extremely helpful. Successful gyroscopic rotation measurements were reported for the 2011 M 9.0 Tohoku-Oki earthquake [153]. Due to the horizontal orientation of the G-ring, the gyroscope is only sensitive to horizontally polarized shear waves, i.e., SH-type waves, which correspond to the

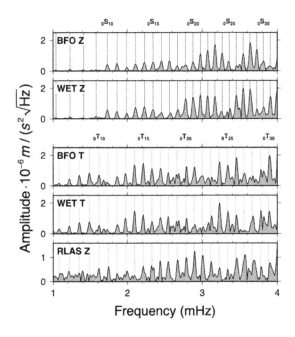

Figure 6.6 Observed power spectral density of a 36-h time window following the Tohoku earthquake on March 11, 2011.The upper two spectra present the vertical component from broadband seismometers, located in the Black Forest Observatory and in Wettzell. The next two spectra show the transverse component and the bottom spectrum is taken from the G ring laser (multiplied by a factor of 10,000). (Reprinted with permission from [153] © (2022) by Springer Nature.)

superposition of toroidal modes. In terms of seismic surface waves, this means that this ring laser can record Love waves but not Rayleigh waves. Spheroidal modes do not cause any rotations around a vertical axis. In Figure 6.6 we show a comparison of the measured spheroidal and toroidal oscillations of the Earth, taken by two seismometers spaced 350 km apart, namely at the Black Forest Observatory in Schiltach and the Geodetic Observatory Wettzell, and compare them to the G ring laser. The vertical lines in the diagram correspond to the expected frequencies of the respective eigenmodes, derived from an Earth model. A good agreement between model and measurement is always achieved, not only in this example but also for all other observed cases [101, 153].

6.3.4 Rotational Signals in the P-Coda

In Section 6.3.2 it was shown that the combined observations from a seismometer and the G-ring allowed the estimation of wave field properties, such as the phase velocity of the seismic signal and the direction of propagation. In a spherically symmetric (i.e., layered) Earth model, we do not expect rotational motions around

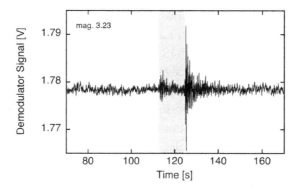

Figure 6.7 Rotation rate around the vertical axis for a local M3.23 earthquake. Assuming a layered medium, there should be no rotational signal in the P-coda, indicated by the gray box. The energy in the ring laser rotational signal is caused by scattering from *p*-waves to *s*-waves.

a vertical axis in an earthquake-induced wave field before the onset of shear waves. Nevertheless, in all earthquake observations, substantial energy is seen almost immediately following the onset of the compressional *P*-wave. This can be explained by the fact that the Earth's subsurface has three-dimensional stuctural heterogeneities and thus scatters waves from *P*- to *S*-type motions. A previous study [159] has shown that *P*-wave-induced tilt – ring laser coupling can be neglected for tele-seismic events because they are too small. An arbitrarily chosen example (Figure 6.7) from an observed small local earthquake by the GEOsensor at the Piñon Flat Observatory illustrates this observation, which is readily obtained from any of our ring lasers, both for local and distant earthquakes. Modeling the data as $P - SH$ scattering and comparing the results to theoretical simulations substantiates this expectation. The observations indicate that the energies of the *P*-coda rotations are predominantly present at high frequencies. They come from many different directions and increase over time more slowly than the energy in the vertical component of translation. Under the model assumption of a random crustal medium, rotational signals in the *P*- coda can be generated efficiently.

6.3.5 Rotations in Microseismic Noise

The excitation of ocean waves from wind activity gives rise to a pressure modulation on the sea floor, which in turn generates a small global seismic noise signal and can be represented in the form of Rayleigh and Love waves. There are two distinct frequency bands, namely the primary microseismic band in the range of 0.05–0.07 Hz and the secondary microseismic band in the range of 0.1–0.4 Hz. The horizontally orientated G ring laser structure, measuring around the vertical axis only, is exclusively sensitive to the Love wave signal. The tiny periodic uplift of the Rayleigh waves does not introduce tilt effects large enough to be detectable. Since the ocean-generated microseismic noise, detected by G, depends on large scale weather patterns on the Atlantic Ocean, the amplitude of the detected rotational signal varies greatly over time. Figure 6.8 illustrates this observation of Love wave signals. It is a common observation that the microseismic activity is greater in winter than in summer. During a calm period at the end of November in 2014, a consistent low noise level was recorded between days 330 and 340. When a storm built up over the Atlantic Ocean around day 344 and then again around day 353, the microseismic background noise level rose by a factor of 4–5. Several years of ring laser observations have shown that this is quite a common behavior [220]. The noise level is usually calculated in the following way. The 2 kHz data stream from the digitizer is binned into segments of 1 min duration. From each segment a frequency estimate of the Sagnac frequency (Earth rate) is established by the "single tone" frequency estimator, as explained in Section 4.3. Every hour the mean value and the standard deviation over the respective 60 data bins is estimated. Since the Earth rotation rate can be considered constant over each time window and geophysical signals, and instrumental perturbations determine the value of the standard deviation, this quantity is a good measure of the instrumental performance and the instantaneous geophysical conditions. Teleseismic rotational signals from earthquakes also show up and enlarge the standard deviation, as the Earth rotation rate is modulated by the ground motion. However, these events are only brief. An example for an earthquake signature appears in the diagram shown in Figure 6.8 on the left side of the diagram at day 330. A more specific classification of the noise sources becomes available when the spectral content of the measurements are taken into account. Figure 6.9 shows a typical result. The spectra on the left side were taken by a broadband STS-2 seismometer, located in the vicinity of the G ring laser in Wettzell, and show the translational accelerations caused by the microseismic activity, while the right-hand diagram presents the spectral power content of the Love waves in the sidebands of the ring laser record more than 80 dB below the main spectral peak, corresponding to the Earth rate. There is a clear correspondence, both for the secondary microseismic band (b) between 0.1 and 0.4 Hz and for the approximately 20 dB weaker primary microseismic band (a) between 0.05 and 0.1 Hz. The detection of the Love wave content in the primary microseismic band currently marks the sensitivity limit of the G ring laser as set by the Q-factor of the cavity. A reduction in thermal coating and substrate noise (Section 3.10.3) is required to lower the detection threshold further. Although not evident from Figure 6.8, the detection limit of G is still defined by the instrument, not yet by the microseismic background noise level. However, it requires a very calm period of low background noise to reach that sensor limit, which rarely occurs during winter.

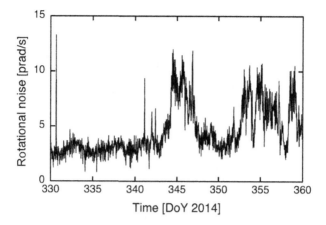

Figure 6.8 Microseismic Love wave activity in the frequency band around 0.1 Hz is variable over time in amplitude. While the instrumental detection limit of G is at 1 prad/s, the lowest obtained level of microseismic induced rotational dither is around 2 prad/s in a calm period. During storms in the Atlantic, the signal level may reach and even exceed 10 prad/s.

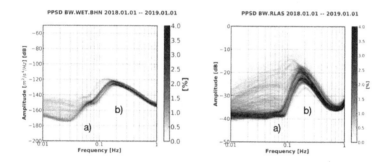

Figure 6.9 The spectral distribution of the recorded microseismic activity for translations (left) obtained from a broadband seismometer, and the rotational Love wave component derived from the G ring laser (right). The primary microseismic band is indicated by (a) and the secondary microseismic band by (b). (Plots by courtesy of C. Hadziionannou.)

Another source of spurious rotational motion on a large ring laser gyroscope is given by local wind: wind loading on the mound of the G ring laser, as well as wind friction exacerbated by torsional forces exerted on local trees, nearby ground obstacles and the air ground interface. Soil covering vegetation under wind loading causes broadband rotational dither, which can make the observations noisy. The pattern of the signals change rapidly in amplitude and in phase in accordance with the appearance of wind gusts outside the laboratory. Therefore, the induced frequencies are significantly higher than the ocean-induced microseismic Love waves. Modeling the effect of wind with a finite element analysis approach [73] reveals

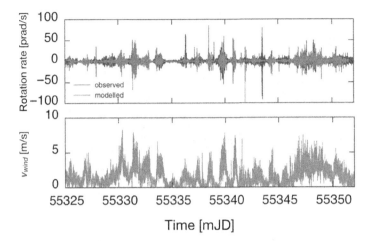

Figure 6.10 Comparison of the high pass filtered rotation rates from the G ring laser with the calculated wind induced perturbations based on a local terrain model. Due to the rather inaccurately measured wind parameters (direction and velocity), only a moderate agreement is obtained. (Adapted with permissions from [73] ©2022 Springer Nature)

that the perturbations from wind load and friction are very local effects. The magnitude of the wind-induced rotations are of the order of 15 prad/s or below, which makes this effect comparable to the ocean-induced microseismic signal. Due to the small size of the wind gust cells, the induced rotational signal excitations quickly become incoherent with increasing distance from the gyro and approach a mean value of about zero quickly, except for some occasional outliers. Figure 6.10 recalls an example for the effect of wind on the ring laser laboratory from the discussion in Section 4.9. While wind increases the noise level of the beat note of the Sagnac interferogram in a characteristic way, no indication was found that wind loading could create a coherent local torsional signal over a larger area or time. Modeled tilts from wind friction turned out to be much smaller than 1 nrad and are too small to perturb a horizontally placed ring laser such as G.

The detection of Love waves with a ring laser gyro is omni-directional. In equation 6.3 we have seen that the acceleration measured by a seismometer and the rotation rate obtained from a gyro have the same signal shape. In contrast, the polarization for the accelerations in the two directions of x and y obtained from the seismometer are not omni-directional. By rotating the frame of reference of the seismometer according to

$$\begin{pmatrix} \tilde{x} \\ \tilde{y} \end{pmatrix} = \begin{pmatrix} \cos\phi & -\sin\phi \\ \sin\phi & \cos\phi \end{pmatrix} \begin{pmatrix} x \\ y \end{pmatrix}, \tag{6.5}$$

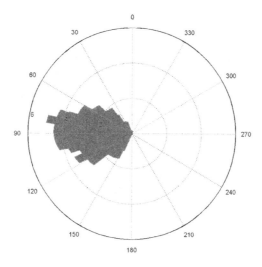

Figure 6.11 The correlation of transverse acceleration from a seismometer and the rotation rate of a gyro allows the direction from a single point of measurement to the signal source to be established. In this example the center of activity of the storm Xynthia (2010) was located west of the ring laser site in Wettzell.

we can obtain the direction between the origin of the signal and our single point of measurement, by finding the maximum correlation between the two types of sensors as a function of the angle ϕ. This concept allows us to track the activity center of a storm. The storm Xynthia, moving along the Atlantic coast of Europe between February 27 and 28 in 2010, provides a good example. Figure 6.11 illustrates the result. The continuously acquired observation data from the ring laser and the seismometer was binned in segments. By rotating the frame of reference of the seismometer from $0° \leq \phi \leq 360°$, the ring laser and seismometer signals were correlated. The repeatedly established angles for maximum correlation then were accumulated in the polar plot of Figure 6.11. For the data shown, the center of activity of the storm was west of Wettzell. Corresponding measurements could be obtained from the evaluation of finite differences between a number of seismometers of an array. This correlated type of measurement has a particular importance for measurement applications where the placement of arrays of sensors is not an option, such as on the ocean floor or in planetary seismology.

6.3.6 Seismic Rotation Sensing at Volcanoes and the Ocean Floor

Seismic activities happen over a very wide dynamic range of more than six orders of magnitude, and measurement scenarios differ from close proximity to long distance between source and sensor. While ring laser gyros today are certainly the instrument of choice for the detection of rotations from teleseismic events and the seismically excited vibrational toroidal eigenmodes of the Earth, fiber optic gyros, on the other hand, perform well in strong motion conditions close to the source. There we find that the correct capture of tilt is one of the most important and at the same time most demanding requirements, since the sensor can be kicked around considerably, changing the orientation of the measurement device constantly and significantly. Under such conditions some percentage of the accelerations in the x, y and z directions is unavoidably mapped into the other channels, and the retrieval of the true ground motion is no longer possible. The additional measurement of the three degrees of freedom of rotations, however, provides this lacking information. Apart from the tilt effects under strong motion conditions, we may also find saturation of the sensor as a complication. Both seismometers and large ring lasers applied to the geosciences are usually operated as open loop sensors in order to have sufficient gain under low signal conditions. Closed loop sensors, such as an FOG from an inertial measurement unit in an aircraft, in contrast, have a much wider dynamic range, however at the price of a very limited resolution for weak signals. The blueSeis-3A is a commercial closed loop rotation sensor based on FOG technology, for the application in geophysics. The self-noise is typically 20 nrad/s/$\sqrt{\text{Hz}}$ over the entire frequency range of 0.01 Hz $\leq f \leq$ 50 Hz [104]. A six degree of freedom measurement of a volcano-induced earthquake has been recorded from the May to August 2018 eruption of the Kilauea volcano in Hawaii [230] with a blueSeis-3A rotation sensor and co-located seismometer (STS-2). In earlier publications, such as [100], we have demonstrated how the measurements of three components of translation augmented with three components of rotation can help to improve the inversion for local (S-wave) velocity, to localize events and to estimate source mechanisms even under the condition of sparse instrumentation and possibly tilt effects. FOGs, like ring laser devices, exploit the properties of light. Hence, they are entirely insensitive to translational motion. This property has the effect of acting like a physical S-wave polarizer, since it only senses the curl of the wave field, namely the SH-type waves on the vertical component and the P-and SV-type waves on the horizontal components. The measurements presented in [230] also confirm the effect of static rotation around the vertical axis, caused by the earthquake activity. The measurement of the additional components of rotations made the interpretation of the effects of this local earthquake induced by a collapsing caldera more complex than expected. It also provided indications of measurement platform dynamics, which would not have shown up on pure translational seismograms. So, while the picture of the observations turned out more difficult than expected, it also revealed potential issues, which otherwise would have gone unnoticed.

An ocean bottom seismometer (OBS) significantly benefits from the addition of an inertial rotation sensor [136]. There can be considerable tilt effects induced on the OBS platform by the ocean surf, causing rotations around the x-axis as well as translations along the y-axis and vice versa. Ocean floor seismometers have extended the application of seismometer applications into the offshore regime. However, since the instrumentation is surrounded by water in motion, interacting with the OBS platform, the horizontal components have a much lower data quality than the vertical component [62, 231]. Surf-induced sensor rotation around the horizontal axis induces a sensitivity to the gravitational acceleration of the Earth, which is not desired. Building on the experience with large ring lasers of successfully measuring seismically induced rotations around the vertical axis [85, 100, 145, 157], it was a natural step to apply rotation sensing for the measurement of OBS platform rotation. The Northrop Grumman LITEF LCG-Demonstrator employs a three component assembly of fiber optic gyroscopes and is suitable to be operated as an inertial measurement unit, strapped to an OBS platform; it resolves the rate of rotation with a sensitivity of $\delta\omega = 200 \times 10^{-9}$ rad/s at a sampling rate of 200 Hz. In this evaluation experiment, the platform was lowered to a depth of 30 m into the North Sea, not far from the island of Heligoland. The experiment was restricted to four days, mainly because of the high power consumption of the FOGs. Figure 6.12a illustrates the setup of the OBS platform. The LCG-Demonstrator and the battery pack are located in the smaller tubes on the right side of the payload, while the seismometer and additional instrumentation were placed near the larger floatation cylinders. The sensitivity of seismometers to tilt is a known issue [80]. The response x of the horizontal x-component of a seismometer can be expressed by

$$\ddot{x} + 2\omega_x D_x \dot{x} + \omega_x^2 x = -\ddot{u}_x + g\Omega_y, \tag{6.6}$$

where y is the horizontal y-component, ω_x and D_x are the natural frequency and fraction of critical damping on the x-oscillator and g is the gravitational

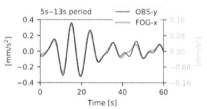

Figure 6.12 Setup of the OBS platform prior to deployment (left). The LCG-Demonstrator and the battery pack are located in the shiny tubes on the right side. A comparison of the obtained rotation rates, converted to horizontal acceleration, and the measured acceleration along the x-axis show almost perfect agreement, indicating the strong tilt contamination of the seismometer signal (right). (Photo and diagram courtesy of H. Igel)

acceleration of the Earth. Equation 6.6 expresses that the x-oscillator is sensitive to the translational acceleration $\ddot{u}_x$ along the x-axis and the rotation Ω_y around the y-axis. The same applies to the y-oscillator with the indices exchanged. This means that the horizontal component of a seismometer consists of a superposition of linear acceleration and angular displacement multiplied by g. Converting the seismometer observations in the frequency band of highest activity between 0.07 and 0.2 Hz to acceleration and the gyro measurements to angle, a sliding window comparison shows a high correlation of 0.77 on average, which increases to 0.94 for periods of higher amplitude. Tilt-induced acceleration contaminating the translational observations reaches amplitudes of 10^{-5}–10^{-4} m/s^2 over the period of the measurement. This translates to 10^{-6}–10^{-5} rad and rad/s in terms of rotation and rotation rate, indicating that OBS observations can be improved by the measurement of the dynamic platform tilt behavior. The diagram of Figure 6.12 illustrates the excellent agreement between acceleration and rotation from a short sequence from the observations. This is the first direct evidence that signals on horizontal seismometer components on the sea floor (indicating horizontal displacements) are almost entirely due to tilting.

6.4 FOGs on Civil Engineering Structures

The applications introduced in this section concentrate on some specific aspects, unique to inertial rotation sensors. There are more potential applications, such as the vibrational frequency analysis of complex structures. However, these are not unique to gyros and can be facilitated by other sensors in a more general and cheaper way.

The effects of rotations have been neglected in studies on the seismic properties of civil engineering structures in the past. This was mainly because their influence was thought to be small, and there were no suitable sensors available to properly measure the system response of buildings to rotations. Only the effects of torsions caused by asymmetries in buildings, where the center of stiffness differs from the center of mass, are known from differential measurements of accelerometers. However, since the inferred rotations are highly dependent on the particular positions of the respective accelerometers, these measurements are tedious and only of limited value. Navigation grade fiber optic gyros with a drift rate of 1–6°/h are compact and small and therefore convenient to use in a strapped-down application to estimate the rotational motion of a building structure during an earthquake. This approach has several advantages, which are:

- FOGs do not require a reference against which the angular excursion is measured, since they measure absolute rotations.

- Regardless of their location of deployment relative to the center of rotation, they measure rotations correctly.
- FOG measurements always contain the known Earth rotation rate. This can be used as a permanent in-situ calibration and for sensor health monitoring.
- When fixed to a wall, they provide the inter-story drift of the building (story) in situ and in realtime, which is not accessible with accelerometers.
- If the inter-story drift during an earthquake exceeds a critical value (usually 4%), the health state of the structure is no longer safe. This can be used as an automatic early warning indicator.

Since the concept of operation is entirely based on optical signals, there are no mechanical moving parts inside the sensor, so that the transfer function is constant and the system works over a very wide range of excitation frequencies (10^{-3} Hz $< f_{FOG} < 2$ kHz). A critical aspect of early warning sensors is the requirement of a permanent in situ check of the sensors' operational integrity. Since navigation grade FOGs are sensing the rate of rotation of the Earth, which in this context is a very stable and well known signal, any deviation from this given value is an indication of sensor failure, when other signal characteristics of earthquakes are absent. We have operated a FOG of the type μFORS-1 on the sixth floor of the eight story "West" building of the University of Canterbury at the Ilam campus in Christchurch (NZ). The shape of the building is a rectangular block. In order to measure the response, we have operated a single axis gyro fixed to a wall, so that the rocking about the short axis of the building can be observed. Figure 6.13 depicts the measured excursions in response to an earthquake of mag. 4.2, 20 km west of Christchurch, which took place shortly before two o'clock in the morning of November 14, 2010. The amplitude of the integrated rotation rate, taken on the sixth floor, reached 4 cm. Most prominently, the earthquake excited the first rocking mode around the short axis of the building at 2.6 Hz.

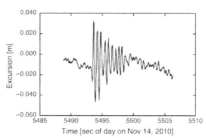

Figure 6.13 Measurements of the rocking of the "West" building in response to a mag. 4.2 earthquake, 20 km west of Christchurch, recorded on November 14, 2010.

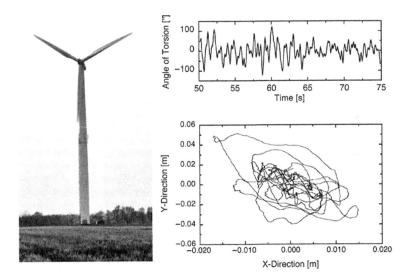

Figure 6.14 Measurements of the torsional motion and the deflection of a wind turbine pylon at a location just below the generator cabin, approximately 70 m above the ground.

When tall structures are subjected to external forces, such as for example wind loads on wind turbines [190], this may cause torsion, and this is not readily quantifiable without the availability of inertial rotation sensing. We have carried out measurements on a roughly 70 m tall windmill in the northern part of Germany. In this experiment the μ-FORS-1 FOG unit was strapped to the tall pylon structure just underneath the generator cabin. Figure 6.14 illustrates the advantages of operating an inertial rotation sensing device on a wind turbine. The left side of the figure gives an impression of the design of the structure that was tested. In the top right panel we represent the torsional excursions as a function of time under moderate gusty wind loads. Rotations of ± 120 seconds of arc were recorded during our measurement sequence. The lower right panel displays the excursions experienced by the generator cabin from the rocking of the entire structure in the east–west (X) and north–south (Y) directions, which exhibits a rather complex pattern of up to 15 cm deflection.

6.4.1 Ring Laser or Fiber Optic Gyro

The question of whether to operate a large ring laser structure or fiber optic gyros in practice answers itself from the perspective of the desired application. If high sensitivity is of importance and there is room for a larger installation, then this will certainly call for a ring laser installation. In the case of a highly dynamic scenario with much larger rotation rates, then this suggests a FOG application. This can be

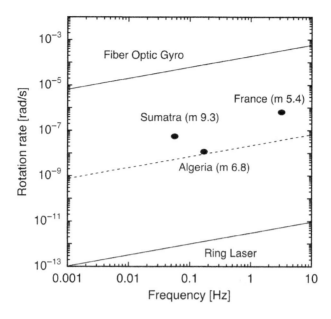

Figure 6.15 Required sensitivity of a rotation sensor for the detection of seismic signals. Reference lines are given for the G ring laser at the bottom, the μFORS-1 FOG at the top. The dashed line in the middle marks the current sensitivity level of the BlueSeis3A seismic rotation sensor.

illustrated by Figure 6.15, which displays the rotation rate versus frequency range in a window suitable for seismic signals. In order to read this diagram properly, we have to keep in mind that the Earth acts as a low pass filter for seismic frequencies. Closer earthquakes have higher frequencies, and they do not need to be so strong in order to generate recordings with a sufficiently good SNR. At the time of writing, it appears to be possible for the BlueSeis3A sensor concept to gain probably another order of magnitude in sensitivity, but that will mark the upper limit of what is feasible with this technique.

6.5 Ring Lasers in Geodesy

Geodesy explores the shape of the Earth and maps it out. The processing includes the geometry, the gravity field and the rotational motion of the Earth. The Earth is not a rigid body, and there are a large number of processes over a wide range of timescales that induce changes [127]. In more recent times and with the help of the measurement techniques of space geodesy, this field of science is also observing variations in the shape of the Earth. Temporal variations of the gravity field, for example, indicate mass transport processes, which may act at very different timescales, reaching from diurnal through seasonal to geological periods. These

are caused by ocean currents, mass loss in the cryosphere or transport phenomena in the Earth's interior. The combined efforts of the geodetic measurement
techniques – namely *Satellite Laser Ranging* (SLR), *Very Long Baseline Interferometry* (VLBI), the *Global Navigation Satellite System* (GNSS) and *Doppler
Orbitography and Radio-Positioning Integrated by Satellite* (DORIS) – are the
celestial and the terrestrial reference frames that ultimately define positions on the
ground, the shape of the Earth and the evolution of the surface of the Earth over
time – i.e., the crustal motion.

These reference frames provide the necessary metrics against which global
change and mass transport phenomena of the fluids of the Earth, namely atmosphere, hydrosphere, cryosphere and the Earth interior, can be quantified [161].
While the terrestrial reference frame is our prime interest, its counterpart, the
celestial reference frame, is important in order to establish link between measurement objects in space (quasars and satellites) and the observing stations on the
Earth's surface. The essential link for the transformation between the two reference
frames is the rate of rotation of the Earth and the orientation of the instantaneous
rotational axis in space. The exact measurement of propagation delays, obtained
interferometrically from a large number of radio sources by a global observation
network, provides polar motion and the variable rotational velocity, known as the
phase angle of Earth rotation UT1. VLBI determines this quantity with a resolution
of 15 µs. The well defined stable positions of the quasars across the sky and the
wide separation of the radio telescopes make this outstanding resolution possible.

Ring lasers are inertial sensors, and they do not require these reference frames.
In order to obtain the rotational velocity of the Earth unambiguously, a high
resolution three-component error-free ring laser structure, strapped down rigidly to
the ground, would be entirely sufficient. The VLBI technique has to operate a full
network of six or more radio telescopes simultaneously, in order to interferometrically establish the connecting baselines between the stations for reconstruction of
the motion of the Earth body. Ring lasers can operate as a self-contained standalone
measurement device. This means that such measurements are available in real time,
continuously and at a high data rate. However, this comes at a price. A single local
strap-down sensor is then used to infer a global measurement quantity. In the case
of VLBI it is the enormous distance covered by the baseline of up to 10,000 km that
makes the system independent from small local ground motion effects. In Chapter 3
we have developed the necessary sensor model to correct the measurements for
systematic errors originating from the laser operation. In Chapter 4 we have looked
at the desired geophysical signals and the sensor orientation. In this section we
are bringing both of these parts together in order to illustrate where Earth rotation
spectroscopy stands today. Figure 6.16 shows a super position of two datasets,
covering a timespan of two months. Apart from some small sections, there is

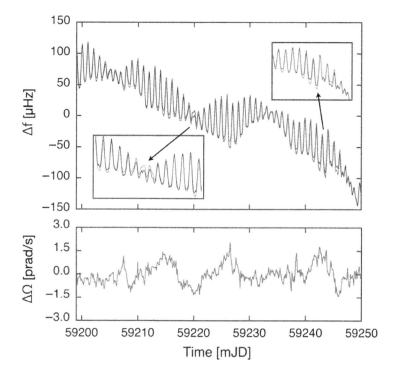

Figure 6.16 Top panel: Comparison of the residuals of the truly experienced rotation rate (gray) of the Earth in the underground laboratory in Bavaria, expressed in terms of the Sagnac beat note and the measurement signal obtained (black). The modeled geophysical signals have been taken from the IERS daily rapid data base and are adjusted to the instrument orientation by adding the local tilt measurements. Bottom panel: The difference between the two curves, expressed in terms of the corresponding rotation rate.

excellent agreement between the actually experienced motion of the rotation sensor (modeled data) and the gyro observation. From the irregular structure of the modeled and the observed rotation rate, one can see that the local tilts play a significant role. In fact, the local tilt-induced variation of the ring laser orientation turns out to be currently a significant obstacle for a single component gyro. The resolution of G is in the femtorad/s regime, and the bumps in the lower panel of Figure 6.16 are not associated with any significant change of the operational laser parameters. More probably, they are either real effects or caused by some subtle variation in the laser dynamics of the plasma that has not yet been identified. Another serious candidate for such kinds of perturbations is the microseismic activity of the Earth, if the broadened scatter does not entirely average out to produce a mean contribution of zero. Excited by wind and ocean waves, this noise source exhibits a pattern whose

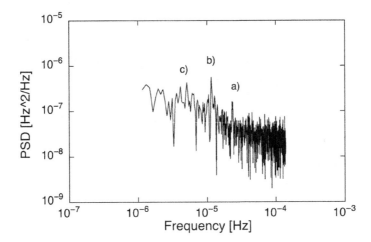

Figure 6.17 The power spectral densities of a 75-day-long dataset of the envelope of the microseismic activity, recorded by the G ring laser between the beginning of March and the end of May, 2021. While the spectrum is flat for periods shorter than half a day, there is a distinct set of frequency bands contained in the data for longer periods, apparently indicating the duration of variable weather patterns. Furthermore, there are some distinct signal peaks visible with periods at exactly half a day (a), one day (b) and around 2.3 days (c).

envelope peaks in the one to several days regime. We have taken beat note estimates from 60 one-minute datasets and computed the standard deviation of the Sagnac frequency for each hour. A spectrum of the observation sequence of 75 days of hourly rms scatter values is depicted in Figure 6.17.

Microseismic activity dithers the rotation rate of the Earth, and the effect on the measured rotation rate is a small perturbation, which acts in proportion to the intensity of the experienced ground vibration. For periods shorter than half a day, the spectrum is flat, while it exhibits several distinct peaks for longer periods within a region of several small frequency bands. We observe two familiar frequency components, with a period of half a solar day at 23.26 µHz (a) and a period of 1 day at 11.56 µHz (a) suggesting effects from atmospheric loading. The third regime is not so well defined in frequency and is thought to be related to the duration of noisy weather patterns. The most significant contributor (c) has a period of 2.3 days. However, there is little if any correlation between the signal residuals, as shown in the lower panel of Figure 6.16, and the envelope of the ground vibration scatter. Therefore, we can either exclude this mechanism as a cause, or it creates one subtle effect on top of other error contributors.

With our analysis approach we transform the Earth orientation parameters from the C04 time series of the IERS augmented by the tidal model [4] and the diurnal polar motion model [36] onto the local ring laser measurements, which are corrected for local tilt effects. This enables a comparison between the sum of all actually acting geophysical signals experienced on a given local spot on the Earth's crust and the measured sensor response. The result is a high resolution instantaneous rotation measurement, or in other words, a continuous monitoring of UT1. The entire goal of the development of large ring lasers for the application in space geodesy is the continuous and unambiguous detection of the polar motion and the variable part of Earth rotation in real time and with high temporal resolution. These signals have magnitudes between 0.1 and 100 ppb and cover a frequency range with periods between several hours and several hundreds of days. VLBI and GNSS perform an analysis over a global network and have between one and several days delay before the rapid or final solutions become available, at a typical rate of one data point per day. The example shown in Figure 6.16 provides one data point every three hours, and this interval could even be reduced to one hour without a significant loss of resolution.

Apart from the complication of the observed intensity modulation (Figure 6.17) caused by the microseismic activity, we find that the uncertainty of the established local tilt is a serious hurdle. With only a single axis-high resolution gyroscope available, the instantaneous orientation critically depends on the measurement of instrumental tilts with respect to the local gravity vector **g**. This is the second likely candidate for the cause of the remaining excursions in the lower panel of Figure 6.16. It underlines that we have to carefully choose the location for the operation of such a device. An installation deep underground, away from the top surface, is preferential. Furthermore, it would be necessary to operate at least a small network of such devices at different locations around the Earth to reliably distinguish global rotations from local effects. A gyroscope array, like ROMY, can recover tilt effects quite well. However, since tilt can be caused by local deformation (including of the gyroscopes themselves), there is still a tight reference required, in order to relate the rotation measurements to the Earth's crust sufficiently well. A gyroscope in a deep underground facility avoids the zone of higher deformation of the top soil, where tilts due to atmospheric pressure loading and ground water interaction are larger. Purely from the viewpoint of the determination of the sensor orientation, we can say that a larger gyro is advantageous, as the orientation of the sensor is better defined by longer interferometer sides. Since the antennas of the VLBI systems are spaced apart at intercontinental distances, this relaxes the requirement for orientation stability of the respective antennas to about 1 cm. Finally, we need to point out that the VLBI technology has the inherent advantage that the referenced quasar positions are very stable in the long term, due to the enormous distances of billions of light years between the sources and the receiver antennas. This means that the celestial reference frame, defined by the quasar positions, does not drift at a perceptible level in our lifetime. In order to balance the stability of the VLBI technology with the high resolution realtime capability

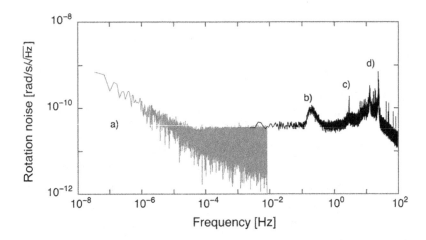

Figure 6.18 The long-term G ring laser performance, taken over a whole year, with all known geophysical signals removed (left side). The part of the spectrum on the right side was obtained from the instantaneous frequency estimates of the Hilbert transform from data of a single day. The mirror coating noise (a) currently limits the long-term resolution. Other remaining features are rotational noise in the microseismic band (b), the adjustment frequency of the piezo actuator (c) and plasma generated noise (d).

of large ring laser gyroscopes, a proper sensor fusion of these two techniques can provide a great advantage. However, it is not yet clear how this could be facilitated ideally with the currently available sensor configuration.

The standalone single component G ring laser system reaches the shot noise limit for periods between 1 s and 1 day at the level of 20 prad/s/ $\sqrt{(Hz)}$ for a dataset that is one year long. The negative slope toward longer periods (a) we believe is due to the currently available IBS mirror coatings in the cavity. With three hours of integration, G resolves rotation rates of 276 frad/s before reaching the current flicker noise floor. The measurements are illustrated in Figure 6.18. The displayed noise spectrum shows the current sensor resolution of the G ring laser over more than nine decades in frequency space, reaching from 50 nHz to 100 Hz, and exceeds the published performance of any other ground based rotation sensor techniques[1] [121] by more than an order of magnitude. In particular the longterm stability of the G ring laser is exceptional in this comparison. There are several distinct features in the noise spectrum that can be identified, in particular at higher frequencies. Brownian mirror coating noise is the likeliest candidate for the $1/f$ signal trend at the low frequency end (a). Rotational noise in the secondary microseismic band causes the

---

[1] We acknowledge the outstanding performance of the Gravity Probe B mechanical gyros, operated in a low earth orbit. They resolved 1.347×10^{-13} rad/s after 10 h of integration [70].

hump around 0.2 Hz (b), while the noise from the primary microseismic band is barely visible at the threshold of the noise floor. The update cycle of the optical frequency stabilization feedback loop around 3 Hz is a fairly sharp and distinct feature (c). The rather fuzzy frequency sideband structure in the range of 12–24 Hz (d) above and below the Sagnac beat note is not yet understood. It could probably be associated with some kind of plasma brightness fluctuations and has been addressed in Section 4.4. Apart from a broad and very diffuse frequency band, two distinct dominant frequencies can be identified, at 12.6 and 22.8 Hz. While the amplitude envelope of the noise band changes in proportion as the laser beam power is varied up and down, these two frequency components appear to be fixed. At this point in time it is also not known if these spurious frequency bands present any limitation for the general sensor operation.

How do we stand with respect to the desired resolution of the LoD signal? The signal known as the variation in "Length of Day" captures the momentum exchange between the solid Earth and the main fluids of the Earth, such as the atmosphere, the cyrosphere and the hydrosphere. As a result of this interaction, the solar day of 86,400 seconds is then either shorter or longer by a fraction of a millisecond. From theory, one would expect a clear systematic seasonal (biannual) trend and a distinct fortnightly periodic signature related to the lunar orbit. So we have to detect a spectral component just below 1 μHz and a somewhat larger signal at around 80 nHz. The latter is currently still outside of the stability range of G (see also Section 4.7.2). Figure 6.19 provides a comparison between the current G ring laser performance and the respective LoD signature. The spectrum of the LoD signal

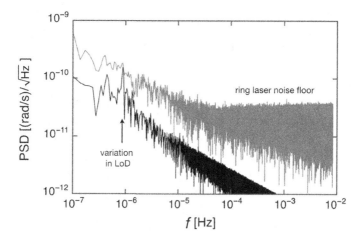

Figure 6.19 Ring laser observation of LoD, based on one year of data (upper curve). The measurements of LoD were taken from the daily GNSS solutions from the IERS archive [168] and were converted to a corresponding rotation rate.

is calculated in units of the corresponding Sagnac frequency from the daily final solutions of the GNSS network, which are posted on the IERS website [168]. The ≈ 1 µHz LoD signal is now just detectable within the sensitivity regime of the G ring laser. For geodetic application, the signal to noise ratio for the sensor noise floor has to be improved at frequencies below 10 µHz. For this matter, the reduction of mirror coating noise by a set of suitable crystalline mirrors in the 633 nm regime could be a game changer and should eventually make reliable high resolution realtime detection of the LoD signal by a single sensor possible.

6.6 Infrasound Detection with Open Cavity Ring Lasers

In Section 4.12 we introduced open ring resonators and their ability to sense infra-sound by modulating the length of the cavity under the influence of significant backscatter pulling. Ring lasers of this specific design have been reported by the group of Robert Dunn at the Hendrix College in Conway (AR) [63, 64]. Infrasound in this context can be considered as a very small periodic atmospheric pressure variation at very distinct frequencies. There are three different types of waves known that can propagate in a non-rotating spherically symmetric atmosphere, namely acoustic waves, Lamb waves and gravity waves [65]. Lamb waves are boundary waves. They travel at the velocity of sound along the ground. As with acoustic waves, compression acts as a restoring force. Gravity waves are similar to Lamb waves with gravity contributing to the restoring force. They are mostly generated on the lee side of orographic obstacles and travel at the speed of the wind. Very low frequencies in the range of several mHz have a wavelength much larger than the size of the ring laser cavity. Furthermore, the air density changes slowly, thus allowing a much larger integration time compared to the rapidly changing higher frequency sources with a wider spectral distribution. Infrasound has long been associated with severe weather patterns like tornados and hurricanes [15]. Since the attenuation of an acoustic frequency in the atmosphere increases with the square of the frequency, infrasound at frequencies of less than 10 mHz experience very little attenuation as they propagate. So they reach very far over long distances and can well be detected more than 2000 km away from their origin. Severe weather conditions have been found to generate pressure amplitudes in the range of 10–100 mPa at distances between several hundred and more than 1000 kilometers [15]. Recalling the esti-mated pressure sensitivity ($\delta p \ll 23$ mPA) of the open cavity ring laser derived in Section 4.12, infrasound from tornados, hurricanes and other severe weather events are clearly in the detectable range of the open ring laser design.

Hurricanes out on the ocean excite low frequency rotation-induced signals, due to an increased microseismic noise level in the range of 80–200 mHz (see Section 6.3.5). This is caused by wind-induced wave activity and is detectable

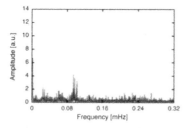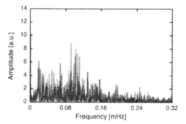

Figure 6.20 The recorded rotational component of the microseismic background noise, caused by the Hurricane Katrina when it was approaching the Louisiana shore on August 28, 2005 (a) and one day later when it hit Louisiana (b). (Adapted with permission from [66] © American Institute of Physics)

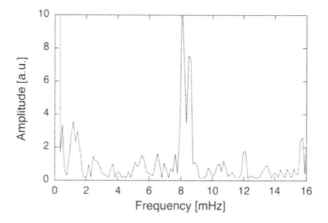

Figure 6.21 Hurricane Wilma (October 21, 2005) generated two spectral lines instead of one, corresponding to a double eyewall with the inner eyewall exhibiting a frequency of 8.5 mHz and the outer eyewall 8.1 mHz. (Adapted with permission from [66] © American Institute of Physics)

with all ring laser gyros of sufficient sensitivity. The origin of this signal is true ground rotation, caused by s-waves. Figure 6.20 depicts a recording induced by Hurricane Katrina on August 28, 2005 (left), when it was about 1000 km away from the sensor. The signal becomes stronger one day later, as the hurricane comes ashore (right). The fact that the signals in the primary microseismic frequency band provide a much stronger signal than the secondary microseismic band indicates that at least some of the recorded signal is caused by infrasound. Under normal conditions, the secondary microseismic activity is typically 20 dB stronger than signals in the primary microseismic band. When hurricanes make landfall, they develop a distinct infrasound signal, which is related to the diameter of the eyewall [66]. Larger eyewall diameters generate lower infrasound frequencies. Figure 6.21 illustrates this typical observation by two examples. While Hurricane Katrina generated a distinct infrasound frequency of about 7.3 mHz when it was making landfall on August 29, 2005, Hurricane Wilma, two months later, is characterized by a double peak with frequencies of around 8.1 and 8.5 mHz, corresponding to a double eyewall feature, as the respective satellite image exhibited [66]. The higher frequency thereby corresponds to the inner eye and the lower frequency to the outer wall. Other sources of infrasound are tornados, which have also been observed by the open ring laser [67]. Since the diameter of the funnel is much smaller than the eye of a hurricane, the infrasound frequencies f_s observed by the Sagnac interferometer in Arkansas are higher, namely in the regime of $0.94 \leq f_s \leq 2.41$ Hz. According to a model of Abdullah [2], the emitted frequency of a fundamental mode of vibration can be approximated by

$$f_s = \frac{v_s}{2\pi r} \sqrt{\frac{4U^2}{v_s^2} + \frac{25\pi^2}{16}}, \tag{6.7}$$

where v_s is the velocity of sound, r the radius of the vortex and U the tangential component of the wind velocity. For a vortex radius of 10 m or less, the vibration frequency enters the regime where it becomes audible. The observed tornados occurred much closer to the ring laser site, and the infrasound frequency components were observed approximately 30 minutes before the funnel reached the ground [67] and disappeared when the funnel lifted up again.

Explosive volcano eruptions have also been identified as a source for stable infrasound signals at distinct frequencies in the range of 3.68–5.14 mHz [234]. The observations have been obtained from the vertical component of long period seismographs, suggesting a loading effect. Unlike the situation for hurricanes and tornados, where, with the exception of double eyewall scenarios, distinct single frequency signals inversely proportional to the size of the respective vortex are generated, infrasound from explosive volcano eruptions show a more complex signal structure. Most of the infrasound energy is emitted in the regime of 0.5–10 Hz [109]. However, after traveling over large distances, only the low frequency components remain detectable. From the Mount. St. Helens eruption in 1980, for example, infrasound frequencies of 2.7, 3.7, 4.76 and 5.55 mHz were detected on the vertical component of a Press–Ewing seismograph at a distance of more than 7000 km [65]. The open cavity ring laser in Arkansas proved also to be very sensitive to this type of signal. Although the infrasound signals populate the same frequency regime as hurricanes, the amplitudes are considerably weaker, and there is more than one distinct frequency detectable. Groups of 4–7 frequencies in the range of 1.2–5.85 mHz were found in ring laser observations of six explosive volcano eruptions in 2013 and 2014 [65]. There are still some more candidates for natural mechanisms for infrasound, but these are much more difficult to isolate and to quantify. It still remains to be seen where an open cavity ring laser gyro could contribute best in the area of geophysical infrasound detection.

6.7 Tests of Non-reciprocal Phenomena

Ring laser setups are very useful for experiments that require an exactly reciprocal beam path. The observation of magnetochiral birefringence [225] is a good example of this, since it is scalar and non-reciprocal, just like the Sagnac effect itself. The design of the experiment is based on a square ring laser design, roughly of the size of the C-II ring laser, with a perimeter of 3.47 m. The gain medium is a discharge tube filled with Ar^+ gas, lasing on $\lambda = 488$ nm. Unlike for a Sagnac interferometer, a high cavity Q is not essential for this experiment. Therefore, it was possible to use an etalon as a mode selection device. The cavity length P was stabilized by locking a frequency controlled HeNe laser to the ring cavity. Similar to our Sagnac interferometers, the two counter-propagating laser beams, leaking through one of the mirrors, were superimposed on a beam splitter to provide the beat note. One of the arms of the ring laser contained two identical sample cells made of fused silica, each 1 cm long, where one of them contained one chiral species and the other its enantiomer, in this case $R(+)$ and $S(-)$ limone. This setup ensures that no rotation of the plane of polarization, caused by natural optical activity, takes place. Faraday rotation is canceled by placing two identical Ni-Fe-B permanent magnets in front of the two cells such that they induce two opposite longitudinal magnetic fields in the beam path. At the same time the contribution of the enantiomers to the observed non-reciprocity adds up and a beat note of

$$\delta\nu = \nu_{cw} - \nu_{ccw} = 4\Delta n \frac{cL}{\lambda P} + \Delta\nu_0 \tag{6.8}$$

is obtained [225], leading to a difference of the refractive index $\Delta n = (10 \pm 1.6) \times 10^{-11}$. In order to check for consistent behavior, the contents of the cells were exchanged, which changed the polarity of the sinusoidal excitation by the rotating magnets. When the cells contained a racemic mixture of the two limonene components, the non-reciprocity vanished as expected. A quantity as small as the non-reciprocal frequency offset induced by magnetochiral birefringence could be unambiguously recovered because the experiment was designed such that the signal of interest produced a well defined periodic oscillation. This made the measurements resilient against drift effects and systematic biases. This advantage is not available for the proposed measurements in fundamental physics (see next section).

6.8 Tests of Fundamental Physics

Predictions based on the theory of general relativity have been demonstrated to be the most precise descriptions for gravitational phenomena and therefore place very stringent requirements on any experimental verification, with a necessity for relative errors to be of the order or significantly smaller than 1 part in 10^9. Measurement methods from modern *Space Geodesy* perform at about this error level. Lunar Laser Ranging (LLR), for example, provides precise round trip optical travel times between a geodetic observatory and cube corner retro-reflectors placed on the moon by the American APOLLO and the Russian LUNA landers. With a long time series of observations and continuous technical improvements, reaching a precision for the measured ranges of several millimeters in recent years, the error margin has reduced to a level of $10^{-9} - 10^{-11}$. As one of many results, according to [150, 151], this led to improved constraints for the gravitational constant and its spatial and temporal variation of $\dot{G}/G = (2\pm 7)\times 10^{-13}$ yr^{-1} and $\ddot{G}/G = (4\pm 5)\times 10^{-15}$ yr^{-2}. If we take the G Ring laser as a reference, this technology has demonstrated a sensor resolution around 3×10^{-9} for geophysically induced signals in the frequency range between 1 minute and about 1 day, which places optical Sagnac interferometry roughly into the same ballpark as LLR in the short term. However, for the long-term stability we are currently limited by $1/f$ noise, as illustrated in Figure 6.18, which constitutes a significant current obstacle but does not present a fundamental limit. Due to the fact that active and passive Sagnac interferometers are both inertial sensors, which measure absolute rotations, it is worthwhile to explore their suitability for tests in fundamental physics. Furthermore it may be possible to constrain theories of gravity thereby [38].

6.8.1 Gyroscopic Tests at a Global Scale

Since general relativity predicts that the stationary field of a rotating body differs from that when it is not rotating, ring lasers may be an instrument of choice for the measurement of this rotational Lense–Thirring frame-dragging effect [34, 56, 133]. Probing the gravito-magnetic effect on Earth with light in an active Sagnac interferometer yields a constant minute difference for the two counter-propagating beams. According to [34], the linear approximation for the time delay for an instrument with its normal vector contained in the local meridian plane is

$$c\delta\tau = \frac{4A}{c}\Omega_\oplus \left[\cos(\theta + \alpha) - 2\frac{GM}{c^2 R}\sin\theta\sin\alpha \right.$$
$$\left. + \frac{GI_\oplus}{c^2 R^3}(2\cos\theta\cos\alpha + \sin\theta\sin\alpha) \right], \qquad (6.9)$$

where A is the area enclosed by the laser beams, α is the angle between the local radial direction and the normal to the plane of the instrument (measured in the meridian plane) and θ is the co-latitude of sensor location. $\Omega_\oplus$ is the rotation rate of the Earth as measured in the local reference frame, which includes the local gravitational time delay. Rewriting the equation such that it contains the flux of an effective angular velocity Ω through the cross section of the sensor yields

$$\delta\tau = \frac{4}{c^2}\mathbf{A} \cdot \mathbf{\Omega}. \tag{6.10}$$

Here $\mathbf{A}$ is the area enclosed by the beams and orientated according to the normal vector of the sensor, $\mathbf{\Omega} = \mathbf{\Omega}_\oplus + \mathbf{\Omega}'$ is composed of $\mathbf{\Omega}_\oplus$, the kinematic Sagnac term due to the rotation of the Earth, and $\mathbf{\Omega}' = \mathbf{\Omega}_G + \mathbf{\Omega}_B + \mathbf{\Omega}_W + \mathbf{\Omega}_T$, which accounts for relativistic effects. In this expression the geodetic precession is $\mathbf{\Omega}_G$, $\mathbf{\Omega}_B$ the Lense–Thirring precession, $\mathbf{\Omega}_W$ results from the preferred frame effect and $\mathbf{\Omega}_T$ is the Thomas precession. For a ring laser in an Earth-bound laboratory, the geodetic and Lense–Thirring terms are both of the order of $\sim 10^{-9}$ relative to the rate of rotation of the Earth; within an order of magnitude of the current G ring laser resolution. Since the Thomas precession is about three orders of magnitude smaller, and the preferred frame term is about two orders of magnitude smaller, they are neglected from this discussion, and we use the reduced expression: $\mathbf{\Omega}' \simeq \mathbf{\Omega}_G + \mathbf{\Omega}_B$. In order to measure $\mathbf{\Omega}'$, a ring laser gyroscope set up horizontally at mid-latitudes requires a sensor resolution of around 0.04 prad/s. This rough order of magnitude estimate ignores the issue of obtaining the true orientation of the effective ring laser plane for the moment.

We recall that the tolerances for the angular alignment are of the order of 1 nrad. In comparison, the average width of the laser beam of the G ring laser is about one million times larger than the tolerances for the tilt angle. A bailout solution for this problem could be a collinear alignment of the ring laser normal vector with respect to the Earth rotation axis. In this orientation, the obtained rotation rate would be at a maximum and could therefore provide a clear criterium for the alignment of the laser plane. The $\cos\alpha$ angular dependence term would also not matter so much, since the condition $\alpha = 0$ has to be met, where this function is very insensitive. Another problem is the reduction of the $1/f$ noise contribution that limits the long-term stability currently, which means that longer averaging no longer delivers higher resolution. At this point in time there are two candidates for the cause of this issue, namely noise from Brownian motion in the mirror substrates and coatings ([155] and Section 3.10.3) and laser gain medium dynamics (see Section 3.9.7). If the former is more dominant, it would require crystalline mirrors to mitigate this issue. However, if the latter cause generates the main contribution to the observed fluctuations, it will require a considerable modeling effort to improve this situation.

For the time being, slowly changing non-reciprocal effects in the sensor operation present a significant obstacle for the required overall sensor stability. This situation is exacerbated by the fact that the Lense–Thirring frame-dragging signal itself corresponds to a very small DC offset contribution to the superimposed rotation rate of the Earth. In the presence of several additional potential mechanisms of instrumental systematic biases, it is particularly difficult to extract a small constant value without ambiguity. Additional offsets with a clear DC characteristic are easily overlooked. We have already discussed the effect of the refractive index of the laser gain medium on the scale factor, the Goos–Hänchen area correction, the effect of waveform curvature (Gouy phase) and the phase change induced by the laser gain medium in Section 3.9.1. All of these effects require corrections that are between one and two orders of magnitude larger than the expected value of the Lense–Thirring contribution. It is reasonable to expect that each of the corrections is accurate to better than 10% of the estimated respective values, but it is not clear if they are also accurate to better than 1%, which would be required as a bare minimum. We also recall the observed offsets caused by the influence of magnetic fields, both at the location of the mirrors and at the gain medium, as discussed in Section 4.11. In this case the resulting effects were rather large, due to the close proximity of a strong magnet to either a mirror or the plasma. Fluctuations of the Earth's magnetic field are much smaller, and possible effects on the Sagnac frequency have not yet been identified by efforts where the magnetic field variations of the Earth were correlated with the observed Sagnac frequency. This, however, may no longer be the case when the domain of nHz variations in the Sagnac frequency is reached in pursuit of the relativistic frame dragging effect. While the sufficiently accurate determination of the real scale factor of the ring laser instrument has many challenges, the overall stability of the ring laser cavity of at least one order of magnitude below the level of the Lense–Thirring contribution has to be ensured, which is another practical challenge, and it may need high quality active cavity stabilization techniques to satisfy this requirement. Currently we expect the accuracy even from our flagship gyro to be only in the range of about 1 ppm.

Apart from actually measuring the Lense–Thirring effect with a ground based gyroscope, high precision tests of metric theories of gravity in the framework of the PPN formalism also come within reach. With $\mathbf{J} = I_\oplus \mathbf{\Omega}_\oplus$ defined according to [34], one obtains

$$\mathbf{\Omega}_G = -(1 + \gamma) \frac{GM}{c^2 R} \sin \vartheta \, \Omega_\oplus \mathbf{u}_\vartheta, \tag{6.11}$$

and

$$\mathbf{\Omega}_B = -\frac{1 + \gamma + \frac{\alpha_1}{4}}{2} \frac{G I_\oplus}{c^2 R^3} [\mathbf{\Omega}_\oplus - 3(\mathbf{\Omega}_\oplus \cdot \mathbf{u}_r)\mathbf{u}_r]. \tag{6.12}$$

In Eqs. 6.11 and 6.12, α_1 and γ represent the PPN parameters that account for the effect of a preferred reference frame and the amount of space curvature produced by a unit rest mass. High precision ring laser measurements should be able to place constraints on α_1 and γ.

6.8.2 Gyroscopic Tests at Laboratory Scale

About one decade ago, Tajmar and de Matos predicted the possibility of a greatly enhanced gravitomagnetic field, which would be measurable in close proximity to a rotating superconductor [216, 217]. The expectations require that the density of the Cooper pairs is sufficiently large, relative to the mass density. The theory was apparently supported by experiments carried out with spinning rings of superconducting lead, niobium and other materials operated close to either laser gyroscopes or accelerometers [218]. According to the theory, any angular acceleration should produce a gravitational field along the ring's surface, which should be detectable with a laser gyroscope or a low-noise accelerometer. Since the expected effect is still very small, the detection devices have to be rigidly fixed to the ground in order to avoid any physical motion. The experiment reported in [218] employed several small FOGs with a resolution of 100 nrad/s and an angular random walk error 2×10^{-5}. The reported test measurements were carried out over a time span of no more than 25 s and indicated a signal response of up to 20 μrad/s. However, independent observations with spinning superconducting lead operated at the UG2 ring laser did not reproduce any evidence for the claimed effect [79]. Since the gyroscope was located outside the dewar compartment, the respective signal must be smaller. The UG2 ring laser at the time of the experiment operated at a sensitivity limit of 8 prad/s in the short-term, which is about six orders of magnitude lower than the reported size of the effect in the literature.

6.9 Future Perspectives

Large ring lasers have come a long way over the last three decades. When proposed in 1993, the construction of C-II was met with a lot of skepticism, regarding the technical feasibility. From an initial sensor stability of only 0.1% with respect to the Earth rotation rate, C-II moved quickly to a performance domain of better than 1 part in 10^6. Based on this success, a concept for an even larger ring laser, the G-ring, was proposed at the end of 1997. At that time it was not known if such a cavity would even allow single mode operation. This led to the construction of the first heterolithic stainless steel ring laser, namely G-0. The only objective for this device was to attain laser action on a single longitudinal cavity mode. This goal was easily achieved, and G-0 continued to operate well as a gyroscopic laser until the devastating Christchurch earthquake in 2011 when the Cracroft Cavern laboratory became unavailable. Once these necessary foundations had been laid, the G-ring laser could be built. It shifted the limit of gyroscopic sensor resolution well beyond 1 part in 10^8. Encouraged by this results, even larger ring lasers were built, demonstrating their viability for applications in this new field of high resolution Sagnac spectroscopy. The sensor stability for the monolithic G has been improved to the point that continuous ultra-stable gyroscopic operation is now achieved over many months.

By building the 834 m^2 UG-2 gyroscope, we have exceeded the limit to which increasing the scale factor by geometric upscaling increases the usable rotational sensitivity of the device. Therefore, earlier proposals to build ever larger ring cavities, such as the Michelson Centenary Ring (MCR) [210], are not currently a priority. They would require a significant advance in active cavity stabilization and novel concepts in sensor design. The multicomponent ROMY ring laser has demonstrated a promising way forward toward a self-consistent stand-alone Earth rotation sensor array. The tetrahedral design overcomes the problem of finding the right orientation on Earth, something that the single component G ring alone cannot provide. Nevertheless, it has to distinguish between local and global rotation. While the Zerodur based gyroscopes C-II and G are by far the most stable instruments, the heterolithic stainless steel structures are much larger in numbers, both realized as active lasers and as passive cavities by working groups in the USA [121], Italy [18, 54], China [137], New Zealand [96] and Germany [74]. Apart from being much more affordable, these types of sensor construction also provide much better accessibility when it comes to modifications, such as mirror substitution. However, due to the limited mechanical and thermal stability, the ring cavity needs a very stable environment and a high performance perimeter control. A deep underground location, such as the *Laboratori Nazionali del Gran Sasso* (LNGS), is very promising due to high temperature stability and few deformation effects from atmospheric and ground water loading.

The above notwithstanding, the current lasers have successfully measured diurnal polar motion and solid Earth tides as well as the repeated observation of the Chandler and Annual wobbles. The latter is a significant achievement, because the key to future advances lies in our ability to detect near-DC signals. We have already come a long way through cavity stabilization, the elimination of undesirable error sources, such as backscatter coupling, the inclusion of the null-shift bias corrections and the reduction of minute cavity bias effects arising (for example) from the gain medium, the drift of the optical frequency and beam steering effects. In Figure 6.18 we have demonstrated that the G ring laser operates close to the expectation. With a current rms noise level of 1.7 μHz over 3 h of averaging, the gyro operates within the limits set by the quantum noise and by the mirror substrate and coating noise, discussed in Section 3.10.3. The remaining dominant $1/f$ noise component in the G output, however, is not easily mitigated and is the focus of our ongoing effort. One contributor to the error margin is microseismic noise. Originally it was hoped that the larger scatter from distant ocean waves and local wind gusts would contribute with a mean value of zero, and hence would average out. The fact that it appears to have a trend that depends on the actual noise amplitude suggests that the noise level might introduce an intensity-related bias. If there is a deterministic relationship behind it, this noise source can probably be removed either by excluding highly affected data from the analysis process or by allowing for an intensity-dependent correction. However, at this point in time, there is no conclusive indication for ambient noise as a fundamental limitation, and this potential error source is not the only one left. Figure 6.17 illustrates that microseismic noise accounts for some but not all of the noise.

What would be the best roads to pursue for the future of large ring laser gyros? Three basic improvements come to mind. A lower noise limit on the bases of crystalline mirrors for visible light give an altogether lower noise level boundary and will already be enough to access the Length of Day signal, which sits just at the current limit of resolution. Although we see evidence of the primary microseismic signal when the activity is moderate, the G ring laser gyroscope is not yet limited in sensor resolution by the Earth itself. This is in contrast to current performance levels of modern seismometers. While a lower noise floor is helpful, it does not remove the principal constraint of rotation sensing, which hinges on a proper estimate of sensor tilts and the effect of noise associated with the intensity of microseismic activity. It may, however, be possible to reverse the analysis approach by using the model predictions of the diurnal polar motion signal and using the observations to solve for the appropriate sensor orientation, before other signals of interest, like the LoD signal, are extracted. This assumes that the model is sufficiently complete and accurate.

An actively stabilized sensor array in the shape of the ROMY installation appears to be a promising approach. At the time of writing, ROMY has shown some significant advantages but still falls far behind the G ring laser, due to the lack of mechanical stability of the heterolithic construction. So far the active cavity stabilization method has not matured sufficiently to be on a par with the monolithic approach based on a large Zerodur slab. Apart from that, we still have to demonstrate that the installation of the four subsystems of ROMY can be referenced well enough with respect to each other over a sufficiently long time, which may sound straightforward but may turn out to be a tedious effort. The ROMY structure may be well suited to overcome the problem of microseismic $1/f$ noise signal contribution. Even if this is the case, it will leave the question open as to whether there are further local or regional signals of geophysical origin in existence that may cause other bias effects on a different timescale. In the end we want to obtain the global effect of Earth rotation and polar motion without ambiguity. Furthermore, we wish to extract the current value of LoD on a daily basis. At the moment we can only provide some evidence that a gyro can access these signals. Therefore, the third future goal would be the installation of several large high performing ring lasers in different locations. While the global Earth rotation signal would be common to all sensors, the micro-seismic and other local or regional signals would differ, and by correlating the sensor output among the various different sensors, the global signal could be cleanly separated from the local effects. Undoubtedly, the operation of several large high quality ring laser gyroscopes at different orientations and longitudes would be desirable for this purpose.

Acronyms

ADEV   Allan deviation. 59, 242, 266
AlGaAs   aluminum–gallium arsenide. 261

BIPM   Bureau International des Poids et Mesures. 180
BMBF   German Ministry of Education and Research. 274

C-I   Canterbury Ring Laser (1st generation). xii, 37
C-II   Canterbury Ring Laser (2nd generation). 41
C04   Time series of earth orientation parameters from the IERS. 292
CM   common mode. 117
cw   continuous wave. 29, 41

DC   Direct component of the measurement signal. 34, 35, 37, 54, 56, 120, 138, 167
DM   differential mode. 117
DORIS   Doppler Orbitography and Radio-positioning Integrated by Satellite. 289
DTG   Dynamically tuned gyroscopes. 265
DWD   Deutscher Wetterdienst. 203

ECMWF   European Centre of Medium Weather Forcast. 203
ELSy   Experimental Laser Gyroscope System. 34
ENSO   El Niño Southern Oscillation. 219
EOM   electro optical modulator. 29
EOP   Earth orientation parameters. 208

FFT	fast Fourier transform. 72, 186
FM	frequency modulation. 73
FOG	fiber optic gyroscope. 240, 242, 244, 245, 264, 265
frad	femto-radian. 293
FSR	free spectral range. xii, 30, 42
FWHM	full width half maximum. 182, 261
G	Grossring, Wettell. 41
G-FORS	'Grosse' Fiber Optic Rotation Sensor. 248
GaAs	gallium arsenide. 152
GH	Goos–Hänchen displacement. 106
GNSS	Global Navigation Satellite System. xi, 180, 209, 222, 289
GOW	Geodetic Observatory Wettzell. xiv
GPS	Global Positioning System. 84
HeNe	helium–neon (laser). xii, 45
HRG	hemispherical resonator gyroscope. 257
IAU	International Astronomical Union. 208
IBS	ion beam sputtering. 24, 50, 152, 238
IERS	International Earth Rotation and Reference System Service. 208, 222, 295
IMU	inertial measurement unit. 72, 265
IR	infrared. 161
IRIS	Incorporated Research Institute for Seismology. 197
LED	light emitting diode. 194
LLR	lunar laser ranging. 299
LNGS	Laboratori Nazionali del Gran Sasso. 303
LoD	length of day. 31, 177, 213, 219, 224, 295
MBE	molecular beam epitaxy. 152
MEMS	microelectromechanical systems. 256, 264, 265
mJD	modified Julian date. 206
mK	milliKelvin. 69
mPa	milliPascal (= 1 milli Nm^2). 295

mrad	milliradian. 249
nHz	nanohertz. 266
NIR	near-infrared regime. 159
nrad	nanoradian. 39
NZ	New Zealand. 161
OBS	ocean bottom seismometer. 283
OFC	optical frequency comb. 95
PDH	Pound–Drever–Hall optical frequency locking technique. 240, 269
PFO	Piñon Flats (geophysical) Observatory. 276
PI	principal investigator. 235
PLL	phase locked loop. 72, 268
ppb	parts per billion. 106
ppm	parts per million. 40
prad	picoradian. 27, 293
PSD	power spectral density. 149, 178
RAM	residual amplitude modulation. 240
RF	radio frequency. 9, 37
rms	root mean square. 304
SLR	satellite laser ranging. 289
SNR	signal to noise ratio. 89, 140, 262
SQUID	superconducting quantum interference device. 252
SSIRU	scalable space inertial reference unit. 258
STS-2	Streckeisen triaxial broadband seismometer. 230, 272, 283
TDEV	time deviation. 161, 182
UHV	ultra high vacuum. 164
ULE	ultra-low-expansion glass. 44, 231
ULISS	cryogenic sapphire microwave oscillators. 268
USGS	United States Geological Survey. 274

UT1   Universal Time corrected for polar motion. 289, 292

UTC   Coordinated Universal Time. 180

VCO   voltage controlled oscillator. 73

VLBI   very long baseline interferometry. xi, 209, 222, 289

YAG   yttrium–aluminium–garnet. 260

zm   zeptometer. 29

References

[1] Abbott, B. P., and the LIGO VIRGO Scientific Collaboration. 2016. Observation of gravitational waves from a binary black hole merger. *Physical Review Letters*, **116**(6), 061102.

[2] Abdullah, A. J. 1966. The "musical" sound emitted by a tornado. *Monthly Weather Review*, **94**(4), 213–220.

[3] Adler, R. 1973. A study of locking phenomena in oscillators. *Proceedings of the IEEE*, **61**(10), 1380–1385.

[4] Agnew, D. C. 1997. NLOADF: A program for computing ocean-tide loading. *Journal of Geophysical Research: Solid Earth*, **102**(B3), 5109–5110.

[5] Ahearn, W. E., and Horstmann, R. E. 1979. Nondestructive analysis for HeNe lasers. *IBM Journal of Research and Development*, **23**(2), 128–131.

[6] Allen, L., and Jones, D. G. C. 1965. The Helium–Neon laser. *Advances in Physics*, **14**(56), 479–519.

[7] Anderson, R. 1994. "Sagnac" effect: A century of Earth-rotated interferometers. *American Journal of Physics*, **62**(11), 975.

[8] Anyi, C. L., Thirkettle, R. J., MacDonald, G. J., Schreiber, K. U., and Wells, J.-P. R. 2019. Gyroscopic operation on the $3s_2 \longrightarrow 2p_{10}$ 543.3 nm transition of neon in a 2.56 m^2 ring cavity. *Optics Letters*, **44**(12), 3074.

[9] Anyi, C. L., Thirkettle, R. J., Zou, D., et al. 2019. The Macek and Davis experiment revisited: A large ring laser interferometer operating on the $2s_2 \longrightarrow 2p_4$ transition of neon. *Applied Optics*, **58**(2), 302–307.

[10] Arissian, L., and Diels, J. C. 2009. Investigation of carrier to envelope phase and repetition rate: Fingerprints of mode-locked laser cavities. *Journal of Physics B: Atomic, Molecular and Optical Physics*, **42**(18), 183001.

[11] Aronowitz, F. 1971. *The Laser Gyro in Laser Applications*. Academic Press.

[12] Aronowitz, F. 1999. Fundamentals of the ring laser gyro. *Report: AGARDograph*, 12, 1–46.

[13] Ash, M., and De Bitetto, P. 1995. Hemispherical resonator gyroscope assessment for space applications. *Charles Stark Draper Laboratory report for Naval Research Laboratory (NRL)*.

[14] Baxter, T. D., Saito, T. T., Shaw, G. L., Evans, R. T., and Motes, R. A. 1983. Mode matching for a passive resonant ring laser gyroscope. *Applied Optics*, **22**(16), 2487–2491.

[15] Bedard, A. J. 2005. Low-frequency atmospheric acoustic energy associated with vortices produced by thunderstorms. *Monthly Weather Review*, **133**(1), 241–263.

[16] Beghi, A., Belfi, J., Beverini, N., et al. 2012. Compensation of the laser parameter fluctuations in large ring-laser gyros: A Kalman filter approach. *Applied Optics*, **51**(31), 7518.

[17] Belfi, J., Beverini, N., Bosi, F., et al. 2010. Rotational sensitivity of the G-Pisa gyrolaser. *IEEE Transactions on Ultrasonics, Ferroelectrics and Frequency Control*, **57**(3), 618–622.

[18] Belfi, J., Beverini, N., Bosi, F., et al. 2012. A 1.82 m^2 ring laser gyroscope for nano-rotational motion sensing. *Applied Physics B: Lasers and Optics*, **106**(2), 271–281.

[19] Belfi, J., Beverini, N., Carelli, G., et al. 2012. Horizontal rotation signals detected by "G-Pisa" ring laser for the Mw = 9.0, March 2011, Japan earthquake. *Journal of Seismology*, **16**(4), 767–776.

[20] Belfi, J., Beverini, N., Bosi, F., et al. 2012. Performance of "G-Pisa" ring laser gyro at the Virgo site. *Journal of Seismology*, **16**(4), 1–10.

[21] Belfi, J., Beverini, N., Carelli, G., et al. 2018. Analysis of 90 day operation of the GINGERINO gyroscope. *Applied Optics*, **57**(20), 5844–5851.

[22] Bergh, R. A., Lefevre, H. C., and Shaw, H. J. 1981. All-single-mode fiber-optic gyroscope. *Optics Letters*, **6**(4), 198.

[23] Bergh, R. A., Lefevre, H. C., and Shaw, H. J. 1981. All-single-mode fiber-optic gyroscope with long-term stability. *Optics Letters*, **6**(10), 502–504.

[24] Bernauer, F., Wassermann, J., Guattari, F., et al. 2018. BlueSeis3A: Full characterization of a 3C broadband rotational seismometer. *Seismological Research Letters*, **89**(2A), 620–629.

[25] Beverini, N., Di Virgilio, A., Belfi, J., et al. 2016. High-accuracy ring laser gyroscopes: Earth rotation rate and relativistic effects. *Journal of Physics: Conference Series*, **723**(07), 012061–012067.

[26] Bilger, H., Wells, P., and Stedman, G. 1994. Origins of fundamental limits for reflection losses at multilayer dielectric mirrors. *Applied Optics*, **33**(11), 7390.

[27] Bilger, H., Stedman, G., Li, Z., Schreiber, K. U., and Schneider, M. 1995. Ring lasers for geodesy. *IEEE Transactions on Instrumentation and Measurement*, **44**(2), 468–470.

[28] Bilger, H. R., and Stedman, G. E. 1987. Stability of planar ring lasers with mirror misalignment. *Applied Optics*, **26**(17), 3710–3716.

[29] Bilger, H. R., Shaw, G. L., and Simmons, B. J. 1984. Calibration of a large passive laser ring. *Physics of Optical Ring Gyros*, 110–113.

[30] Bilger, H. R., Stedman, G. E., and Wells, P. V. 1990. Geometrical dependence of polarisation in near-planar ring lasers. *Optics Communications*, **80**(2), 133–137.

[31] Birch, K. P. 1991. Precise determination of refractometric parameters for atmospheric gases. *JOSA A*, **8**(4), 647–651.

[32] Bishof, M., Zhang, X., Martin, M. J., and Ye, J. 2013. Optical spectrum analyzer with quantum-limited noise floor. *Physical Review Letters*, **111**(9), 093604.

[33] Black, E. D. 2000. An introduction to Pound-Drever-Hall laser frequency stabilization. *American Association of Physics Teachers*, **69**(1), 79–87.

[34] Bosi, F., Cella, G., Di Virgilio, A., et al. 2011. Measuring gravitomagnetic effects by a multi-ring-laser gyroscope. *Physical Review D*, **84**(12), 122002.

[35] Bryan, G. H. 1890. On the beats in the vibrations of a revolving cylinder or bell. *Proceeding of the Cambridge Philosophical Society*, **7**, 101–114.

[36] Brzeziński, A. 1986. Contribution to the theory of polar motion for an elastic earth with liquid core. *Manuscripta Geodaetica*, **11**, 226–241.

[37] Burrell, G. J., Hetherington, A., and Moss, T. S. 1968. Faraday rotation resulting from negative absorption. *Journal of Physics B: Atomic and Molecular Physics (1968-1987)*, **1**(4), 692.

[38] Capozziello, S., Altucci, C., Bajardi, F., et al. 2021. Constraining theories of gravity by GINGER experiment. *The European Physical Journal Plus*, **136**(4), 1–21.

[39] Chow, W., Hambenne, J., Hutchings, T., et al. 1980. Multioscillator laser gyros. *IEEE Journal of Quantum Electronics*, **16**(9), 918–936.

[40] Chow, W., Gea-Banacloche, J., Pedrotti, L., et al. 1985. The ring laser gyro. *Reviews of Modern Physics*, **57**(1), 61–104.

[41] Chugani, A., Samant, A. R., and Cerna, M. 1998. *Labview signal processing*. Prentice Hall.

[42] Clivati, C., Calonico, D., Costanzo, G. A., et al. 2013. Large-area fiber-optic gyroscope on a multiplexed fiber network. *Optics Letters*, **38**(7), 1092–1094.

[43] Cochard, A., Igel, H., Schuberth, B., et al. 2006. *Rotational motions in seismology: Theory, observation, simulation*. Springer. Pages 391–411.

[44] Cole, G. D., Zhang, W., Martin, M. J., Ye, J., and Aspelmeyer, M. 2013. Tenfold reduction of Brownian noise in high-reflectivity optical coatings. *Nature Photonics*, **7**(8), 644–650.

[45] Cole, G. D., Zhang, W., Bjork, B. J., et al. 2016. High-performance near- and mid-infrared crystalline coatings. *Optica*, **3**(6), 647–656.

[46] Collaboration, The LIGO Scientific. 2015. Advanced LIGO. *Classical and Quantum Gravity*, **32**(7), 074001.

[47] Cook, A. K., Foster, D. H., and Nöckel, J. U. 2007. Goos-Hänchen induced vector eigenmodes in a dome cavity. *Optics Letters*, **32**(12), 1764–1766.

[48] Cordover, R H, Jaseja, T S, and Javan, A. 1965. Isotope Shift Measurement For 6328 Å He—Ne Laser Transition. *Applied Physics Letters*, **7**(12), 322–324.

[49] Currie, B. E., Stedman, G. E., and Dunn, R. W. 2002. Laser stability and beam steering in a nonregular polygonal cavity. *Applied Optics*, **41**(9), 1689.

[50] D., Dickson L., and A., Good T. 2004. *Holographic barcode scanners: Applications, performance and design in handbook of optical and laser scanning*. CRC Press.

[51] Davis, J. L., and Ezekiel, S. 1981. Closed-loop, low-noise fiber-optic rotation sensor. *Optics Letters*, **6**(10), 505–507.

[52] Dehnen, H. 1967. Zur Prüfung allgemein-relativistischer Rotationseffekte mittels eines Ringlasers. *Zeitschrift für Naturforschung A*, **22**(5), 816–821.

[53] Di Virgilio, A. D. V., Belfi, J., Ni, W.-T., et al. 2017. GINGER: A feasibility study. *European Physical Journal Plus*, **132**(4), 157.

[54] Di Virgilio, A. D. V., Beverini, N., Carelli, G., et al. 2019. Analysis of ring laser gyroscopes including laser dynamics. *The European Physical Journal D*, **79**(7), 573. https://doi.org/10.1140/epjc/s10052-019-7089-5.

[55] Di Virgilio, A. D. V., Beverini, N., Carelli, G., et al. 2020. Identification and correction of Sagnac frequency variations: An implementation for the GINGERINO data analysis. *The European Physical Journal D*, **80**(2), 163.

[56] Di Virgilio, Angela, Schreiber, K. U., Gebauer, A., et al. 2010. A laser gyroscope system to detect the gravito–magnetic effect on Earth. *International Journal Of ModernPhysics D*, **19**(1), 2331–2343.

[57] Dobrowolski, J. A. (ed). 1982. *Handbook of Optics*, 2nd ed. McGraw-Hill.

[58] Donner, S., Lin, C. J., Hadziioannou, C., Gebauer, A., et al. 2017. Comparing direct observation of strain, rotation, and displacement with array estimates at Piñon Flat Observatory, California. *Seismological Research Letters*, **88**(4), 1107–1116.

[59] Dorschner, T., Haus, H., Holz, M., Smith, I., and Statz, H. 1980. Laser gyro at quantum limit. *IEEE Journal of Quantum Electronics*, **16**(12), 1376–1379.

[60] Dotsenko, A. V., Kornienko, L. S., Kravtsov, N. V., et al. 1986. Use of a feedback loop for the stabilization of a beat regime in a solid-state ring laser. *Journal of Physics B*, **16**(1), 58–63.

[61] Dresden, M., and Yang, C. N. 1979. Phase shift in a rotating neutron or optical interferometer. *Physical Review D*, **20**(8), 1846–1848.

[62] Duennebier, F. K., and Sutton, G. H. 1995. Fidelity of ocean bottom seismic observations. *Marine Geophysical Researches*, **17**(6), 535–555.

[63] Dunn, R. W. 1989. Multimode ring laser lock-in. *Applied Optics*, **28**(13), 2584–2587.

[64] Dunn, R. W. 1998. Design of a triangular active ring laser 13 m on a side. *Applied Optics*, **37**(27), 6405–6409.

[65] Dunn, R. W., and Hosman, A. R. 2014. Detection of volcanic infrasound with a ring laser interferometer. *Journal of Applied Physics*, **116**(1), 173109.

[66] Dunn, R. W., Slaton, W. V., and Kendall, L. M. 2012. Detection of low frequency hurricane emissions using a ring laser interferometer. *Journal of Applied Physics*, **112**(7), 073110.

[67] Dunn, R. W., Meredith, J. A., Lamb, A. B., and Kessler, E. G. 2016. Detection of atmospheric infrasound with a ring laser interferometer. *Journal of Applied Physics*, **120**(12), 123109.

[68] Marion, J. E., and Weber, M. J. 1991. Phosphate laser glasses. *European Journal of Solid State Inorganic Chemistry*, **28**, 271–287.

[69] Everitt, C. W. F. 1988. *The Stanford Gryoscope Experiment (A): History and Overview*. W. H. Freeman & Company.

[70] Everitt, C. W. F., DeBra, D. B., Parkinson, B. W., et al. 2011. Gravity Probe B: Final results of a space experiment to test general relativity. *Physical Review Letters*, **106**(22), 221101.

[71] Ezekiel, S., and Balsamo, S. R. 1977. Passive ring resonator laser gyroscope. *Applied Physics Letters*, **30**(9), 478–480.

[72] Franco-Anaya, R., Carr, A. J., and Schreiber, K. U. 2008. Qualification of fibre-optic gyroscopes for civil engineering applications. *2008 NZSEE Conference*, 04, 8.

[73] Gebauer, A., Schreiber, K. U., Klügel, T., Schön, N., and Ulbrich, U. 2012. High-frequency noise caused by wind in large ring laser gyroscope data. *Journal of Seismology*, **16**(4), 777–786.

[74] Gebauer, A., Tercjak, M., Schreiber, K. U., et al. 2020. Reconstruction of the instantaneous earth rotation vector with sub-arcsecond resolution using a large scale ring laser array. *Physical Review Letters*, **125**(3), 033605.

[75] Gerstenberger, D. C., Drobshoff, A., and Sheng, S. C. 1988. Isotope shift of the 543.3 nm laser transition of neon. *IEEE Journal of Quantum Electronics*, **24**(3), 501–502.

[76] Giordano, V., Grop, S., Dubois, B., et al. 2012. New-generation of cryogenic sapphire microwave oscillators for space, metrology, and scientific applications. *Review of Scientific Instruments*, **83**(8), 085113.

[77] GmbH, Northrop Grumman Litef. 2007. Datasheet of μFORS-1. Northrop-Grumman-Litef GmbH.

[78] Graham, R. D. 2010. *New Concepts for Operating Ring Laser Gyroscopes*. Dissertation, University of Canterbury, Christchurch, NZ, 08, 1–294.

[79] Graham, R. D., Hurst, R. B., Thirkettle, R. J., Rowe, C. H., and Butler, P. H. 2008. Experiment to detect frame dragging in a lead superconductor. *Physica C: Superconductivity*, **468**(5), 383–387.

[80] Graizer, V. 2009. The response to complex ground motions of seismometers with galperin sensor configuration. *Bulletin of the Seismological Society of America*, **99**(2B), 1366–1377.

[81] Gray, B. S., Latimer, I. D., and Spoor, S. P. 1996. Gain measurements at 543 nm in helium neon laser discharges. *Journal of Physics D: Applied Physics*, **29**(1), 50–56.

[82] Gross, R. 2006. Earth rotation variations – long period. *Treatise on Geophysics*, **3**(11), 1–50.

[83] Gustavson, T. L., Bouyer, P., and Kasevich, M. 1997. Precision rotation measurements with an atom interferometer gyroscope. *Physical Review Letters*, **78**(11), 2046.

[84] Press, W. H., Teukolsky, S. A., Vetterling, W. T., and Flannery, B. P. 1995. *Numerical recipes in Fortran 77*. Cambridge University Press.

[85] Hadziioannou, C., Gaebler, P., Schreiber, K. U., Wassermann, J., and Igel, H. 2012. Examining ambient noise using colocated measurements of rotational and translational motion. *Journal of Seismology*, **16**(4), 787–796.

[86] Harrison, J. C. 2012. Cavity and topographic effects in tilt and strain measurement. *Journal of Geophysical Research*, **81**(2), 319–328.

[87] Harry, G., Bodiya, T. P., and DeSalvo, R. (eds). 1982. *Optical Coatings and Thermal Noise in Precision Measurement*. Cambridge University Press.

[88] Hecht, J. 1992. Helium neon lasers flourish in face of diode-laser competition. *Laser Focus World*, 00, 1–5.

[89] Heer, C. V. 1982. Laser gyro history. *Physics Today*, **35**(5), 134.

[90] Heer, C. V. 1984. History of The Laser Gyro. In: Jacobs, S. F., Killpatrick, J. E., Sanders, V. E., Murray, S. III, Marlan, S. O., and Simpson, J. H. (eds.). Vol. 0487. *Physics of Optical Ring Gyros* (pp. 2–12).

[91] Hochuli, U. E., Haldemann, P., and Li, H. A. 1974. Factors influencing the relative frequency stability of He–Ne laser structures. *Review of Scientific Instruments*, **45**(11), 1378.

[92] Holdaway, J., Hurst, R. B., Graham, R. D., Rabeendran, N., and Schreiber, K. U. 2012. Self-locked operation of large He–Ne ring laser gyroscopes. *Metrologia*, **49**(3), 209–212.

[93] Höling, B., Leuchs, G., Ruder, H., and Schneider, M. 1992. An argon ion ring laser as a gyroscope. *Applied Physics B*, **55**(1), 46–50.

[94] Hurst, R. B., Dunn, R. W., Schreiber, K. U., Thirkettle, R. J., and MacDonald, G. K. 2004. Mode behavior in ultralarge ring lasers. *Applied Optics*, **43**(11), 2337–2346.

[95] Hurst, R. B., Wells, J.-P. R., and Stedman, G. 2007. An elementary proof of the geometrical dependence of the Sagnac effect. *Journal of Optics A: Pure and Applied Optics*, **9**(10), 838–841.

[96] Hurst, R. B., Stedman, G., Schreiber, K. U., et al. 2009. Experiments with an 834 m^2 ring laser interferometer. *Journal of Applied Physics*, **105**(1), 3115.

[97] Hurst, R. B., Rabeendran, N., Schreiber, K. U., and Wells, J.-P. R. 2014. Correction of backscatter-induced systematic errors in ring laser gyroscopes. *Applied Optics*, **53**(31), 7610–7618.

[98] Hurst, R. B., Mayerbacher, M., Gebauer, A., Schreiber, K. U., and Wells, J.-P. R. 2017. High-accuracy absolute rotation rate measurements with a large ring laser gyro: Establishing the scale factor. *Applied Optics*, **56**(4), 1124–1130.

[99] Igel, H., Schreiber, K. U., Flaws, A., et al. 2005. Rotational motions induced by the M 8.1 Tokachi-oki earthquake, September 25, 2003. *Geophysical Research Letters*, **32**, L08309.

[100] Igel, H., Cochard, A., Wassermann, J., Flaws, A., et al. 2007. Broad-band observations of earthquake-induced rotational ground motions. *Geophysical Journal*, **168**(1), 182–196.

[101] Igel, H., Nader, M. F., Kurrle, D., et al. 2011. Observations of Earth's toroidal free oscillations with a rotation sensor: The 2011 magnitude 9.0 Tohoku-Oki earthquake. *Geophysical Research Letters*, **38**(2), L21303.

[102] Igel, H., Bernauer, M., Wassermann, J., and U., Schreiber K. 2015. *Rotational seismology: Theory, instrumentation, observations, applications*. Encyclopedia of Complexity and System Science. Springer, New York.

[103] Igel, H., Schreiber, K. U., Gebauer, A., et al. 2021. ROMY: A multicomponent ring laser for geodesy and geophysics. *Geophysical Journal International*, **225**(01), 684–698.

[104] iXblue Photonics, Saint Germain, France. 2021. blueSeis-3A datasheet. *iXBlue*.

[105] Jacobs, S. F., and Zanoni, R. 1998. Laser ring gyro of arbitrary shape and rotation axis. *American Association of Physics Teachers*, **50**(7), 659–660.

[106] Jamal, R., and Pichlik, H. 1999. *Labview: Programmiersprache der vierten Generation*. Prentice Hall.

[107] Javan, A. 2005. *Encyclopedia of modern optics: The Helium-Neon laser*. Elsevier Academic Press.

[108] Javan, A., Bennett, W. R., Jr., and Herriott, D. R. 1961. Population inversion and continuous optical maser oscillation in a gas discharge containing a He–Ne mixture. *Physical Review Letters*, **6**(3), 106–110.

[109] Johnson, J. B. 2003. Generation and propagation of infrasonic airwaves from volcanic explosions. *Journal of Volcanology and Geothermal Research*, **121**(1), 1–14.

[110] Jones, D. G. C., Sayers, M. D., and Allen, L. 1968. Mode self-locking in gas lasers. *Journal of Physics A: General Physics*, **2**, 95–101.

[111] Kaula, W. M. 1964. Tidal dissipation by solid friction and the resulting orbital evolution. *Reviews of Geophysics*, **2**(4), 661–685.

[112] Kay, S. M. 1993. *Fundamentals of Statistical Signal Processing: Volume 1, Estimation Theory*. Prentice Hall.

[113] Kay, S. M., and Marple, S. L. 1981. Spectrum analysis – A modern perspective. *Proceedings of the IEEE*, **69**(11), 1380–1419.

[114] Kessler, T., Legero, T., and Sterr, U. 2012. Thermal noise in optical cavities revisited. *Journal of the Optical Society of America B*, **29**(1), 178–184.

[115] King, B. T. 2000. Application of superresolution techniques to ring laser gyroscopes: Exploring the quantum limit. *Applied Optics*, **39**(33), 6151–6157.

[116] Klochan, E. L., Kornienko, L. S., Kravtsov, N. V., Lariontsev, E. G., and Shelaev, A. N. 1974. Laser operation without spikes in a ruby ring. *Radio Engineering and Electronic Physics*, **19**, 58–64.

[117] Klochan, E. L., Kornienko, L. S., Kravtsov, N. V., Lariontsev, E. G., and Shelaev, A. N. 1974. Oscillation regimes in a rotating solid state ring laser. *Soviet Physics*, **38**, 669–673.

[118] Klügel, T., and Wziontek, H. 2009. Correcting gravimeters and tiltmeters for atmospheric mass attraction using operational weather models. *Journal of Geodynamics*, **48**(3–5), 204–210.

[119] Klügel, T., Schlüter, W., Schreiber, K. U., and Schneider, M. 2005. Grossringlaser zur kontinuierlichen Beobachtung der Erdrotation. *Zeitschrift für das Vermessungswesen*, 02, 10.

[120] Knopf, O. 1920. Die Versuche von F. Harreß über die Geschwindigkeit des Lichtes in bewegten Körpern. *Naturwissenschaften*, **8**(42), 815–821.

[121] Korth, W. Z., Heptonstall, A., Hall, E. D., et al. 2016. Passive, free-space heterodyne laser gyroscope. *Classical and Quantum Gravity*, **33**(3), 035004.

[122] Kravtsov, N. V., Lariontsev, E. G., and Shelaev, A. N. 1993. Oscillation regimes of ring solid-state lasers and possibilities for their stabilization. *Laser Physics*, **3**, 22–62.

[123] Kurrle, D., Igel, H., Ferreira, A. M. G., Wassermann, J., and Schreiber, K. U. 2010. Can we estimate local Love wave dispersion properties from collocated amplitude measurements of translations and rotations? *Geophysical Research Letters*, **37**(4), 3281.

[124] Lai, M., Diels, J.-C., and Dennis, M. L. 1992. Nonreciprocal measurements in femtosecond ring lasers. *Optics Letters*, **17**(21), 1535.

[125] Lalezari, R. 1987. Putting the non-red HeNe to Work. *Photonics Spectra*, 117.

[126] Lamb, W. E. 1964. Theory of an optical maser. *Physical Review*, **134**(6A), 1429–1450.

[127] Lambeck, K. 1980. *The Earth's Variable Rotation*. Cambridge University Press.

[128] Lawrence, A. 1998. *Modern Inertial Technology: Navigation, Guidance, and Control*. Springer.

[129] Lefevre, H. C. 2014. *The Fiber-optic Gyroscope*, second edition. Artech House Applied Photonics Series. Artech House Publishers.

[130] Lefevre, H. C., Bergh, R. A., and Shaw, H. J. 1982. All-fiber gyroscope with inertial-navigation short-term sensitivity. *Optics Letters*, **7**(9), 454–456.

[131] Léger, P. 1996. Quapason – a new low-cost vibrating gyroscope. *Proceedings of the Symposium Gyro Technology, Stuttgart Germany*.

[132] Lenef, A., Hammond, T. D., Smith, E. T., et al. 1997. Rotation sensing with an atom interferometer. *Physical Review Letters*, **78**(5), 760.

[133] Lense, J., and Thirring, H. 1918. Über den Einfluß der Eigenrotation des Zentralkörpers auf die Bewegung der Planeten und Monde nach der Einsteinschen Gravitationstheorie. *Physikalische Zeitschrift*, 01, 156.

[134] Levin, Y. 2008. Fluctuation–dissipation theorem for thermo-refractive noise. *Physics Letters A*, **372**(12), 1941–1944.

[135] Li, H.-N., Sun, L.-Y., and Wang, S.-Y. 2002. Frequency dispersion characteristics of phase velocities in surface wave for rotational components of seismic motion. *Journal of Sound and Vibration*, **258**(5), 815–827.

[136] Lindner, F., Wassermann, J., Schmidt-Aursch, M. C., Schreiber, K. U., and Igel, H. 2017. Seafloor ground rotation observations: Potential for improving signal to noise ratio on horizontal OBS components. *Seismological Research Letters*, **88**(1), 32–38.

[137] Liu, K., Zhang, F. L., Li, Z. Y., Feng, X. H., Li, K., Lu, Z. H., Schreiber, K. U., Luo, J., and Zhang, J. 2019. Large-scale passive laser gyroscope for earth rotation sensing. *Optics Letters*, **44**(11), 2732.

[138] Liu, K., Zhang, F., Li, Z., et al. 2020. Noise analysis of a passive resonant laser gyroscope. *Sensors (Basel)*, **20**(18), 5369.

[139] Lodge, O. J. 1893. Aberration problems – a discussion concerning the motion of the ether near the Earth, and concerning the connection between the ether and gross matter; with some new experiments. *Philosophical Transactions A* 1897, **189**, 149–166.

[140] Lodge, O. J. 1897. Experiments on the Absence of mechanical connexion between ether and matter. *Proceedings of the Royal Society of London*, **61**, 31–32.

[141] Loukianov, D., Rodloff, R., Sorg, H., and Stieler, B. (eds). 1999. *Optical gyros and their application*. North Atlantic Treaty Organization.

[142] Macek, W. M., and Davis, D. T. M. Jr. 1963. Rotation rate sensing with traveling-wave ring lasers. *Applied Physics Letters*, **2**(3), 67–68.

[143] MacKenzie, D. 1996. *Knowing Machines: Essays on Technical Change*. MIT Press.

[144] Maiman, T. H. 1960. Stimulated optical radiation in Ruby. *Nature*, **187**(4736), 493–494.

[145] McLeod, D. P., Stedman, G. E., Webb, T. H., and Schreiber, K. U. 1998. Comparison of standard and ring laser rotational seismograms. *Bulletin of the Seismological Society of America*, **88**(6), 1495–1503.

[146] Meyer, R. E., Ezekiel, S., Stowe, D. W., and Tekippe, V. J. 1983. Passive fiber-optic ring resonator for rotation sensing. *Optics Letters*, **8**(12), 644–646.

[147] Michelson, A. A. 1904. Relative motion of Earth and Aether. *Philosophical Magazine*, **8**(48), 716–719.

[148] Michelson, A. A., and Gale, H. G. 1925. The effect of the Earth's rotation on the velocity of light, Part II. *The Astrophysical Journal*, **61**(140-145), 2.

[149] Milonni, P. W., and Eberly, J. H. 1988. *Lasers*. Wiley & Sons.

[150] Müller, J., and Biskupek, L. 2007. Variations of the gravitational constant from lunar laser ranging data. *Classical and Quantum Gravity*, **24**(17), 4533–4538.

[151] Müller, J., Murphy, T. Jr, Schreiber, K U., et al. 2019. Lunar laser ranging – A tool for general relativity, lunar geophysics and earth science. *Journal of Geodesy*, 01, 1–19.

[152] Murugesan, S., and Goel, P. S. 1989. Autonomous fault-tolerant attitude reference system using DTGs in symmetrically skewed configuration. *IEEE Transactions on Aerospace and Electronic Systems*, **25**(2), 302–307.

[153] Nader, M. F., Igel, H., Ferreira, A. M. G., et al. 2012. Toroidal free oscillations of the Earth observed by a ring laser system: A comparative study. *Journal of Seismology*, **16**(4), 745–755.

[154] Notcutt, M., Ma, L.-S., Ludlow, A. D., Foreman, S. M., Ye, J., and Hall, J. L. 2006. Contribution of thermal noise to frequency stability of rigid optical cavity via Hertz-linewidth lasers. *Physical Review A*, **73**(3), 033033–4.

[155] Numata, K., Kemery, A., and Camp, J. 2004. Thermal-noise limit in the frequency stabilization of lasers with rigid cavities. *Physical Review Letters*, **93**(25), 250602.

[156] Packard, R. E., and Vitale, S. 1992. Principles of superfluid-helium gyroscopes. *Physical Review B*, **46**(6), 3540.

[157] Pancha, A., Webb, T. H., Stedman, G., McLeod, D., and Schreiber, K. U. 2000. Ring laser detection of rotations from teleseismic waves. *Geophysical Research Letters*, **27**(2), 3553–3556.

[158] Park, J., Song, T.-R. A., Tromp, J., et al. 2005. Earth's free oscillations excited by the 26 December 2004 Sumatra-Andaman earthquake. *Science (New York, N.Y.)*, **308**(5725), 1139–44.

[159] Pham, N. D., Igel, H., Wassermann, J., Cochard, A., and Schreiber, K. U. 2009. The effects of tilt on interferometric rotation sensors. *Bulletin of the Seismological Society of America*, **99**(2B), 1352–1365.

[160] Pircher, G., and Hepner, G. 1967. Perfectionnement aux dispositifs du type gyrometre interferometrique a laser. *French Patent: 1.563.720*.

[161] Plag, H.-P., and Pearlman, M. (eds.). 2009. *Global Geodetic Observing System-meeting the Requirements of a Global Society on a Changing Planet in 2020*. Springer.

[162] Post, E. J. 1967. Sagnac effect. *Reviews of Modern Physics*, **39**(2), 475–493.

[163] Pritsch, B., Schreiber, K. U., Velikoseltsev, A., and Wells, J.-P. R. 2007. Scale-factor corrections in large ring lasers. *Applied Physics Letters*, **91**(6), 061115.

[164] Qiao, W., Xiaojun, Z., Zongsen, L., et al. 2014. A simple method of optical ring cavity design and its applications. *Optics Express*, **22**(12), 14782–14791.

[165] Rabeendran, N. 2005. New approaches to gyroscopic lasers. *Dissertation, University of Canterbury, New Zealand*, 07.

[166] Rodloff, R. 1990. Concept for a high precision experimental laser gyroscope system 'ELSy' (laser gyro goniometry). *Physica C*, **12**(2), 67–74.

[167] Rosenthal, A. H. 1962. Regenerative circulatory multiple-beam interferometry for the study of light-propagation effects. *Journal of the Optical Society of America*, **52**(10), 1143–1148.

[168] Rotation, International Earth, and Service, Reference System. 2022. *International Earth Rotation and Reference System Service*. https://www.iers.org/IERS/EN/Home/home_node.html. Accessed: February 14, 2022.

[169] Rotge, J. R., Simmons, B. J., Kroncke, G. T., and Stech, D. J. 1986. *Final report on optical rotation sensor*. https://apps.dtic.mil/sti/pdfs/ADA169357.pdf. Accessed: February 14, 2022.

[170] Rotge, J. R., Simmons, B. J., Kroncke, G. T., and Stech, D. J. 1986. Optical rotation sensors – final report. *Final Report on Project 2301-F1-68. Air Force System Command, United States Air Force*.

[171] Rowe, C. H., Schreiber, K. U., Cooper, S. J., et al. 1999. Design and operation of a very large ring laser gyroscope. *Applied Optics LP*, **38**(12), 2516.

[172] Rozelle, D. M. 2009. The hemispherical resonator gyro: From wineglass to the planets. *Proc. 19th AAS AIAA Space Flight Mechanics*.

[173] Sagnac, G. 1913. L'ether lumineux demontre par l'effet du vent relatif d'ether dans un interferometre en rotation uniforme. *Comptes rendus de l'Académie des Sciences*, **157**, 708–710.

[174] Sagnac, G. 1913. Sur la preuve de la réalité de l'éther lumineux par l'experiénce de l'interférographe tournant. *Comptes rendus de l'Académie des Sciences*, **157**, 1410–1413.

[175] Sanders, G. A., Prentiss, M. G., and Ezekiel, S. 1981. Passive ring resonator method for sensitive inertial rotation measurements in geophysics and relativity. *Optics Letters*, **6**(11), 569–571.

[176] Santagata, R. 2015 (11). *Sub-nanometer length metrology for ultra-stable ring laser gyroscopes*. Ph.D. thesis, University of Siena, Italy.

[177] Santagata, R., Beghi, A., Belfi, J., Beverini, N., Cuccato, D., Di Virgilio, A., Ortolan, A., Porzio, A., and Solimeno, S. 2015. Optimization of the geometrical stability in square ring laser gyroscopes. *Classical and Quantum Gravity*, **32**(5), 055013.

[178] Sato, Y., and Packard, R. E. 2012. Superfluid helium quantum interference devices: Physics and applications. *Reports on Progress in Physics*, **75**(1), 016401.

[179] Schiff, L. I. 1960. Possible new experimental test of general relativity theory. *Physical Review Letters*, **4**(5), 215–217.

[180] Schmelzbach, C., Donner, S., Igel, H., et al. 2018. Advances in 6C seismology: Applications of combined translational and rotational motion measurements in global and exploration seismology. *GEOPHYSICS*, **83**(3), WC53–WC69.

[181] Schreiber, K. U. 1999 (02). *Ringlaser für die Geodäsie*. Technical University Munich.

[182] Schreiber, K. U., and Wells, J.-P. R. 2013. Invited review article: Large ring lasers for rotation sensing. *Review of Scientific Instruments*, **84**(4), 041101–041126.

[183] Schreiber, K. U., Rowe, C. H., Wright, D. N., Cooper, S. J., and Stedman, G. 1998. Precision stabilization of the optical frequency in a large ring laser gyroscope. *Applied Optics LP*, **37**(36), 8371–8381.

[184] Schreiber, K. U., Klügel, T., and Stedman, G. 2003. Earth tide and tilt detection by a ring laser gyroscope. *Journal of Geophysical Research: Solid Earth*, **108**(B), 2132.

[185] Schreiber, K. U., Velikoseltsev, A., Rothacher, M., Klügel, T., and Stedman, G. 2004. Direct measurement of diurnal polar motion by ring laser gyroscopes. *Journal of Geophysical Research: Solid Earth*, **109**(B), 6405.

[186] Schreiber, K. U., Velikoseltsev, A., Stedman, G. E., and Hurst, R. B. 2004. Large ring laser gyros as high resolution sensors for applications in geoscience. *Proceedings of the 11th International Conference on Integrated Navigation Systems, St. Petersburg*, 326.

[187] Schreiber, K. U., Stedman, G. E., Igel, H., and Flaws, A. 2006. *Ring laser gyroscopes as rotation sensors for seismic wave studies*. Springer Berlin Heidelberg (pp. 377–390).

[188] Schreiber, K. U., Igel, H., Velikoseltsev, A., et al. 2006. *The GEOsensor project: Rotations – a new observable for seismology*. Springer Monograph: "Observation of the Earth System from Space".

[189] Schreiber, K. U., Hautmann, J., Velikoseltsev, A., et al. 2009. Ring laser measurements of ground rotations for seismology. *Bulletin of the Seismological Society of America*, **99**(2B), 1190–1198.

[190] Schreiber, K. U., Velikoseltsev, A., Carr, A. J., and Franco-Anaya, R. 2009. The application of fiber optic gyroscopes for the measurement of rotations in structural engineering. *Bulletin of the Seismological Society of America*, **99**(2B), 1207–1214.

[191] Schreiber, K. U., Klügel, T., Velikoseltsev, A., et al. 2009. The large ring laser G for continuous earth rotation monitoring. *Pure and Applied Geophysics*, **166**(8–9), 1485–1498.

[192] Schreiber, K. U., Klügel, T., Wells, J.-P. R., Hurst, R. B, and Gebauer, A. 2011. How to detect the chandler and the annual wobble of the Earth with a large ring laser gyroscope. *Physical Review Letters*, **107**(10), 173904.

[193] Schreiber, K. U., Klügel, T, Wells, J.-P. R., Holdaway, J., Gebauer, A., and Velikoseltsev, A. 2012. Enhanced ring lasers: a new measurement tool for Earth sciences. *Quantum Electronics*, **42**(11), 1045–1050.

[194] Schreiber, K. U., Gebauer, A., and Wells, J.-P. R. 2012. Long-term frequency stabilization of a 16 m^2 ring laser gyroscope. *Optics Letters*, **37**(11), 1925–1927.

[195] Schreiber, K. U., Gebauer, A., and Wells, J.-P. R. 2013. Closed-loop locking of an optical frequency comb to a large ring laser. *Optics Letters*, **38**(18), 3574–3577.

[196] Schreiber, K. U., Thirkettle, R. J., Hurst, R. B., et al. 2015. Sensing Earth's rotation with a Helium–Neon ring laser operating at 1.15 µm. *Optics Letters*, **40**(8), 1705.

[197] Schulz-DuBois, E. 1966. Alternative interpretation of rotation rate sensing by ring laser. *Quantum Electronics, IEEE Journal of*, **2**(8), 299–305.

[198] Schwartz, S., Gutty, F., Feugnet, G., Loil, E., and Pocholle, J.-P. 2009. Solid-state ring laser gyro behaving like its helium-neon counterpart at low rotation rates. *Optics Letters*, **34**(24), 3884.

[199] Schwarzschild, B. M. 1981. Sensitive fiber optic gyroscopes. *Physics Today*, **34**(10), 20–22.

[200] Seitz, F., and Schmidt, M. 2005. Atmospheric and oceanic contributions to Chandler wobble excitation determined by wavelet filtering. *Journal of Geophysical Research*, **110**(B11), 25–11.

[201] Shankland, R. S. 1974. Michelson and his interferometer. *Physics Today*, **27**(4), 37–43.

[202] Shaw, G. L., and Simmons, B. J. 1984. A 58 m^2 passive resonant ring laser gyroscope. *Fiber Optic and Laser Sensors II*, 117–121.

[203] Shaw, G. L., Rotge, J., and Simmons, B. J. 1986. Progress on 58 m² passive resonant ring laser gyroscope. *Proceedings of the SPIE*, **566**(01), 84–89.

[204] Siegman, A. E. 1986. *Lasers*. University Science Books.

[205] Sollberger, D., Igel, H., Schmelzbach, C., et al. 2020. Seismological processing of six degree-of-freedom ground-motion data. *Sensors*, **20**(23), 6904.

[206] Spreeuw, R. J. C., Centeno Neelen, R., van Druten, N. J., Eliel, E. R., and Woerdman, J. P. 1990. Mode coupling in a He–Ne ring laser with backscattering. *Physical Review A*, **42**(7), 4315–4324.

[207] Stedman, G., Bilger, H., Li, Z., et al. 1993. Canterbury ring laser and tests for nonreciprocal phenomena. *Australian Journal of Physics*, **46**(00), 87–101.

[208] Stedman, G. E. 1997. Ring-laser tests of fundamental physics and geophysics. *Reports on Progress in Physics*, **60**(6), 615.

[209] Stedman, G. E., Li, Z., and Bilger, H. R. 1995. Sideband analysis and seismic detection in a large ring laser. *Applied Optics LP*, **34**(2), 5375.

[210] Stedman, G. E., Schreiber, K. U., and Bilger, H. R. 2003. On the detectability of the Lense-Thirring field from rotating laboratory masses using ring laser gyroscope interferometers. *Classical and Quantum Gravity*, **20**(13), 2527–2540.

[211] Stokes, L. F., Chodorow, M., and Shaw, H. J. 1982. All-single-mode fiber resonator. *Optics Letters*, **7**(6), 288.

[212] Stone, J. A., and Stejskal, A. 2004. Using helium as a standard of refractive index: Correcting errors in a gas refractometer. *Metrologia*, **41**(3), 189–197.

[213] Suryanto, W., Igel, H., Wassermann, J., et al. 2006. First comparison of array-derived rotational ground motions with direct ring laser measurements. *Bulletin of the Seismological Society of America*, **96**(6), 2059–2071.

[214] Szöke, A., and Javan, A. 1963. Isotope shift and saturation behavior of the 1.15-μ transition of Ne. *Physical Review Letters*, **10**(12), 521–524.

[215] Tackmann, G., Berg, P., Schubert, C., et al. 2012. Self-alignment of a compact large-area atomic Sagnac interferometer. *New Journal of Physics*, **14**(1), 015002.

[216] Tajmar, M., and de Matos, C. J. 2005. Extended analysis of gravitomagnetic fields in rotating superconductors and superfluids. *Physica C: Superconductivity*, **420**(1), 56–60.

[217] Tajmar, M., and de Matos, Clovis J. 2003. Gravitomagnetic field of a rotating superconductor and of a rotating superfluid. *Physica C: Superconductivity*, **385**(4), 551–554.

[218] Tajmar, M., Plesescu, F., Seifert, B., Schnitzer, R., and Vasiljevich, I. 2008. Search for frame-dragging-like signals close to spinning superconductors. *arXiv.org*, 00.

[219] Tang, C. L., Statz, H., and de Mars, G. 1963. Spectral output and spiking behavior of solid-state lasers. *Journal of Applied Physics*, **34**(8), 2289–2295.

[220] Tanimoto, T., Hadziioannou, C., Igel, H., et al. 2016. Seasonal variations in the Rayleigh-to-Love wave ratio in the secondary microseism from colocated ring laser and seismograph. *Journal of Geophysical Research: Solid Earth*, **121**(4), 2447–2459.

[221] Tartaglia, A., Di Virgilio, A., Belfi, J., Beverini, N., and Ruggiero, M. L. 2017. Testing general relativity by means of ring lasers. Ring lasers and relativity. *The European Physical Journal Plus*, **132**(2), 73.

[222] Tian, W. 2014. On tidal tilt corrections to large ring laser gyroscope observations. *Geophysical Journal International*, **196**(1), 189–193.

[223] Vali, V., and Shorthill, R. W. 1976. Fiber ring interferometer. *Applied Optics*, **15**(5), 1099–100.

[224] Vali, V., and Shorthill, R. W. 1977. Ring interferometer 950 m long. *Applied Optics*, **16**(2), 290–291.

[225] Vallet, M., Ghosh, R., Le Floch, A., et al. 2001. Observation of magnetochiral birefringence. *Physical Review Letters*, **87**(18), 183003.

[226] Velikoseltsev, A. 2005. *The development of a sensor model for Large Ring Lasers and their application in seismic studies*. Ph.D. thesis, Technical University Munich.

[227] Verdeyen, J. T. 2000. *Laser electronics*. Prentice Hall Series in Solid State Physical Electronics. Prentice Hall.

[228] von Laue, M. 1907. Die mitführung des lichtes durch bewegte Körper nach dem relativitätsprinzip. *Annalen der Physik*, **328**(1), 989–990.

[229] Walsh, P., and Kemeny, G. 1963. Laser operation without spikes in a ruby ring. *Journal of Applied Physics*, **34**(4), 956–957.

[230] Wassermann, J., Bernauer, F., Shiro, B., et al. 2020. Six-axis ground motion measurements of Caldera collapse at Kilauea Volcano, Hawai'i–more data, more puzzles? *Geophysical Research Letters*, **47**(5), e2019GL085999.

[231] Webb, S. C. 1988. Long-period acoustic and seismic measurements and ocean floor currents. *IEEE Journal of Oceanic Engineering*, **13**(4), 263–270.

[232] Wei, D. T., and W., Louderback A. 1979. Method for fabricating multi-layer optical films. *Patents*.

[233] White, A. D., Gordon, EI, and Rigden, JD. 1963. Output power of the 6328-Å gas maser. *Applied Physics Letters*, **2**(5), 91.

[234] Widmer, R., and Zürn, W. 1992. Bichromatic excitation of long-period Rayleigh and air waves by the Mount Pinatubo and El Chichon volcanic eruptions. *Geophysical Research Letters*, **19**(8), 765–768.

[235] Widmer-Schnidrig, R., and Zürn, W. 2009. Perspectives for ring laser gyroscopes in low-frequency seismology. *Bulletin of the Seismological Society of America*, **99**(2B), 1199–1206.

[236] Wilkinson, J. R. 1987. Ring lasers. *Progress in Quantum Electronics*, **11**(1), 1–103.

[237] Zarinetchi, F., and Ezekiel, S. 1986. Observation of lock-in behavior in a passive resonator gyroscope. *Optics Letters*, **11**(6), 401–403.

[238] Zou, D., Anyi, C. L., Thirkettle, R. J., Schreiber, K. U., and Wells, J.-P. R. 2019. Sensing Earth rotation with a helium-neon laser operating on three transitions in the visible region. *Applied Optics*, **58**(10), 7884.

[239] Zürn, W., and Widmer-Schnidrig, R. 2002. Globale Eigenschwingungen der Erde. *Physik Journal*, **1**(10), 1–7.

Subject Index

Printed in the United States
by Baker & Taylor Publisher Services